Tales of the Quantum

Tales of the Quantum

Understanding Physics' Most Fundamental Theory

ART HOBSON

OXFORD
UNIVERSITY PRESS

Oxford University Press is a department of the University of Oxford. It furthers
the University's objective of excellence in research, scholarship, and education
by publishing worldwide. Oxford is a registered trade mark of Oxford University
Press in the UK and certain other countries.

Published in the United States of America by Oxford University Press
198 Madison Avenue, New York, NY 10016, United States of America.

© Oxford University Press 2017

All rights reserved. No part of this publication may be reproduced, stored in
a retrieval system, or transmitted, in any form or by any means, without the
prior permission in writing of Oxford University Press, or as expressly permitted
by law, by license, or under terms agreed with the appropriate reproduction
rights organization. Inquiries concerning reproduction outside the scope of the
above should be sent to the Rights Department, Oxford University Press, at the
address above.

You must not circulate this work in any other form
and you must impose this same condition on any acquirer.

Library of Congress Cataloging-in-Publication Data
Names: Hobson, Art, 1934– author.
Title: Tales of the quantum : understanding physics' most fundamental theory / Art Hobson.
Description: New York, NY : Oxford University Press, [2017] |
Includes bibliographical references and index.
Identifiers: LCCN 2016018154 | ISBN 9780190679637
Subjects: LCSH: Quantum theory.
Classification: LCC QC174.12 .H53 2016 | DDC 530.12—dc23
LC record available at https://lccn.loc.gov/2016018154

1 3 5 7 9 8 6 4 2

Printed by Sheridan Books, Inc., United States of America

Also by Art Hobson

Concepts in Statistical Mechanics
Physics and Human Affairs
The Future of Land-Based Strategic Missiles (co-editor, co-author)
Physics: Concepts & Connections

For my brother Richard Hobson

CONTENTS

List of Illustrations ix
Preface xi
Acknowledgments xv

1. Introduction: The Tale of the Quantum in the Window   1

PART 1   THE UNIVERSE IS MADE OF QUANTA

2. What Is Quantum Physics About?   17

3. Particles and Classical Mechanics   38

4. Fields and Classical Electromagnetism   59

5. What Is a Quantum?   77

PART 2   HOW QUANTA BEHAVE

6. Perfect Randomness   109

7. Quantum States and How They Change   131

8. Superpositions and Macroscopic Quanta   149

9. An Entangled, Nonlocal Universe   166

PART 3   GETTING BACK TO THE NORMAL WORLD

10. Schrödinger's Cat and "Measurement"   191

11. The Environment as Monitor   217

Notes   235
Glossary   261
Index   277

LIST OF ILLUSTRATIONS

1.1 Mach-Zehnder interferometer   3
2.1 Water wave interference   23
2.2 Double-slit experiment with light   24
2.3 Experimental result of the double-slit experiment with light   25
2.4 Double-slit experiment with light: experimental setup   25
3.1 Galileo's experiment with a ball on an incline   39
4.1 Creating electromagnetic waves   66
4.2 The electromagnetic spectrum   67
5.1 Double-slit experiment with dim light and time-lapse photography   81
5.2 A photo emerges from individual particle-like impacts   81
5.3 Single-slit experiment with light   82
5.4 Buildup of an interference pattern in the double-slit experiment using electrons   87
5.5 Unifications during the history of physics   94
6.1 Diffraction when a wave passes through a single slit   116
6.2 Diffraction of light passing through a slit   116
6.3 Five "snapshots" of one electron passing through single-slit diffraction   118
6.4 Double-slit trajectories of 100 point particles, in Bohm's model   128
7.1 Graph of a typical quantum state of one electron at one instant   133
7.2 A string between two fixed points, in several states of vibration   135
7.3 Portraits of an electron in a hydrogen atom   137
7.4 The lowest five energy levels for the electron in a hydrogen atom   142
8.1 Mach-Zehnder interferometer   150
8.2 Evidence of quantum superposition   152
8.3 Schematic diagram of a superconducting quantum interference device   161
8.4 Evidence for quantum superposition of a macroscopic object   165
9.1 When two quanta interact and then separate, they typically remain entangled   169

9.2 The RTO photon entanglement experiment as a double-slit experiment 170
9.3 RTO experiment using Mach-Zehnder interferometers for entangled photons 173
9.4 The interference of correlations 174
9.5 The detector effect 185
11.1 A variation on the Stern–Gerlach experiment 219
11.2 The experiment of Haroche and colleagues 230

Table

5.1 Summary of the Standard Model of Fundamental Fields and Quanta, a theory of the fundamental fields 95

PREFACE

From Dr. Suess to the Hardy Boys to Tom Sawyer, the tales told in books were my most dependable boyhood buddies. I first encountered physics many years later, following a degree in music and a stint in the Army. As a student of physics at Kansas State University in Manhattan, I was delighted to find much of what I learned in textbooks and lectures to be fascinating, often strange, and always mind-expanding, stories. But now they were true stories, about a natural world that increasingly intrigued me. An account of the forces acting on an airplane, an elegant proof that Earth follows an elliptical path around the sun, or an account of the famous experiment demonstrating that light is a wave—all were fascinating adventures of the mind. Fifty-five years later, my love affair with nature persists.

This book tells tales of our quantum universe, in words accessible to everyone, without mathematics or unnecessary technicalities. My protagonist is the quantum, arguably *the* central actor on the cosmic stage. Although most popular books about quantum physics follow the subject's history, *Tales of the Quantum* follows the phenomena: wave–particle duality, fundamental randomness, being in two places at once, and quantum jumps, to name a few. It presents history and people only to the extent that they illuminate the phenomena. Nevertheless, I hope *Tales of the Quantum* has a storylike quality that will be meaningful to both nonscientists and scientists. It's written for all who would like to better fathom, before they depart this mortal coil, what makes the universe tick.

Tales of the Quantum has a central message: quantum physics is fine and healthy just as it is. From the 1920s through today, the radical nature of the theory has prompted many, including Albert Einstein, to find fault with one or the other fundamental quantum precept and to try to fix it. But quantum physics doesn't need fixing. The theory may be strange, but it's not a mystery. We'll discover that all its supposed paradoxes are resolvable and can be explained consistently in ordinary English without invoking algebra, technicalities, or supernatural powers.

The book's most original feature is a suggested solution of the measurement problem, aka Schrödinger's cat. Chapter 10 argues that the solution arises from a suggestion first made in 1968 and rediscovered by many physicists, including me, since that time. Working from this suggestion, my own research has shown that Schrödinger's famous cat is not the outrageous "quantum superposition" of a dead cat and an alive cat that it at first appears to be, but is instead an entirely non-paradoxical "superposition of correlations" as I'll explain in Chapter 10. In my view, measurement remains the only significant quantum foundational issue still in dispute. Regarded by some as unresolved, by others as resolved, and by still others as a mere pseudo-problem that needs no resolution, the problem splits the experts and is compounded by its title. By calling it the *measurement problem*, we suggest that quantum foundations have something to do with the human beings who make scientific measurements, leading some to conclude that quantum physics reinstates human minds at the center of physics for the first time since Copernicus. It's even been seriously suggested that human consciousness is required for normal physical reality to emerge from the quantum world.

The work of Wojciech Zurek and many others, although not entirely solving the measurement problem, has clarified that measurements are indeed crucial to understanding how quantum physics leads to the world of our experience. But these so-called "measurements" have nothing necessarily to do with humans; they are conducted constantly, all over the universe, by the environment. Human consciousness has no essential role in the foundations of quantum physics.

Another distinctive facet is this book's take on the famous issue of wave–particle duality. Is the universe made of waves in spatially extended "fields," or of tiny particles, or both? As the title of my 2013 *American Journal of Physics* article states, "there are no particles, there are only fields." The universe is made entirely of fields, such as Earth's gravitational field and the magnetic fields you've probably experienced when playing with magnets. With the notable exception of Richard Feynman, most quantum field theorists—physicists who integrate quantum physics with Einstein's theory of relativity—have taken this viewpoint, but somehow it hasn't filtered through to the broader ranks of physicists, other scientists, and the public.

Tales of the Quantum contains no mathematics beyond a few numbers. Mathematical physicist Paul Dirac stated, "Mathematics is only a tool and one should learn to hold the physical ideas in one's mind without reference to the mathematical form."[1] But some physicists are convinced that any accurate presentation of quantum physics must be mathematical and thus can't be understood by nonscientists, and that any popularly understandable presentation must be inaccurate. I heartily disagree! A plethora of good physics books, written in nontechnical language for nonscientists, disproves this notion. Books such as Albert Einstein and Leopold Infeld's classic *The Evolution of Physics: From Early Concepts to Relativity and Quanta* (Simon & Schuster, New York, 1938 and 1966), Brian Greene's *The Elegant Universe: Superstrings, Hidden Dimensions, and the Quest for the Ultimate Theory* (W. W. Norton & Company, 1999), and Louisa Gilder's *The Age of Entanglement: When Quantum Physics Was Reborn* (Alfred A. Knopf, 2008), illustrate that one can explain physics, including quantum physics, accurately and nontechnically.

In aiming for a book that is suitable for both nonscientists and scientists, I have relegated many details to numbered endnotes that provide additional commentary, supporting references, and some more technical discussion. These endnotes can be safely omitted without disturbing the flow of the text. There is also an extensive glossary.

At least since the days of the early Greeks, philosophical people have wanted to know the ultimate constituents of the universe. What is the stuff of reality and how does it behave? One popular answer, that it's made of atoms, is outdated and incorrect. We've known for a few decades that most of the universe is actually not made of the chemical atoms. However, atoms and everything else are made of things more fundamental and even more intriguing than atoms, namely "fields" that are bundled into "quanta." This book proceeds from these two key concepts, fields and quanta. They take getting used to, but you can grasp them. The quantum, like an eccentric friend, requires a modicum of time to be understood. *Tales of the Quantum* will spiral in, introducing our protagonist in general terms in Chapters 1 through 5 before zooming in on the quantum's details in Chapters 6 through 11.

Following the presentation of a new quantum theory of the elementary particles at Columbia University in 1958 by physicist Wolfgang Pauli, Niels Bohr—one of quantum physics' founding fathers—said to Pauli, "We are all agreed that your theory is crazy. The question that divides us is whether it is crazy enough to have a chance of being correct."[2] Nature is far more inventive than is human imagination, and the microscopic world is not what Niels Bohr or anyone else could have guessed. Quantum physics is indeed strange, and some have rejected some aspects of it on grounds of this strangeness, but strangeness alone is not a compelling reason to reject a scientific theory.

Modern physics does take some getting used to. Albert Einstein, upon hearing Werner Heisenberg's 1927 lecture announcing his indeterminacy principle, stated "Marvelous, what ideas the young people have these days. But I don't believe a word of it."[3] Learning contemporary physics might be like a baby first confronting life outside the womb. It's not what you expected, but you can make sense of it by maintaining an open mind.

Working on this book has made me appreciate quantum physics as the most fascinating array of ideas I could ever hope to encounter. I hope you enjoy reading *Tales of the Quantum* as much as I enjoyed writing it.

ACKNOWLEDGMENTS

This book, like all science, is collaborative. Many people gave their time and attention. My brother Richard Hobson and my dear friend Ulrich Harms read an embarrassingly rough early draft and provided feedback from their perspectives—Richard as a writer, mathematician, and teacher, and Ulrich as a physicist and teacher. Their comments, provided at an early stage of my writing, guided me throughout the remainder of the project and helped shape the book's final form.

The University of Arkansas, especially its Department of Physics, was enormously supportive. I interrupted the work of colleagues Claud Lacy and Suren Singh several times each week with questions about everything from the Big Bang to interferometry. I received valuable input from colleagues Julio GeaBanacloche, Daniel Kennefick, Michael Lieber, William Oliver, Paul Thibado, and Reeta Vyas. My university and my department did me the huge favor of providing this retired professor with the perfect ivory tower, an office in the physics building. I had valuable conversations about writing and grammar with James Bennett of the English Department, about finding a good publisher with Peter Ungar of the Anthropology Department, and about quantum foundations with Barry Ward of the Philosophy Department.

I thank Maximilian Schlosshauer for advice and for his wonderful book *Decoherence and the Quantum-to-Classical Transition*, which remained on my desk or in my hands rather than on my shelves throughout this project. My friend Rodney Brooks and his book *Fields of Color* was an important inspiration. James Malley provided important encouragement and advice.

I received significant ideas, advice, and inspiration from Stephen Adler, Nathan Argaman, Casey Blood, Jeffrey Bub, Shelley Buonaiuto, Howard Carmichael, Massimilliano de Bianchi, Dennis Dieks, Edward Gerjuoy, Nicolas Gisin, David Green, Daniel Greenberger, Nick Herbert, Paul Hewitt, David Jackson, Ruth Kastner, Gary Knack, Peter Lewis, Ruby Lord, Tim Maudlin, David Mermin, Peter Milonni, Michael Nauenberg, Roland Omnes, Philip Pearle, Mario Rabinowitz, A. R. P. Rau, Stefan Rinner, Greg Rohm, Marlan Scully, Marc

Sher, Anwar Shiekh, Sjaak van Dijk, Steven van Enk, Frank Wilczek, Dieter Zeh, and Wojciech Zurek.

Professional organizations, and their administrators and staff, are essential to science. The American Association of Physics Teachers has provided lifelong support to me via their meetings, publications, knowledge, and opportunities for social interactions. I especially thank the many people who have been active within this Association in connection with physics and society education—in particular, Gordon Aubrecht, Al Bartlett (now deceased), Harvey Leff, John Roeder, and Steven Shropshire. The American Physical Society, the APS Forum on Physics and Society, and the American Association for the Advancement of Science also provided lifelong organizational support.

I could not do my work without assistance from support staff. I thank my helpers at the University of Arkansas, especially Dianne Melahn and her physics office staff, Kathleen Lehman and Stephanie Freedle and their physics library staff, the University's Department of Information Technology Services, and other university staff.

I was overjoyed when Oxford University Press agreed to publish this book. My thanks and good wishes go out to my editor Jeremy Lewis, assistant editor Erik Hane, editorial assistant Anna Langley, marketing manager Michelle Kelly, production manager Prabhu Chinnasamy of Newgen Publishing, copy editor Cat Ohala, and others without whom this book would have been written but not necessarily read.

My dear friend James Bennett was supportive throughout the project. My eternal love and appreciation go to my wife, Marie Riley, and my children Ziva Branstetter, David Hobson, and Justin Riley. I couldn't have written this book without their love and support. I thank my wife—a member of the university's Department of Information Technology Services—also for her computer expertise. I thank my departed Mom and Dad. Marjorie Breitweg Hobson, who taught me the joy of reading and writing, and Leland Hobson, who taught me the joy of science.

1

Introduction

The Tale of the Quantum in the Window

> Quantum theory, in particular, is the most seminal change in viewpoint since the early Greeks gave up mythology to initiate the search for a rational understanding of the universe.
> —Leon Lederman, Nobel Laureate and Director Emeritus of Fermilab, and Christopher Hill, in *Quantum Physics for Poets*

There was more mystery in the shop window on that Saturday morning than Alice could have expected. The old city square in Fayetteville was in its springtime bloom of native flowers, and the farmers' market spilled bouquets, food, paintings, crafts, music, conversation, dogs, and people over the sidewalks and surrounding streets. As Alice strolled the far side of Block Avenue, shopping and sipping local coffee from a street-side booth, her reflection was visible in a plate glass shop window. Being currently enrolled in a physics-for-nonscientists course at the nearby University of Arkansas, Alice understood that for her to view her own image, sunlight reflected from her body must travel to the window, reflect from the glass, and travel back to her eye. But there was more to it than that.

Alice waved to her friend Bob, a clerk working inside the shop. Bob returned the gesture. Noting that he must have also seen Alice's visible image, she concluded the window not only reflected her image back to her, but also transmitted her image through the window. So there were two images, one reflected and one transmitted.

This is food for thought, especially when one stops to consider that, as I'll explain throughout this book, all light and all of everything else appears to be made of tiny, highly unified bundles of energy. Call them *quanta*. They are this book's leading characters. Our pondering about Alice that morning on the square will quickly survey several fundamental principles, giving you a feel for the quantum world. Some details might escape you for now, but I'll return to them later, so don't worry about it.

The light that carries Alice's image is transmitted by the rapid motions of an electromagnetic field similar to the familiar magnetic field that surrounds every magnet and exerts forces on nearby iron objects. The word *field*, as used by physicists, is similar to its ordinary usage in phrases such as "baseball field" or "field of wheat." I'll define this important word in more detail later, but for now a field is a region of space that has certain physical properties. The electromagnetic field fills space the way smoke can fill a room. It's one of the universe's several *quantized fields*—meaning, its energy comes in the form of highly unified bundles or *quanta*, each carrying a *quantum* (from the Latin "quantus" or "quantity") of energy. In the case of light, these quanta are called *photons*. So the light that carries Alice's image is made of zillions (my word for a really large number) of photons.

The key point about quanta is that they are indivisible.[1] So because both Alice and Bob see Alice's image, some entire photons must be detected by Bob inside the shop, whereas the remainder are detected by Alice outside on the sidewalk. The fractions detected inside and outside are determined, as you might expect, by the properties of the glass window.

But there's some strangeness here: *What happens when one photon reaches the window?* Because it can't split into two, it must either be detected by Alice (indicating reflection) or be detected by Bob (indicating transmission).[2] What determines which way one quantum will go? The surprising answer: nothing determines this. The unified nature of the quantum entails, and experiments show, that *nature is fundamentally undetermined or "random."* In careful experiments in which laser light is directed at a high-quality partially reflecting glass plate, scientists can ensure all the photons are identical and the plate is perfectly uniform. Because every photon is then subjected to precisely the same conditions, you might expect them to behave identically when they hit the plate; yet, some transmit through the plate while others reflect back from the plate. Nobody, not even nature, knows which photons will go which way. Nature is fundamentally random.[3]

Before 1900, scientists thought identical conditions led to identical results. Well, they don't. Nature contradicts this plausible principle predicted by "classical physics" throughout the period 1650 to 1900, according to which the universe is precisely predictable the way an accurate clock is predictable. But nature is not like a clock. Nature herself doesn't know the future as she actively creates what happens at every instant to every quantum in the universe.

This is sufficiently odd that you might want to ask: How do we know? *This is always an excellent question.* It's basic to all of science. How do we know? Scientific knowledge, and odd scientific claims in particular, must be based on evidence.

Here's evidence for quantum randomness. A *Mach-Zehnder interferometer* (Figure 1.1) is a device for bringing reflected and transmitted light beams back

Introduction

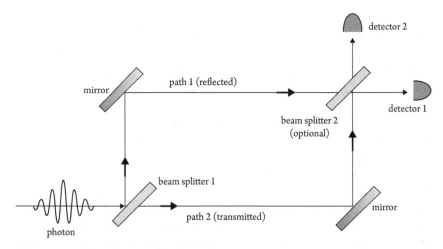

Figure 1.1 Mach-Zehnder interferometer. The length of either of the two paths between the beam splitters can be varied by e.g. slightly repositioning either mirror. This simple device can demonstrate most of the characteristic quantum phenomena.

together. In the figure, beam splitter 1 operates like the storefront window. It's a glass plate, seen edge-on in the figure, that reflects 50% of the light hitting it and transmits 50%. But instead of being perpendicular to the path of the incoming photons, this plate is positioned at a 45-degree angle to the incoming photons so that a reflected photon veers vertically along path 1 and a transmitted photon moves horizontally along path 2.[4] Each path is a meter (m) or two long and is laid out carefully on a lab table. Mirrors bring the light beams to a crossing point, where experimenters can, if they wish, place a second beam splitter.

Suppose a single photon enters at the lower left and beam splitter 2 is absent. What happens when the photon reaches the first beam splitter? Is it reflected, transmitted, or both? The experiment shows *it is always detected as a single, whole photon at either detector 1 or detector 2*. Detectors 1 and 2 could be the eyes of two different observers,[5] or both could be lab detectors that click, or give some other indication, when they detect light. So this experiment, without beam splitter 2, mimics the reflected and transmitted photons of Alice's and Bob's shop window observations. *The experiment verifies the existence of indivisible quanta*, because the detectors always detect either one photon or no photon, never some fraction of a photon. *The experiment also verifies quantum randomness*: when many photons are injected one by one at the lower left, the statistics of the impacts are absolutely random—meaning, the individual outcomes (photon impacts detector 1 or photon impacts detector 2) are as unpredictable as it's possible for them to be. You could watch photons hit one or the other detector all day long, but still have no idea of which detector the next photon will hit. Yet, all photons are prepared identically and subjected to identical conditions.

You might object that this single experiment does not prove absolutely that the "which path" choice is random. Perhaps a different beam splitter or a different experimental layout would produce a predictable outcome. Such objections are quite reasonable. In fact, experiments can never prove absolutely any general scientific principle, because you can't prove a general statement logically on the basis of individual cases. Experiments can only "verify"—support but not prove absolutely—a general principle, although a single compelling experiment that *contradicts* the general principle can *disprove* it. Science operates on the basis of what is reasonable in light of the observable evidence. Chapter 6 presents further evidence for the absolute unpredictability of this experiment.

With beam splitter 2 in place, something new happens. Beam splitter 2 mixes the two paths. If the photon was on path 1, it is (with 50–50 probabilities) either transmitted to detector 1 or reflected to detector 2; if the photon was on path 2, it is (also with 50–50 probabilities) either reflected to detector 1 or transmitted to detector 2. So no matter which path the photon takes, it ends up in either detector 1 or detector 2 with 50–50 probabilities. Thus, *the detectors now provide no clue regarding which path the photon took*. The mathematics of quantum physics imply and nature confirms that, whenever an experiment provides no information regarding which path a photon, or any other quantum, takes, *the photon takes all possible paths simultaneously*. Does this seem odd to you? If so, you're not alone.

This means each photon moves along both of the two separated paths! That's hard to believe. Only one photon was injected, and the detectors always detect precisely one photon, and photons never break into pieces, and yet I'm telling you the one photon is present on both paths? Come on! How can one photon go in two directions, and be in two places, at the same time?

Furthermore, this going-in-two-directions-at-once must happen regardless of the second beam splitter's presence or absence. After all, as the photon arrives at the first beam splitter, there is no way for it to "know" whether a second beam splitter lies "downstream." This means all the photons in the Fayetteville shop window went both ways, to Alice and also to Bob, but somehow each one is *detected* by only one of them, either Alice or Bob.

It would be premature pedagogically for me to present at this point the substantial evidence supporting all of this. That comes in Chapter 8. There's a brief technical explanation in the endnotes.[6]

This going-in-two-directions, called *superposition*, makes us question the whole notion, presented in an earlier paragraph, that light is made of unified bundles.[7] Perhaps each photon splits into two pieces at beam splitter 1, with one piece going each way. But the experiment provides evidence against such a suggestion. If light simply split at the first beam splitter, with part of it going each way, both detectors should click simultaneously when beam splitter 2 is absent. But both detectors never click, implying that only one object, not two, is present in the device at any time. There is no evidence that the photon ever splits.

Beware: Although photons and other quanta impact detectors one at a time like tiny particles, they are not particles. As you'll learn throughout this book, all quanta are spatially extended objects that *sometimes* behave like small particles. This notion helps us understand how one photon can take both paths through an interferometer: Each photon is an extended object, analogous to a puff of smoke, that spreads along both paths whenever both paths are available.

If you find this eccentric, I agree. Because science always tries to find the least eccentric explanation for phenomena (a simplicity principle known as *Occam's razor*), it took decades of experiments, discussions, and theorizing before these odd notions were accepted; in fact, some quantum fundamentals are still debated.

Werner Heisenberg, for example, was frustrated by the quantum's nonintuitive and apparently contradictory behavior. He writes, "I remember discussions with Bohr which went through many hours till very late at night and ended almost in despair; and when at the end of the discussion I went alone for a walk in the neighboring park I repeated to myself again and again the question: Can nature possibly be so absurd as it seemed to us in these atomic experiments?" On another occasion, Heisenberg was reduced to tears during the course of an intense conversation with Bohr.[8]

Eventually, odd explanations such as a single photon moving along two different paths were accepted because *every simpler explanation turned out to violate the observed evidence.* We scientists don't compete with each other to see who can come up with the oddest theory. However, evidence and logic often force us to such conclusions even though the process might be painful and even reduce us to tears.

Our tale now takes another devious twist. The behavior described here applies not only to light, but also to everything we know, including material objects—objects that, unlike light, have weight. Consider, for example, *atoms*—the smallest constituents of the chemical elements that still retain the characteristics of that element (an iron atom is the smallest piece of iron that still behaves like iron). Unlike photons, atoms and molecules (connected groups of atoms) are material objects. Yet, they do the same quantum tricks that photons do, tricks such as superposition and fundamental randomness. In fact, quantum physics also applies to macroscopic (visible without a microscope) objects such as baseballs and tables, although it's extremely difficult to create observable quantum effects in such large objects.[9] So far as we know, everything obeys quantum physics.

The material objects around you are made of zillions of atoms. Each atom has a really small nucleus at its center, with really, *really* small electrons near the nucleus, and each nucleus containing things called *protons* and *neutrons*. Because of its resemblance to the sun and planets, this simplified view of atomic structure is known as the *planetary model of the atom* (a "model" is a simplified explanation). Atoms are held together by electric forces; each proton is "positively charged," each electron is "negatively charged," and opposite electrical charges attract. So the nucleus attracts the electrons, holding them into their locations near the nucleus. Neutrons, on the other hand, are uncharged (i.e., electrically neutral).

I'll say it again. As far as we know, all these material objects—electrons, protons, neutrons, nuclei, atoms, molecules, and even baseballs and tables—obey the quantum rules, just as light does.

The most fundamental quantum feature of light and other electromagnetic radiation is that it is divided up or *quantized* into individual bundles called *photons*. Matter is also quantized. There are several kinds of material quanta. You've heard of some of them: electrons, protons, atoms, and molecules. Because we've all heard that matter is made of atoms, this notion of the quantization or bundling of matter is not surprising. The surprising part is that *these material quanta obey the same odd principles just discussed for photons*.

For example, University of Vienna physicist Anton Zeilinger has managed to coax molecules made of 60 or more carbon atoms into superpositions of following two paths simultaneously, like the photons that follow both paths through a Mach-Zehnder interferometer. By atomic standards, such a molecule is a sizeable hunk of matter. Each of its 60 atoms contains six protons, six neutrons, and six electrons, for a total of more than a thousand subatomic quanta. It's surprising that such a large and complex object can perform the trick of being in two places simultaneously. After all, scientists have never actually observed a baseball, or even the tiniest dust particle, to occupy two positions at once. Yet, this happens regularly in the microscopic world, and quantum physics predicts that it *can* happen, albeit with great experimental difficulty, even with objects as large as baseballs. In Chapter 8, we encounter an object sufficiently large to be seen (barely) by the naked eye that has been put into a superposition of moving in two distinct ways at the same time—the largest quantum superposition yet achieved.

There's a simple but surprising explanation (details in Chapter 5) for this odd phenomenon. *Reality is made of waves in unseen fields.* Quanta such as photons, electrons, atoms, and molecules are not "things." They are, instead, waves in fields, much as water ripples on a pond are waves in water. Superposition is common for waves. As a classical (nonquantum) analogy, imagine an unruffled pond with an eastern half covered by a thin layer of oil and a western half that is oil free. Imagine dropping a small stone into the western half, creating ripples that move outward (please note that water ripples are not a quantum phenomenon; they are a classical phenomenon that happens to mimic some aspects of quanta). It probably wouldn't surprise you if these ripples partially bounced off of the water-oil boundary and headed back westward, and also partially crossed over the boundary into the oil and continued heading eastward. The original ripple is now two ripples, in two places. This kind of thing happens all the time with waves, and it's neither surprising nor counterintuitive when you realize that we're talking about a water wave.

Our interferometer experiment is like this. *Each photon* (think of it as a ripple) *goes both ways upon reaching beam splitter 1*. What's more surprising and uniquely quantum is that, unlike water ripples, these ripples remain a *single* photon, but in two places. In the case of the water, the original set of ripples creates two sets of ripples traveling in two directions, as can be confirmed by observers on both the east and west sides of the pond. But in the interferometer experiment, we always detect only one photon, never two. A photon is a single thing that really can be in two places at once.

And matter behaves the same way. A material molecule is also a wave in a field and can be in two places at once. In fact, any object can, in principle, be in two places at once.

So there's a lot you can think about the next time you walk past a shop window.

The Amazing Quantum

Quantum theory is arguably the most wide-ranging, highly accurate, and economically rewarding theory of all time.

Regarding the first claim, although quantum physics deals primarily with small things such as photons and atoms, it has major implications for big things such as the universe. Physicists agree that the large-scale evolution and structure of the universe are correctly described by Albert Einstein's general theory of relativity. They also agree that the small-scale structure of matter and energy is governed by quantum physics. This leads to a problem. When scientists apply the general theory of relativity to certain phenomena involving small regions of space, they run into obvious errors related to the fact that general relativity does not incorporate the principles of quantum physics.

Black holes are an example. Stars are made mostly of the simplest types of atoms—namely, hydrogen and helium. They get their energy to produce heat and light from nuclear fusion, which is a process that converts hydrogen into helium. But stars eventually run out of their hydrogen fuel, leading to stellar death throes that can have several possible end points depending on the overall mass (weight) of the star. This end point is always a collapsed state resulting from the force of gravity squashing the star into a tiny fraction of its original volume. For the most massive stars, this collapsed state is a black hole, a state in which the general theory of relativity predicts that the star quickly collapses to zero volume. Does "zero" really mean absolutely zero? Yes, that's the prediction.

But this prediction turns out to contradict quantum physics, which has significant things to say about the state of the star after it's collapsed to atomic dimensions. So physicists try to incorporate quantum principles into the general theory of relativity. As you might imagine, when science is confronted with two such powerful theories, one geared toward the cosmic and the other toward the microscopic, the effort to combine them leads to amazing theories and speculations—theories that I won't delve into here.

A similar conundrum occurs when one ponders the origin of the universe. It all started with the Big Bang, a quantum event occurring 13.798 ± 0.037 billion years ago. The afterglow of that fiery origin is still observable by sophisticated devices that search far across the cosmos and, thus, far back in time (we're viewing light that's 14 billion years old, after all) to detect rays received on Earth from the Big Bang's aftermath. These rays indicate the Big Bang had an extremely microscopic origin dominated by quantum physics. But because the Big Bang involved the creation of enormous amounts of matter and energy, the principles of general relativity also played a huge role. So we again run into the difficult problem of combining general relativity and quantum physics.

There is a general consensus among physicists that we will eventually develop a single theory that combines general relativity with quantum physics consistently. After all, nature cannot logically contradict herself and the two theories, as they stand today, contradict each other when applied to the zillions of black holes in the universe and to the origin of the universe. There has got to be a way to combine these theories into one theory that approximates general relativity at the large scale and quantum physics at the small scale.

Such a "theory of everything" would be the ultimate tale of the quantum. It's expected to be a quantum theory—a theory that incorporates the main principles of quantum physics. If and when we discover it, it will, in principle, describe every physical thing—from the quarks that constitute protons and neutrons, to the cosmos.

Regarding the second claim in my list of superlatives, the accuracy of quantum physics is astonishing. Consider, for example, the light that comes from atoms. The quantum predictions are especially accurate for light from nature's simplest and most ubiquitous atom: hydrogen. Nearly every hydrogen atom has only a proton in its central nucleus, with one electron moving around it. Quantum theory predicts that all atoms emit light by transitioning between "atomic states," emitting a photon in the process. Here's what this means.

The quantum rules prescribe that the electron in a hydrogen atom must move only in certain "states of motion." There are many such *atomic states* for the electron, each of them having a particular and predictable quantity of energy, somewhat the way an electric fan can be in a state of running fast, running medium, or running slow.[10] (A small aside: I've used the familiar word *energy* several times without defining it, because most people have a general feel for energy that is sufficient for now. It's physics' most important word and requires a clear definition, which shows up in Chapter 3.) According to the theory, atoms emit light and other radiation by transitioning from a higher energy state to a lower energy state, emitting one photon in the process. Because the overall energy is always conserved (Chapter 3), the photon's energy must equal the energy lost by the atom in "dropping" to the lower energy state. It turns out (Chapter 5) that a photon's energy determines the wavelength (the distance from one crest of the wave to the next crest) of the wave of radiation emitted by the atom.

That's how quantum physics can predict the wavelengths of all the different colors of visible light and other radiations that can be emitted by any hydrogen atom. This assortment of wavelengths is called the visible *spectrum* (this word means "range," and usually refers to a range of wavelengths) of hydrogen. If you heat a box of hydrogen atoms until it glows, you "excite" many atoms into higher energy states, and these excited atoms continually release their energy by emitting photons with wavelengths that belong to the hydrogen spectrum.[11] That's why hot gas, such as the sun or a neon sign, glows.[12] If you then use a prism or other device to bend and separate the different colors (different wavelengths) of light, the amount of bending tells you the wavelength of each color. These wavelengths can be predicted, and measured, to many figures. For example, the wavelength emitted by a hydrogen atom in transitioning from its next-to-lowest to its lowest possible energy is predicted to be 0.000,000,121,568 meter. Atomic wavelengths are typically tiny, as you might expect for waves generated within a small object. This particular wavelength of hydrogen happens to be shorter than the range of wavelengths for visible light, and lies in the *ultraviolet* (higher than violet, more energetic than violet light) range. The measured and predicted wavelengths agree, to within experimental error. Such six-figure accuracy is typical of atomic physics.

The study of spectra is one of science's most fruitful endeavors, yielding reams of highly accurate information used in all sorts of sciences and technologies. This would be impossible without quantum physics.

The tale of history's most accurately verified scientific prediction comes from a field called *quantum electrodynamics*.

Every electron creates electric and magnetic effects in the space around it. One cause of the magnetic effects is that every electron has an intrinsic spin: just as Earth spins around a north–south axis, electrons always spin around some axis. It's a basic fact of nature that this spin is fixed in magnitude, the same for every electron, and can be neither shut off nor altered. Such motions of electrically charged objects always create magnetic effects (Chapter 4). The strength of a spinning electron's magnetic effects is expressed quantitatively by something called the electron's *magnetic moment*, a quantity that turns out to be easy to calculate except for one key factor called the electron's *g-factor*. Without worrying about the meaning of this g-factor, what's important for us is that it can be precisely predicted and also precisely measured, so it's an important tool for comparing quantum theory with experimental reality. It happens to be a dimensionless pure number (i.e., the same number in every system of measurement). To measure this property of the electron, experimenters must *suspend a single electron for months* in a strong magnetic field, quite an accomplishment in itself. It's most recent measured value is 0.001,159,652,180,73(28), where the last two digits, in parentheses, are uncertain.[13] So the electron's g-factor has been measured to 12 figures, and less certainly to 14 figures. And quantum theorists are actually able to calculate the electron's g-factor to 12 figures. The calculation involves tens of thousands of daunting individual calculations that have been carried out with increasing accuracy over a period of six decades. The result of this calculation is: precisely the same 12 digits!

It's uncanny. Humans can predict a particular 12-figure number. They can tell other people, "If you go into a lab and do such-and-such, you're going to discover this 12-digit number." And when people do such-and-such, they find that Mother Nature verifies the prediction. Thousands of years ago, some cultures were able to predict, roughly, the timing and the location along the horizon of the rising of the sun on the longest day of the year, as well as other celestial events. Stone Age structures such as Stonehenge in England, constructed in part as an observatory for such events,[14] testify that these predictions inspired great awe. The far more accurate and detailed predictions of quantum physics should be regarded in the same light, as awe-inspiring examples of the ability of humans to connect with and understand nature. I hope the tales told in this book inspire in you a similar astonished wonder.

As for being financially rewarding, today's entire world economy is linked to our understanding of the quantum. Electronic computers, transistors, lasers, and even the World Wide Web were all invented by physicists whose research lay mostly in the quantum realm.[15] Virtually the entire world economy is linked to these technologies. It's hard to put a price tag on all this, but it was said to be 30% of the US gross national product back in 2001,[16] and today surely reaches many trillions of dollars annually in the United States and tens of trillions of dollars annually in the world.

Quantum physics lies behind such applications as the transistor (which forms the basis of the information revolution), tunnel diodes, lasers, masers, fiber optics, X-ray machines, spintronics, synchrotron light sources, radioactive tracers, scanning tunneling microscopes, superconducting magnets, electron microscopes, positron emission tomography (PET) scans, X-ray computed axial tomography (CAT scans), magnetic resonance imaging (MRI), superfluid liquids, nuclear reactors, nuclear bombs, nuclear magnetic resonance, nuclear medicine, radioactive tracers, microchips, lasers, and semiconductors—just to name a few.

The Quandary about Quantum Fundamentals

It's surprising that, more than a century after the quantum's birth, quantum fundamentals are still in dispute. At the 2011 "Quantum Physics and the Nature of Reality" conference organized by Anton Zeilinger, 27 physicists, 5 philosophers, and 3 mathematicians responded to a prepared questionnaire with 16 multiple-choice questions covering major issues in quantum foundations. Conference participants disagreed widely about several fundamental principles. Meeting organizers summed up the poll by saying, "There is still no consensus in the scientific community regarding the interpretation of the theory's foundational building blocks. Our poll is an urgent reminder of this peculiar situation."[17]

The problem is, it's not easy to figure out what the equations and words of quantum theory actually mean. The formal mathematical theory is more abstract, and more difficult to interpret concretely, than other physical theories. Many scientists even question whether the theory describes the real world at all or is, instead, simply a useful mathematical prescription for predicting experimental outcomes. As you'll see, there have been many suggested alterations and interpretations of the theory.

Today the disputation centers around at least three issues: wave–particle duality (Chapter 5), nonlocality (Chapter 9), and the measurement problem (Chapter 10). A few decades ago, quantum randomness would have been on this list of still-controversial topics; Einstein and many others thought the indeterminacy of radioactive decay and other quantum phenomena was only apparent and that there was a yet-undiscovered theory, beyond quantum physics, that would show the future to be fully determined by the present. Although a few physicists continue searching for such a deterministic explanation of nature, most accept the enormous evidence for indeterminacy (Chapter 6).

Regarding wave–particle duality, you'll see that the successes of quantum field theory—the theory that unites quantum physics with Einstein's ideas about space and time as stated in his special theory of relativity—has convinced such leading quantum theorists as Steven Weinberg and Frank Wilczek that the universe is made of waves in fields rather than particles in empty space; both the scientific consensus and the evidence now point in that direction.[18]

Regarding nonlocality, Albert Einstein and coworkers, in 1935, were the first to point out that quantum physics predicts this phenomenon.[19] But nonlocality was, in Einstein's view, so "spooky" that he concluded quantum physics must be superseded by some more "complete" theory that was not spooky—not nonlocal. You'll see in Chapter 9 that experiments since 1970 show that nature actually exhibits the nonlocality predicted by quantum physics. Thus there is increasing acceptance, approaching a consensus, of quantum nonlocality.[20]

The measurement problem, aka "Schrödinger's cat," remains a central issue, in my view the only significant outstanding quantum foundational issue. Some experts regard it as unresolved, others as resolved, and still others as a pseudo-problem that needs no resolution. Most agree that physicists dearly need to find consensus about this. We tackle it in Chapter 10, where I present new arguments supporting a solution first suggested in 1968 by Josef Jauch and developed by others since that time. Developments since the 1960s, especially our deepened understanding of the nonlocal phenomenon called "entanglement" (Chapter 9), shed new light on the true meaning of Schrödinger's cat.

Quantum physics is mostly about things we can't see. Its basic principles are surprising, to say the least, and, during 115 years of quantum history, there has been an unending succession of scientific disagreements about its fundamentals. So it's not surprising that a lot of quantum-inspired *pseudoscience*—misleading distortion of the scientific process presented so as to appear scientific although it lacks supporting evidence and rational plausibility—has raised its ugly head. In a world in dire need of scientific literacy, pseudoscience is exactly what we don't need.

Let's look at a few examples. A popular 2004 film *What the Bleep Do We Know!?* grossed $10 million and won several film awards. Its central tenet is that we create our own reality through consciousness and quantum physics. It purports to show how thoughts change the structure of ice crystals. It interviews a 35,000-year-old spirit "channeled" by a psychic, and features physicists saying things like, "The material world around us is nothing but possible movements of consciousness."[21]

The popular television physician Deepak Chopra informs us that quantum healing can cure our ills by the application of mental power.[22] Chopra's New York Times Bestseller *Ageless Body, Timeless Mind: The Quantum Alternative to Growing Old* sold more than two million copies worldwide.

One of the many quantum interpretations put forward by physicists seeking an escape from the measurement problem is that human consciousness explains how a quantum state can be finally "collapsed" (Chapters 9 and 10). But there is no evidence that human consciousness plays the slightest role in this process. This notion that consciousness can control external physical events is cut from the same cloth as the self-proclaimed psychic Uri Geller's claim to bend spoons through sheer mental power. It begs the question of how quantum states collapse in the rest of the universe where there are no humans, or how quantum states could collapse before humans existed, or whether a baby or a smart ape can collapse quantum states. Eugene Wigner, an excellent physicist who should have known better, held this view for awhile, as did the great mathematician and quantum foundations analyst John von Neumann. Wigner abandoned his view in 1970, 10 years after adopting it.[23]

Nevertheless, the discredited "consciousness" view of quantum physics keeps coming up. A widely used textbook published in 2006 and used in liberal arts physics courses at the University of California and elsewhere bears the title *Quantum Enigma: Physics Encounters Consciousness*.[24] The book continues to use Wigner's outdated views as evidence for the consciousness interpretation of quantum physics, despite Wigner's rejection of this view.[25] Although rational thought should dismiss this notion on its face, in Chapter 10 I'll describe a published experiment that takes the trouble to directly and convincingly disprove the consciousness interpretation.

Quantum-inspired pseudoscience supports a variety of amazing but questionable claims from extrasensory perception to alternative medicine.[26] It's not a good day for science when book store managers and librarians wonder whether a particular book should be shelved under new age, religion, or quantum physics.

Tales of the Quantum presents an optimistic outlook—namely, that the overall quantum framework is fine as it is and that the apparent paradoxes are resolved or resolvable within that framework. We should certainly reject pseudo-scientific distortions; furthermore there's no apparent need for fundamental revisions or eccentric interpretations of any kind. Science's most fundamental theory is in better shape than its detractors imagine.

PART 1

THE UNIVERSE IS MADE OF QUANTA

2

What Is Quantum Physics About?

This chapter describes the general nature of *quantum physics*, often described as the science of matter and energy on the smallest scales. Although most quantum physics is done at molecular, atomic, and subatomic scales, this size-based definition is oversimplified because quanta aren't necessarily small; they can, in fact, be as big as Earth (Chapter 5). And two or more "entangled" quanta (Chapter 9) can be considered a single composite quantum, the parts of which are intimately unified and capable of influencing each other instantaneously even if they inhabit different galaxies. So this book proceeds from the following definition:

As far as we know, everything in the universe is made entirely of quanta. Quantum physics is about the nature and behavior of these fundamental constituents of the universe.

What, then, is a quantum? It's not an easy question. Quanta are subtle, counterintuitive, and not entirely understood. We do, however, understand this question better than we did in 1951, when Albert Einstein wrote to a colleague: "All these fifty years of conscious brooding have brought me no nearer to the answer to the question, 'What are light quanta?' Nowadays every Tom, Dick, and Harry thinks he knows it, but he is mistaken."[1]

Having written a groundbreaking paper in 1905 that, for the first time, recognized the idea of the quantum, Einstein knew whereof he spoke. His paper concerned a simple phenomenon called the *photoelectric effect*, in which light falling on a metal causes the metal's surface to eject electrons. By assuming the light is made of small bundles or "particles," Einstein was able to explain the quantitative details of this phenomenon. The first mathematical hint of quanta came from Max Planck in 1900, but Einstein's paper was the first to present the quantum as a physical object. Einstein's 1921 Nobel Prize was awarded for this 1905 paper rather than for his more significant theory of special relativity (1905) and theory of general relativity (1915). By 1921, it was obvious that Einstein deserved a Nobel Prize, but his work on relativity was still controversial so the Nobel Committee instead awarded him the prize for his widely accepted work on the photoelectric effect. He remained involved with quantum physics all his life, mostly as a perennial, sharp-witted critic who provided invaluable commentary on problematic issues.[2]

The word *quantum* derives from "quantity." As a working definition, *a quantum is a highly unified, spatially extended, specific quantity of field energy.* This book is devoted to understanding this mouthful and fleshing out its ramifications. Photons, electrons, protons, atoms, and molecules are examples of quanta. Think of a quantum as a bundle of energy. But the term *bundle* can be misleading, because a single quantum can be spread thinly over a region many kilometers across and can even comprise tenuously connected pieces separated by cosmological distances—the way two pancakes in a frying pan can be connected by a thin line of batter. Quanta are "digital" in the sense that a quantum, even if it's many kilometers wide, either exists entirely or doesn't exist at all. You can't have a part of a quantum, and you can't create a quantum gradually or destroy a quantum gradually.[3] A quantum always comes into existence, or vanishes, or changes its configuration, instantaneously everywhere. When you turn on a light, each spatially extended light quantum (each photon) pops into existence all at once, and when light vanishes, each entire quantum pops out of existence all at once, although the photon might be kilometers wide. Furthermore, quanta sometimes jump instantaneously[4] from one configuration to an entirely different configuration. This all-or-nothing quality has much to do with the quantum's considerable strangeness.[5]

Quantum physics answers the question "What are quanta and how do they behave?" Chapters 2 through 5 discuss the first question and Chapters 6 through 9 discuss the second. Chapters 10 and 11 discuss how these counterintuitive quanta lead to our normal world.

The Universe Is Made of Quanta

Some people say everything is made of atoms, but today we know that's far from true. Many things—cabbages, kings, your foot—are made of atoms. "Atoms" are the smallest recognizable parts of the roughly 100 fundamental chemical "elements" that you may have noticed listed in charts posted in chemistry classrooms. For instance, the element hydrogen is a gaseous substance with its smallest recognizable (i.e., having the properties associated with hydrogen) parts being hydrogen atoms.

But many things aren't made of atoms. Light is one example. Radio, infrared, X-rays, and other unseen *radiations* (so called because they are emitted outward, along a *radius*, from a central source), similar to light but invisible, are not made of atoms. Other things not made of atoms include electric currents, the magnetic field surrounding every magnet, and Earth's gravitational field. Protons, neutrons, and electrons are not made of atoms, although atoms are made of them. Inside each proton and each neutron lie three quarks that are not made of atoms.

The famous Higgs bosons, discovered in 2012 at the Large Hadron Collider in Geneva, are not made of atoms.[6] The trillions of neutrinos coursing harmlessly through your body during this second (some of which are by now, 2 seconds later, out beyond the orbit of the moon, even if they had to pass through our entire planet to get there) are not made of atoms.[7]

Quantitatively more important than any of these are the invisible *dark matter* and *dark energy* comprising 95% of the universe's energy. Although this dark sector is nearly undetectable, it's all around you and it fills the universe. We didn't know it existed until the late 20th century, although a few astronomers suspected dark matter's existence as early as the 1930s. We still don't know what either one is, although we might be closing in on dark matter. We do know that neither dark matter nor dark energy is made of atoms. Because they are probably made of quanta, and because most of the universe is made of them, we'll ponder them a little while.

But hold on. Ninety-five percent of the energy of the universe is not made of atoms, and it's all around us, and we can't see it? How do you know I'm not making this up? Such an odd assertion needs evidence. You should be asking (and if you were, my hat's off to you): Whaaat? How do we know? What's the evidence? Scientists are born skeptics.

To talk about this, we need some new concepts. *Matter* refers to material that has weight. Normal matter (cabbages, kings, your foot) is made of atoms and molecules. Atoms and molecules have an important similarity to cabbages: You can put them in front of you, at rest, subject only to inevitable quantum random vibrational motion (Chapter 6). A concept related to weight—namely, mass—clarifies this. The *mass* of a material object is a measure of the force (the push) needed to accelerate it (to speed it up) starting from rest. It's measured in kilograms. So a 2-kilogram object requires twice as much force to accelerate (by some given amount) as a 1-kilogram object. This difficulty-of-accelerating an object is also known as the object's *inertia*. Surprisingly, light and other radiation have no mass, no inertia. A quantum of radiation, a photon, is infinitely easy to accelerate. No force is required to get it moving. We know this because every photon ever observed was moving at light speed. We never see a photon speed up from zero to light speed. The instant a photon is created by, for example, switching on a light bulb, it's moving at light speed.[8] So the distinction between matter and radiation is that the first has mass and the second doesn't.

The most direct evidence for dark matter comes from observations of galaxies.[9] Galaxies, including our own Milky Way galaxy, are huge collections of stars held together by gravity, usually shaped either like potatoes or like thin pancakes. There are hundreds of billions of galaxies in the observable universe, each typically containing hundreds of billions of stars. Early evidence for dark matter came during the 1930s from the study of clusters of galaxies bound together by the attractive pull of gravity. Some clusters were held together more tightly than astronomers could explain by estimating the gravitational pull of all the glowing, visible matter in the galaxies. Astronomers hypothesized that a new, nonglowing form of matter helped hold the clusters together.

This hypothesis of an invisible form of matter seems radical, but the alternative is even more radical. If there were not some new nonglowing form of matter, then not only Newton's law of gravity but also Einstein's widely accepted general theory of relativity would have to be wrong. This is an important lesson in the scientific process: Scientists don't dream up odd ideas on a whim or to be stylishly radical. The observed facts force these ideas on scientists, because the alternative would be even more radical.

Astronomer Vera Rubin improved on this evidence during the 1970s by studying individual galaxies. Disk-shaped galaxies spin around their center, like a phonograph record, and are held together by the gravitational attraction between their stars. Rubin found, however, that some galaxies were spinning so fast that the stars should have spun out of their roughly circular orbits around the galaxy's center, the way that a speeding car can fly off the highway while negotiating a tight curve. Something in addition to gravity from the galaxy's visible stars must be holding stars in their orbits. Again, nonglowing matter seems to be the best answer.[10]

The best evidence for dark energy comes from observation of the expansion of the universe. Ever since the universe's Big Bang origin 13.8 billion years ago, everything has continued flying outward. The galaxies are moving (on average) away from each other, away from our galaxy, and away from every other galaxy.[11] This expansion of the universe has been verified by numerous observations that measure the speeds at which galaxies move outward, and how far away they are. This book is not the place to go into the fantastic ways astronomers have of measuring these things (see any astronomy textbook).

During the 1980s, two teams of astronomers used a new way of measuring the expansion speeds and distances of very distant galaxies based on the light received from exploding stars, called *supernovas*, within those galaxies. Occurring only once every few decades in a typical galaxy, each supernova outshines its entire galaxy for a few days or weeks after its explosion and is visible across, essentially, the entire known universe. This work determined distances and recession speeds of the supernovas, and therefore of the galaxies that contained them, at huge distances away from us. Anything that's a long distance away is seen as it appeared a long time ago, because it takes a long time for the light to get here from there. For example, we see the sun as it was 8 minutes ago because that's how long it takes sunlight to get here. It could have blown up 7 minutes ago for all we know. So a sufficiently distant supernova, one for which light requires say 12 billion years to get here from there, would be viewed as it appeared a mere 2 billion years after the Big Bang, when the universe was young. The universe is a kind of one-way time machine; we're always peering into the past.

A shocking new result emerged from the two teams' supernova measurements. Not only has the universe been expanding during all of its preceding 13.8 billion years, the expansion is currently accelerating! Everybody had known since 1929 that the Big Bang caused the universe to expand. Everybody also "knew" that this expansion must be slowing down for the same reason that a rock thrown upward must slow down as it rises as a result of the gravitational force pulling backward. The only question seemed to be: How rapidly is the expansion slowing? But science is risky. New observations can throw a monkey wrench into old theories. There was a new fact. The expansion is speeding up.

This was revolutionary. I recall reading about it in 1998 when *Science* magazine announced it. Although earlier hints had suggested the new result, I was astonished, as was the entire scientific world. How could the universe be accelerating, when nothing seems to be pushing on it?

But scientists immediately took the accelerating universe seriously because it was supported by real evidence. The acceleration had been measured painstakingly for several years by two experienced groups working independently of each other. The new result percolated quickly through the scientific community, becoming *Science's* story-of-the-year. Within 2 years, astronomers had accepted the revolutionary new fact.[12]

There's a lesson here. Because scientific conclusions can be inconvenient for one's cherished beliefs, a segment of the US population is skeptical of science. Some skeptics buttress their arguments with the notion that it's scientists, rather than those who cling to nonscientific beliefs—who are stodgy and slow to change when new facts emerge. For example, "creationists" have convinced themselves that the theory of evolution survives only because biologists cannot bear the thought of changing their cherished 150-year-old ideas. But scientists accept evolution because there continues to be massive evidence for it and no evidence against it. True, there have been cases when scientists were irrationally slow to accept new evidence,[13] but the story of the accelerating universe shows that science can accept new theories quickly when the evidence warrants it. The scientific process doesn't work perfectly, but it works.[14]

A universe that slows as it expands requires an initial burst of outward-pushing energy, such as the Big Bang, to set it into motion. When it gets moving, however, it can continue to expand without further pushing, for the same reason that a baseball continues flying toward the outfield even after the batter has finished slamming it.[15] But because the universe is speeding up rather than slowing down, well-established physics implies something must push on it. Although we don't know what this something is, we gave it a name: *dark energy*.[16]

Both dark matter and dark energy are encoded in the high-energy radiation that comes to us from the Big Bang. Very little of this radiation has been absorbed, so it still fills the universe. Where else could it go? This radiation is still all around us, but it's now stretched (because of the expansion of the universe) into long-wavelength microwaves similar to, but much less intense than, the radiation that warms your food in a microwave oven. Using sophisticated microwave detectors aboard orbiting satellites, astronomers have recorded the intensity of this radiation in different directions and throughout the entire microwave spectrum. These waves carry information about the Big Bang's hot spots in some directions and cold spots in other directions, information telling us the universe was stretched and squeezed during the Big Bang in amounts that could be accounted for only if the universe's energy is apportioned into 68% dark energy, 27% dark matter, and only 5% normal matter—numbers that are widely accepted today.[17] Normal matter is not so normal after all!

Finally, this section's title, "The Universe Is Made of Quanta," is far more accurate than "The Universe Is Made of Atoms," but it needs two qualifications. First, there is a good chance the dark energy is the so-called *quantum vacuum*. As we'll see in Chapter 5, although the quantum vacuum forms an essential part of quantum physics, and although it contains energy, it cannot itself be said to be "made of quanta." It's the other way around: quanta are ripples in the quantum vacuum field that fills the universe. Second, it's possible that quantum physics will need modification to account for dark matter and dark energy. The notion that dark matter and dark energy fit into the quantum format is entirely theoretical at this point: Everything else obeys quantum physics, so why not dark matter and dark energy too? I'd place big odds that quantum physics will meet this challenge, but as philosopher and Yankee catcher Yogi Berra put it, "it's tough to make predictions, especially about the future." After all, most of us (me too) also thought the universe's expansion was slowing down.

On the Trail of the Quantum: The Double-Slit Experiment

The quantum is simple but subtle. I've mentioned several examples of quanta: photons, electrons, atoms, molecules, and Higgs bosons. But what are these things? Are they simply little particles, like a tiny pea or dust particle only smaller? There's an easy answer: No. *Particle* is a popular but poor word choice for these things and I try not to use it. Most physicists, despite their frequent use of the word *particle*, know this. Quanta are quite unlike tiny peas or dust particles. Quanta sometimes act like particles, but mostly they don't.

Light furnishes a familiar example. It's all around us, but it's hard to say what it is. Plato thought our eyes emitted invisible rays that moved toward the objects we see.[18] Isaac Newton thought illuminated objects sent out streams of particles that enter our eyes,[19] whereas Newton's contemporary Christian Huygens thought illuminated objects sent out waves that enter our eyes. The nature of light is one of science's oldest questions; the quest to understand it led to the two pillars of modern physics—namely, Einstein's relativity and quantum physics.[20]

Huygens was right: light is a wave. Plato was far wrong. Newton was partly right: light waves are, indeed, made of small quanta called photons. However, these photons are not the small isolated particles, similar to tiny bullets, that Newton had in mind. Quanta are more subtle than that. We're going to approach them step by step. Here, I focus mostly on the wave aspects of light, saving the quantum details for Chapter 5.

Here is the tale of how we know light is a wave.

Figure 2.1 is a photograph looking down from above at the water surface in a so-called *ripple tank experiment*. The photo shows small water waves illuminated from the side so that wave crests appear bright and valleys appear dark. From the left, long straight waves approach a barrier with two small openings. Waves pass through the openings and emerge on the right side of the barrier.

Two phenomena, characteristic of all waves and crucial in quantum physics, are noteworthy. First, on passing through the openings, waves spread out on the far side of each opening into what might be considered the shadow region above and below each slit. This spreading, called *diffraction*, turns out to be more pronounced for narrower openings.

Second, waves from the two openings overlap in such a way as to *interfere* beyond the openings. That is, the two sets of ripples add up to make large ripples at certain places, and cancel each other to make no ripples at other places. As you can see, the places where ripples cancel form straight lines of nearly flat water leading outward from the region of the two openings, and the places where ripples add form lines of large waves leading outward from the same center.

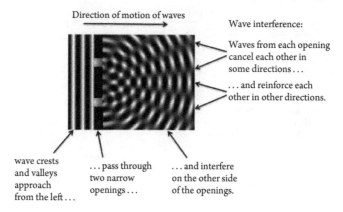

Figure 2.1 Water wave interference, looking down on a ripple tank experiment with waves passing through two openings.

Here's why this happens. Each opening, acting separately, sends out a semicircular set of ripples; the ripples from the two openings pass through (or ride over) each other without disturbance, so the water is raised doubly when two crests meet, lowered doubly when two valleys meet, and both raised and lowered for a net zero change (no raising or lowering) when a crest meets a valley. This phenomenon, in which two sets of waves reinforce or cancel each other alternately, is called *wave interference* or, simply, *interference*. It can't happen without two or more sets of waves and it is guaranteed evidence of the simultaneous presence of at least two sets of waves on the water's surface. Interference proves to be a key quantum phenomenon. It's also the key to demonstrating that light is a wave.

One of the most beautiful physics experiments of all time was first performed by Thomas Young in 1801.[21] It reappears throughout *Tales of the Quantum* in guises he couldn't have imagined. Young showed that light from two sources exhibits wave interference. Figure 2.2 pictures the experiment, but slightly simplified, as explained in the endnotes.[22] Light shines through two long narrow slits cut in an opaque partition. A viewing screen detects the light after it has passed through the slits.

Figure 2.3 shows the image made by the light on the viewing screen. This photo was made by placing photographic film at the position of the receiving screen, as shown in Figure 2.4. We see many long bright lines, interspersed with many long dark lines. To explain these lines, let's return to Figure 2.1 and imagine an observer looking only at the far right boundary of the ripple tank. This observer sees large waves arriving at several points on the boundary, interspersed with points on the boundary where no waves arrive. Imagine this ripple tank experiment extended into the third dimension, coming out of the

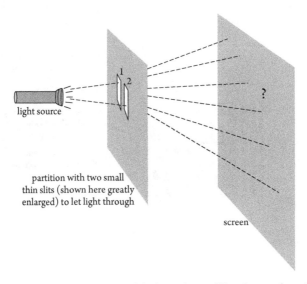

Figure 2.2 Double-slit experiment with light. What will be observed on the viewing screen?

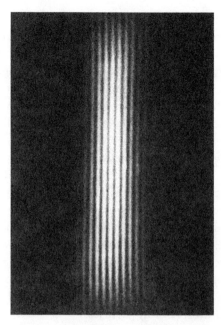

Figure 2.3 Experimental result of the double-slit experiment with light.

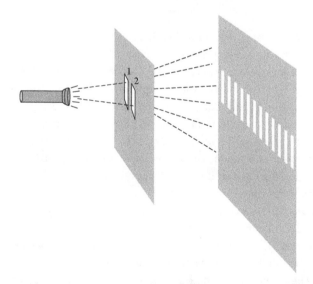

Figure 2.4 Double-slit experiment with light: experimental setup and result.

page so the two openings in Figure 2.1 become the two slits of Figure 2.4. The interference pattern of large waves coming into the boundary interspersed with no waves is analogous to Figure 2.3, with bright lines that are zones where big light waves arrive at the viewing screen, and with dark lines that are zones where no waves arrive.

In other words, Figure 2.3 is the three-dimensional analog of the two-dimensional interference of surface waves in water shown in Figure 2.1. It's exactly what you would expect to see on the viewing screen if waves of light were coming through the two slits. Each long line of light in Figure 2.3 is a region of "constructive interference," where crests coming from one slit meet crests coming from the other slit, and valleys meet valleys; each long dark line is a region of "destructive interference," where crests from one slit meet valleys from the other.

Recall that, before Young's experiment, Newton and Huygens had disagreed with regard to whether light was made of waves or particles. If light were a stream of particles flying out of the light source in Figure 2.2, some particles would pass through slit 1 and a roughly equal number would pass through slit 2. Those passing through slit 1 would proceed straight ahead and pile up on the viewing screen directly behind slit 1, whereas those passing through slit 2 would pile up behind slit 2, so we would see two long bright bands of light, with each band centered directly behind one of the slits. That's not what we see in Figure 2.3. The evidence shows a spread-out interference pattern, and this convinced scientists that light is made of extended waves, not tiny particles.

And that, my friend, is how we know light is a wave.

Science depends on careful observation. Let's ponder Figure 2.1. *Although waves (crests and valleys) are moving toward the right, the water surface itself is simply vibrating up and down.* Please check this in a bathtub or some other tub of water. First, allow a single drop to fall into completely still water. The circular ripple that results is a good example of a wave. Now, float a small cork or other object in still water and allow one drop to fall nearby. The cork rides up and down, remaining in one spot as the ripple passes underneath. As this demonstrates, the water merely vibrates up and down, while the wave moves outward from the drop's impact point. The water itself does not move outward.

So what is a water wave?

(This is a pause, for pondering.)

Here's one decent definition: *a water wave is a disturbance*—an alteration of the normally flat condition of the water surface—*that travels through the water*. The disturbance (the shape—a series of crests and valleys) travels along the water surface, but the water itself remains in place, simply vibrating up and down.

All waves, such as waves sent down a rope, "the wave" in a sports stadium, compression waves sent along a slinky toy, and sound waves, behave like the water wave. Every one is a disturbance traveling through some substance that does not itself travel with the wave. This substance is called the *medium* for the wave. Remember that the motion of the medium is quite different from the motion of the wave. The wave travels but the medium stays in place and merely vibrates.

Ponder this: What is the medium for a rope wave? A sports wave? Slinky waves? Sound waves?

(*More time for pondering.*)

Because you can't have a disturbance without something to disturb, each of these waves must have a medium. For a rope wave, it's the rope. For a sports wave, it's the bodies of the fans that "vibrate" just one time by standing and then sitting. For a Slinky wave, it's the coils of the Slinky; note that this one differs a little from the others, because the coils' vibrations are back and forth, in line with the direction of motion of the wave, instead of up and down, perpendicular to the wave's motion. And for sound waves, the medium is air. Although you might know this fact about sound from some science course, it's not so obvious from observations. There's elegant evidence for it in many science museums that exhibit a bell clanging loudly from within a glass-enclosed chamber from which the air is then removed gradually. As the air thins, the sound quiets, eventually turning to silence although you can see the clapper still vigorously, but silently, hitting the bell.

What, then, is the medium for light waves? There was a lot of debate about this after Young's experiment. The medium cannot be air, because light travels through outer space, from the sun to Earth for example, where there is no air.

The answer leads us to a new kind of physical object, one that is the key to most of modern physics: *The universe is filled with unseen but quite real entities called "fields."* Chapter 4 is all about fields. A familiar example is the magnetic field that fills the space around every magnet. A magnet's field exists at any place where another magnet (or any material, such as iron, that is magnetized easily) would, if put at that place, feel a force from the first magnet. Another example is Earth's gravitational field that causes rocks and other objects to fall when dropped. It exists at any place, including places far from Earth, where a dropped rock would fall toward Earth.[23] You can't see magnetic or gravitational fields, but you know they exist because of the effect they can have on objects placed in the field. Physicists think of a field as a property of space. For example, the space near our planet's surface has at least two field properties: Earth's magnetic field (as evidenced by the operation of magnetic compasses) and Earth's gravitational field (as evidenced by falling rocks).

As we see in Chapter 4, magnetic fields are best thought of as a component of a more comprehensive field called the *electromagnetic, or EM, field*. This EM field, like all fundamental fields, fills the universe. *The universal EM field is the medium for light waves.* Light is very much a quantum phenomenon, and the EM field strictly obeys the principles of quantum physics. So light forms a prime example throughout this book.

The most fundamental principle of quantum physics is that *there are just a few different fundamental fields, each of them filling the entire universe, and all of them are quantized.*[24] That's a mouthful, to be sorted out in Chapters 4 and 5. The two key concepts needed for quantum physics are fields and quanta. Quantized fields combine both concepts. For now, a *quantized field* (or quantum field) is a field that is made of indivisible bundles, or *quanta*, of field energy.

So Huygens was right. Light is a wave, as demonstrated by Young's double-slit experiment. Evidence presented in Chapter 4 shows that light is a wave in the EM field. Newton was partly right, too: because the EM field is quantized, light comes in bundles called *quanta*. But as we see next, quanta are nothing like the small, isolated, bulletlike "particles" Newton had in mind. Although they sometimes behave like particles, quanta are not particles at all.

Quanta Versus Small Particles

Let's travel through space–time to ancient Greece, where the foundations for many of the western world's great ideas were laid. One Greek, Democritus (460–370 BCE), was a born physicist, although neither the word nor the field of study had yet been invented. He was also a born philosopher, a term that still means lover of wisdom. His tale represents, in a very general way, the beginning of quantum physics. He invented the world's first known "thought experiment," an imagined experiment that is simple in principle but might be difficult in practice. Imagine cutting a chunk of gold in half, then cutting those halves in half, then cutting *those* halves in half, and so forth. Could you continue this process forever, or would you eventually come to a point where no further cutting was possible? Is matter divisible without limit or is it made of indivisible parts?

To Democritus, the first alternative seemed impossible. If subdividing goes on forever, we eventually arrive at pieces of gold so small as to be imperceptible—in other words, indistinguishable from nothingness. The gold would have ceased to be perceptible, which seemed absurd. The cutting process must, he thought, eventually end at indivisible objects. He called them *a-tomos*, a Greek word meaning "not able to be cut" or "without parts."

But he didn't leave it at that bit of intuitive philosophizing; he offered physical evidence based on his own experience, making his work not only philosophical but also scientific. His personal philosophy was that everything that happens is caused by matter and the motions of matter, a principle known as *philosophical materialism*. So Democritus' evidence took the form of material causation. Consider, he suggested, a loaf of bread baking in an oven. Even from a considerable distance and out of sight of the bread, a person can detect fresh-baked bread by its smell. For Democritus, such a phenomenon must have a physical cause that starts within the bread. Some of the bread's a-tomos must emerge from the baking process and drift through the air to people's noses. Bread must give off a-tomos that are characteristic of bread; other a-tomos must emerge from violets to give them their characteristic odor, and similarly for all odorous objects. When these a-tomos get into our noses, said Democritus, they cause us to perceive the appropriate odor. Even from our modern vantage point, this was a decent explanation of odors. Democritus was probably the first recorded thinker to provide clear evidence of this sort for scientific theories.

We use the word *atom* somewhat differently today from the Greek word *a-tomos*. Belying its Greek name, the modern atom is actually made of several parts—namely, a nucleus and orbiting electrons. There are roughly 100 different kinds of atoms, known as the *chemical elements,* such as hydrogen, helium, gold, and uranium; they are listed in the periodic table seen in textbooks and classrooms and are often represented by standard abbreviations such as H for hydrogen, He for helium, Au for gold (*aurum* is Latin for gold), and U for uranium. These different elements are distinguished by having different numbers of protons in their nuclei, with the number of electrons usually equal to the number of protons. For example, all hydrogen atoms have 1 proton in the nucleus, all helium atoms have 2, all gold atoms have 79, and all uranium atoms have 92. Each element has different chemical properties because it has different numbers of electrons, and it's these electrons that actually determine an element's chemical properties. Odor is a chemical property, but it's too complicated to be concocted from a single atom. Most odors are specific "compounds," with their smallest parts or *molecules* made of tens of hydrogen, carbon, and oxygen atoms.[25] Thus, molecules are composites of two or more atoms. Specific molecules are indicated by the elements they comprise and the number of atoms of each element. Thus, H_2O is the water molecule, made of two atoms of the element hydrogen and one atom of the element oxygen.

Democritus had it about right in some respects. His big idea, that the universe is made of small pieces that cannot be divided endlessly into smaller pieces, is correct today according to quantum physics. But today's ultimate pieces, now called *fundamental quanta*, are not much like Democritus' a-tomos. They are neither indestructible nor unchangeable as Democritus and, later, Isaac Newton thought. However, they retain a key quality that Democritus insisted on: a fundamental quantum is not made of separable parts and always acts as a single unit.

Electrons, photons, and other quanta, such as the six kinds of quarks, seem to qualify as basic indivisible constituents of the universe. As far as we know, we can't subdivide an electron, a photon, or a quark. It must be noted, however, that any quantum can be destroyed and its energy used to create one or more other quanta; but these other quanta are not parts of the original quantum. For example, a photon can vanish and create in its stead a pair of material particles such as an electron and a so-called *antielectron*, but the electron and antielectron are new things, not parts of the original photon.

Everything appears to be made from just a few kinds of fundamental quanta. For instance, all protons and all neutrons are composite objects, made of three quarks. All atoms are composites made of quarks and electrons. In Chapter 5 there is a table listing all the known fundamental quanta, from which every other known thing is made. (Dark matter and dark energy are not yet "known things.")

Since Democritus, there has been a debate about whether the universe is made ultimately of indivisible particles moving separately through otherwise empty space or made of one or more "fields" that fill all space. Is space almost empty or is it full? Democritus was definitely a particles guy, as you can tell from this renowned quotation:

> By convention sweet is sweet, by convention bitter is bitter, by convention hot is hot, by convention cold is cold, by convention color is color. But in reality there are a-tomos and the void [empty space]. That is, the objects of sense are supposed to be real and it is customary to regard them as such, but in truth they are not. Only the a-tomos and the void are real.[26]

From our modern perspective, there's some truth to both views but more truth to the fields view. Chapter 5 provides evidence that everything is made of quanta that can be described consistently as waves in unseen fields that fill the universe. There is no known consistent way to describe the experimental evidence in terms of a theory that is based entirely on particles. A theory based on both fields and particles might be logically possible, but it goes against the principle of Occam's razor: Why should the universe be built from two such different kinds of fundamental entities?

Quanta are not much like Democritus' particles. For one thing, Democritus' a-tomos, and Newton's presumed particles as well, were permanent and unalterable. But quanta are quite impermanent and changeable. All quanta can be created instantaneously from other forms of energy, and destroyed instantaneously. For another thing, all quanta can take on a variety of configurations, called *quantum states* (Chapter 7).

For another big thing, contrary to Democritus, *there is no empty space*. Even in the far reaches of the universe, places like the enormous regions between galactic clusters where atoms seldom tread, the gravitational field, EM field, and other fundamental fields all fill every cubic millimeter simultaneously. All these fields are always present everywhere.

This sounds like a confusing plethora of different fields, but the ultimate hope is that all these fields really amount to a single, unified field present everywhere, with a nature and behavior that explains everything we see around us. Such a unified field was certainly Einstein's hope, and it's alive and well today in the vision of a unified quantum field theory of everything.[27]

Regions such as the space between galactic clusters contain mostly vacuum plus just a few actual quanta. But surprisingly, *in modern physics, the word* vacuum *no longer means nothingness*. In modern physics, a physical vacuum means a region devoid of all quanta, but this does not amount to nothingness. In fact, *every quantum is simply a wave, or disturbance, within physically real vacuum fields that contain energy and are not "nothing."* Thus, vacuum fields are the wellspring of everything else, for without them quanta could not exist. We'll see (Chapters 5 and 6) that quantum physics requires real vacuum fields even in regions where there are no quanta at all. "Nothing" is impossible!

We'll never find a region of space with nothing in it. If we managed to remove all the fields, including even the vacuum fields, from some region of space, that region would (according to quantum physics) simply have to vanish from the physical universe. This is because the quantum notion known as Heisenberg's principle demands that every quantum field be present everywhere, at least in the form of the quantum vacuum. If any fundamental field were absent from some region, then this field would have a value—a strength or magnitude—of zero throughout this region, and this value has no quantum randomness (uncertainty), contrary to Heisenberg's principle. Thus, at least the minimal field known as *vacuum fluctuations*—random fluctuations around zero—must be present everywhere. Quantum physics entails that every nook and cranny of the universe is filled with quantum fields.

We will delve further into fields versus particles and the nature of quanta in Chapter 5.

The Quantum Idea: Nature Is Digital

In 1894, prominent American physicist Albert Michelson remarked there were no more fundamental discoveries to be made. Quoting a contemporary physicist, Lord Kelvin, he stated, "An eminent physicist remarked that the future truths of physical science are to be looked for in the sixth place of decimals."[28] Many scientists shared Michelson's and Kelvin's confidence that the known grand principles—namely, Newton's laws, the principles of thermodynamics (energy, "heat," and temperature), and the principles of electromagnetism—were now known and would not change.

But just 6 years later, in December 1900, such matters began to change. Radically new physics was discovered, although it wasn't recognized at that time as radical or even new. At a meeting of the German Physical Society, Max Planck described his analysis of the radiations given off by heated objects. To appreciate Planck's idea, some background is needed.

Microscopic matter never rests. Even within solid materials that seem to be at rest, individual atoms and molecules jiggle incessantly and "randomly"—in a disorganized, unpredictable fashion. These microscopic motions are a reflection of the material's temperature; in fact, "random motion of atoms and molecules" is what we mean, at the microscopic level, by temperature. Although this connection between temperature and molecular motion seems surprising when you think about it, it's been verified by many experiments. One such experiment is observations of tiny pollen or dust particles suspended in water or some other liquid. In a microscope, the pollen particles can be seen to dart this way and that way ceaselessly as a result of impacts caused by the unseen submicroscopic motions of water molecules. As scientists increase the water's temperature, this microscopically visible darting of pollen particles increases in a manner that can be predicted quantitatively from the assumed random motions of the underlying molecules.[29] This random molecular motion resulting from temperature is called *thermal motion*.

As we'll see in Chapter 4, it's a fundamental principle of electromagnetism that vibrating electrified objects must radiate (emit in many directions) EM energy. Because molecules contain electrified protons and electrons, and because molecules are in constant back-and-forth thermal motion in all materials, we conclude that every material radiates EM energy constantly, and that this *thermal radiation* generally increases with temperature.

A hot plate used for cooking, for example, glows brighter and brighter as its temperature increase. Also, its color changes from dark red to brighter red and, eventually, would become white hot if heated to sufficiently high temperatures. This white light contains all the visible colors, the entire spectrum of visible wavelengths of light (more about this in Chapter 4). Stars, fires, and incandescent light bulbs all glow for the same reason. In fact, cooler objects such as this book and your body "glow" in a similar manner, but with a lower energy kind of EM radiation the eye cannot detect called *infrared*. All normal matter emits such thermal radiation. It's a general and fundamental physical phenomenon, so one would expect that its details would be explainable quantitatively in terms of basic physics principles.

Many physicists during the 19th century searched for such an explanation. Using the known principles of thermodynamics and electromagnetism, they hoped to predict correctly the amount of energy radiated at each wavelength by an object at any particular temperature. But all explanations came up short. One attempted explanation, based on 19th-century principles of thermodynamics and electromagnetism, predicted that the amount of radiated energy increases enormously for shorter wavelengths. It's true, as we will see in Chapter 4, that shorter wavelength radiation has higher energy, but the predicted energy was so large that short-wavelength, high-energy thermal radiation should blind us every time we look at a fire or an incandescent bulb![30] Because this doesn't happen, that particular theory must be wrong.

Enter Max Planck. Like others, he assumed a glowing object's emitted radiation comes from vibrations of its atoms.[31] Desperate to solve a certain equation that followed from this assumption, he found the mathematics was easier if he assumed the amount of energy emitted by an atom to be restricted to certain, specific possible amounts rather than being allowed to have any arbitrary value from a continuous range of energies. To put this another way, Planck specified that the amount of energy radiated by an atom could range only over a "discrete set of values" (analogous to 1, 2, 3, . . .) rather than a "continuous set of values" (analogous to "any number between 0 and 1"). He tried this only as a mathematical ploy, planning to solve the equation and then to allow the increments between allowed values to shrink to zero to find the solution for the continuous case. His plan worked in part. Using energy increments, he got a solution that approximated the experimental results, but when he allowed the increments to shrink, the solution again predicted ridiculously large energies at short wavelengths. Despite his best efforts, he couldn't make these energy increments go away. He was able to derive the correct experimental results only by choosing the increments to have a particular size.

The resulting formula predicted correctly the amount of energy radiated at every wavelength by a glowing object at any particular temperature. Planck's increments proved essential—in fact, revolutionary. Today, we say the energy emitted by an atom is *quantized*—meaning, *restricted to a discrete set of values*. This changed everything.

Here is an analogy to clarify the notion of quantized energies. The analogy goes beyond what Planck actually stated and lays out the full picture of quantization as we understand it today.

Imagine swinging in a child's swing. If you keep pumping your legs to move the swing, you will swing higher and higher on each oscillation. If you then stop pumping you "die down" gradually and continuously (my Dad called it "letting the old cat die" when he pushed my childhood swing) to lower and lower maximum heights. When you swing higher, each back-and-forth oscillation has more energy; for example, you move faster when you come back through the center point. When you swing lower you have less energy. You can have any amount of energy from zero up to some upper limit beyond which swinging becomes dangerous.

But suppose your swinging energy was *quantized*—restricted to just a few specific values. Let's suppose there are just five such values and they allow you to reach maximum (end point) heights of 2 meters (abbreviated m), 1.5 m, 1 m, 0.5 m, or 0 m (a condition of rest) above the lowest point. How would the cat die in this case?

(*Please ponder....*)

It would be jerky. Our imaginary quantized swing would have to reduce its energy in sudden, in fact, *instantaneous* (happening in zero time) jerks. You would swing for a few oscillations at a 2-m maximum height without losing any energy, then, instantaneously, switch to smaller oscillations at a 1.5-m height, where you would continue for several more oscillations, then switch instantaneously to a 1-m height, and so forth. After several such "quantum jumps" to lower and lower energy values, you would find yourself at rest at the lowest point. *The jumps between allowed energy values would have to be instantaneous*, because our assumed quantization rule says the energy of the swing is restricted to just the allowed values. If your transition from one energy value to the other took any time at all (0.001 second, for example), your energy during the transition would lie between two of the allowed values, and our assumed quantization rule doesn't allow this.

As you might have heard, energy is never created or destroyed (Chapter 3). So when the swing reduces its height instantaneously from 2 m to 1.5 m, the corresponding loss in energy must show up somewhere else as a gain in energy. In other words, the swing must emit an instantaneous burst of energy. So as the cat dies in instantaneous jumps, the quantized swing releases energy in instantaneous bursts. This large-scale example of quantization is mostly fantasy, although physicists have observed quantum jumps in a tiny but macroscopic vibrating "diving board" (Chapter 8). Quantum jumps are common in the microscopic world. The bursts of energy are called *quanta*. The swing's energy, and the energies of the bursts, are said to be *quantized*.

Quantum physics applies, so far as we know, to everything, including even playground swings. But the quanta for a real playground swing are much smaller (in energy) than I have portrayed them. They are so small that you'll never notice your swing reducing its energy in tiny increments. For a real-life swing, the old cat dies smoothly, and quantum physics makes no practical difference, because the energy of one quantum is so tiny relative to the energy of any macroscopic object.

For an atom, on the other hand, these small bundles of energy are comparable with the energy of the atom itself and they make a big difference. Planck's assumption of quantized energies meant that atoms in a heated object would not radiate continually but would, instead, radiate tiny bursts of energy intermittently.

It's an odd concept. Why would energies be restricted in this manner? Planck had no idea. We still have no idea.[32] It's the way the universe is—that is, it's a fundamental principle of physics that we simply have to assume without further explanation. Planck himself was skeptical of this assumption, thinking he was just fudging the mathematics to come up with the right answer. Little attention was paid to this detail for the next 5 years. Who cares if energy is emitted continually or in a stream of tiny bursts? But just as one small move can bring down a house of cards, you can revolutionize physics by slightly changing a small but fundamental assumption. Planck's tiny bursts of radiated energy are now called *photons*. Planck's hypothesis revolutionized physics so profoundly that physicists now use the word *classical* to mean "before quantum physics."

The basic idea of quantum physics is simple. Energy is created, transferred, altered, and destroyed only in unified bundles—quanta—rather than gradually. Such changes must occur in instantaneous tiny energy increments. In the language of information technology, a quantity that can take on only specific values is *digital*, because the different possible values can be counted (using the 10 *digits* 0, 1, 2, . . . 9 and their combinations), as contrasted with a continuously variable or *analog* quantity.

It's a common misconception to believe that quantization implies energy is distributed spatially as small, isolated bundles separated by empty space. This is wrong. Energy fills space smoothly and continuously, but when energy is added, transferred, or removed from a region of space, or when the form or "state" of the energy within some region is altered, one or more entire quanta of energy must be added, transferred, altered, or removed, and this must happen instantaneously. Energy is digital, not analog.

The Fragility of Quanta, and Measurement

It stands to reason that the microworld is easily disturbed. It's made of tiny things having tiny masses, so a tiny push can alter them radically. To get a feel for this, compare a block of iron with a block of plastic foam. A small hammer tap hardly affects the iron, but it sends the foam sailing. Find a chunk of foam filler and try it. What you're experiencing here is a difference in the inertia of the two objects, measured quantitatively as *mass*. An atom or other quantum is touchy, difficult to observe accurately, because mere observation can send it sailing. We could observe an atom by, for instance, shining light on it; but light is made of photons and each photon's interaction with the atom can alter the atom. Because experimentation is central to all science, objects that jump around in response to every observation are likely to prove difficult to investigate scientifically. Humans intervening in the quantum world are like bulls in a china shop: everything we do disturbs the pottery.

Although *measurement*, *observation*, and *detection* are recurring words in this book, they can be misleading. These words can give the misimpression that quantum physics is simply about laboratory measurements rather than about the universe. In standard English usage, a "measurement" is the human process of gathering data about some object. If we follow this usage, then a "quantum measurement" would have to mean the human process of gathering data about some *quantum* object. Historically, the term was used in quantum physics for many decades in just this way, but this turns out to be a far too narrow definition. Without delving into the history at this point, it turns out that this definition of "measurement" as a *human* process led to the misconception that humans have something fundamental to do with the quantum principles of the universe. It might have been best simply to stop using the word *measurement* and find some other term instead, but this word is now so widely used that it's impossible to remove it.

So the best resolution is to broaden the meaning of the term. By a *quantum measurement*, we mean *any process in which a quantum phenomenon causes a macroscopic change*, regardless of whether a human is involved in causing or observing the change. The term *measurement* is perhaps appropriate here because any human who *did* observe such a "measurement" *could,* by observing the macroscopic change, learn something about the quantum phenomenon that caused it.

Here are three examples of measurements: a proton from space strikes and moves a sand grain on Mars, a high-energy photon from a distant supernova melts an ice grain in Antarctica, and a photon passes through a Mach-Zehnder interferometer (Figure 1.1) and one of the detectors clicks. In all three cases, a quantum phenomenon (a proton from space, a photon from space, a photon traversing an interferometer) causes a macroscopic change (a sand grain moves, an ice grain melts, a detector clicks). It matters not whether a human observes it.

Just a few months before his untimely death in 1990, quantum physicist John Bell wrote an article titled "Against Measurement," advocating banning the term *measurement* because it carries the false connotation that quantum physics is connected in some special way to laboratories and humans. Bell knew quantum physics has no more to do with humans than do, say, the laws of gravity. Although I'm not banning the term from this book, I use *measurement* only in the nonanthropocentric sense stated here. I can only hope Bell would have approved.

Measurement, especially the related problem known as "Schrödinger's cat," is often considered paradoxical. However, we'll see in Chapters 10 and 11 that measurements are not paradoxical at all. They're just subtle.

Our notion of scientific objectivity is dependent on the permanence of the objects measured by scientists. Science is considered "objective" or "the same for all observers" as distinguished from "subjective" or "observer dependent," because any number of scientists can observe any particular phenomenon and their observations will be, if not identical, at least consistent. If Alice and Bob measure the mass of a large, motionless rock, they will get consistent results unless somebody made a mistake. One will not find the rock's mass correctly to be about 2 kilograms whereas the other finds it correctly to be about 5 kilograms. Even if Alice alters the rock, she could tell Bob about the alteration and he could take this into account.

But quanta are delicate and are changed easily in unknown ways. If Alice and then Bob study a particular quantum object (an object for which quantum effects cannot be ignored), Alice might alter it significantly without knowing how, or even if, it was altered. It's difficult, then, for Bob to verify that his observation is consistent with Alice's, or even that he observes the same object. Thus, it's difficult for measurements to determine the nature of quantum objects. In fact, it's been difficult for scientists to agree about how the quantum measurement process even works.

However, one aspect of quantum measurement leads to scientifically objective agreement. Whenever a quantum object makes an impression on the macroscopic world, that macroscopic impression can be observed and agreed on by all. For example, the audible click of a detector or a visible flash on a screen is an objective scientific fact, regardless of the fragility of the quantum world. Thus, our point of departure for comprehending the microscopic world must be the impressions quanta make on the macroscopic world, and the experiments that demonstrate these impressions. Such experiments keep discussions of quantum physics grounded in real facts, in contrast to meaningless abstractions. In this book I try to hew closely to such experiments.

Quanta are fragile in a more important respect than size or mass. They are highly unified, so that a small interaction (a small exchange of energy) at one location can alter instantaneously and radically an entire extended quantum. Despite being extended spatially, a quantum is a single thing, not made of parts.[33] You cannot alter a quantum at just one place. Whatever happens to it happens to the entire quantum. It's as though the bull in the china shop necessarily shattered instantaneously every pot it touched. *This unity of the single quantum is the ultimate source of quantum oddness.*

Planck's idea of energy quantization was the seed from which quantum physics flowered. By 1930, Max Planck, Albert Einstein, Erwin Schrödinger, Werner Heisenberg, Niels Bohr, Louis de Broglie, and Paul Dirac, to name only a few, had developed this seed into a rigorous but enigmatic theory. This theory continues making precise predictions and passing every experimental test, but it also continues to be enigmatic and is questioned frequently on grounds of logical consistency and because of its counterintuitive implications. The elders of quantum physics, as pioneers of the post-Newtonian realm, were strangers in a strange land. Chapter 3 discusses the Newtonian physics that formed the scientific background for these pioneers, which remains an essential background for understanding the quantum.

3

Particles and Classical Mechanics

Physics as we know it today began during the 17th and 18th centuries with Galileo and Newton. They initiated *classical physics*, or prequantum physics, comprising Newtonian mechanics, thermodynamics, electromagnetism, and related topics. It presumes Democritus' much older idea that matter is made of zillions of small, separate, solid, changeless particles moving through empty space, but by the 19th century the empty space becomes filled with an extended entity called the *electromagnetic field*. Classical physics never really goes out of style because it's roughly correct in typical macroscopic situations. For instance, it's widely used for constructing large engineering structures such as bridges and buildings. I'm devoting this and the next chapter to classical physics because this background makes the quantum easier to grasp.

This chapter deals with Newton's principles of matter, motion, force, and gravity; the principles of energy and thermodynamics that appeared considerably after Newton; and aspects of Einstein's special theory of relativity, including the famous relationship between mass and energy. Although Einstein's special and general theories of relativity are central to modern physics, the quantum makes no appearance in these theories.

Newtonian Physics

Galileo Galilei (1564–1642) was an independent and revolutionary thinker. His sarcastic critiques of Aristotle's respected (by the Catholic Church, in particular) ancient Greek physics, and his support for sun-centered Copernican astronomy against the Earth–centered view, got him in big trouble with the Pope. Science, because it accepts only evidence and reason, goes frequently against the conventional grain. Science does not accept authority as sufficient reason for reaching conclusions. So beware: Science can be dangerous to your beliefs. And, as Galileo discovered, this can, in turn, be dangerous for your health, especially if you live in a time or place of fervent irrational belief.

Galileo fathered modern experimental science. Experiment, understood broadly as observation or experience, is what makes science revolutionary.

Galileo was the first to experiment systematically in search of new principles. "Try it and see." It's a simple rule, and it's the heart of science.

Galileo performed a series of experiments in which a ball rolled along level surfaces or down smooth, straight inclines. These experiments taught him to appreciate friction and air resistance. He asked: What would happen to a ball rolling on a level surface if friction and air resistance were absent entirely? Although he couldn't obtain such perfect conditions, he found an experiment that provided the answer. He rolled a highly spherical ball down a smooth, straight incline and then back up a second equally sloped incline (Figure 3.1). He noted the ball rolled nearly as far uphill as it had rolled downhill, and could see that if friction and air resistance were removed these two distances would be exactly equal. Thus, in the absence of friction and air resistance, the uphill and downhill motions were exactly symmetric. This symmetry allowed Galileo to extrapolate mentally to the case of motion along a straight, level track with no slope at all, as explained in the caption to Figure 3.1. When a ball starts moving along a level track with no air resistance or friction, it would have to continue moving while neither speeding up nor slowing down. It would move at unchanging speed in a straight line, forever![1]

As shown in Figure 3.1, a ball released at the top of the incline C will roll up incline D just as fast and far as it rolled down C. As we consider longer and longer inclines,[2] we approach the situation shown in E and F, where the "incline" is not inclined at all and the ball, once started, would have to keep moving forever at an unchanging speed.

The notion that *an object will keep moving unless there's something to stop it* was on the minds of other intellectuals at the time. It was quite contrary to Aristotle's physics, according to which heavy objects moving horizontally naturally slow down and stop. Although Galileo's experiments clinched the matter, others such as philosophers René Descartes and Thomas Hobbes also formulated this principle, now known as the *law of inertia* or, more boringly (and besides, Newton didn't discover it) as Newton's first law.

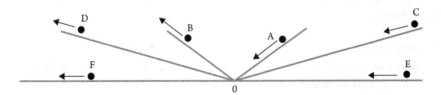

Figure 3.1 Galileo's experiment with a ball on an incline. Assuming no friction or air resistance and no loss of speed when the direction changes, a ball released at the top of incline A will roll up the equally-sloped incline B just as fast and as far as it rolled down A. A ball released at the top of the incline C will roll up incline D just as fast and far as it rolled down C. As we consider longer and longer inclines, we approach the situation shown in E and F, where the "incline" is not inclined at all and the ball, once started, would have to keep moving forever at an unchanging speed.

A digression into the scientific process: *law* is a poor word choice. The annoying habit of calling scientific principles "laws" originated during the 16th and 17th centuries and continued through the 19th century. The word is too rigid. The ideas of science are *principles* or *theories* or, if they are highly tentative, *hypotheses*. *Law* seems to connote an idea set in stone. Scientific ideas are definitely not set in stone. They are true only to the extent they are verified by experience and rational thought, and they always risk being disproved by new experiments. But because nobody refers to "Newton's principles," I'll go along with the crowd and stick with *law* in this chapter.

The law of inertia was a profound insight. It's hard today to appreciate the difficulty, in Galileo's day, of conceiving this notion. An educated person would have responded, "What? It keeps moving steadily in a straight line? All by itself? That's impossible! What would keep it moving?"

And people couldn't conceive of the absence of gravity. It was thought that falling objects were free of all external influences, and they fell because they had an internal natural tendency, a "desire," to seek out Earth's center. This was Aristotelian physics, and it felt right to people. It took a Newton to understand gravity as an external influence exerted by Earth on every object. After Newton, humans began to regard gravity as an external influence rather than an internal tendency—an influence from which one could escape by going sufficiently far from Earth. Thus, for Newton, a sufficiently distant object would not fall toward Earth; it would instead continue moving in a straight line at unchanging speed.

The law of inertia is the key to Newtonian physics and a key to the postmedieval Age of Enlightenment during the 17th and 18th centuries. Although it came well after the death of Copernicus, the 16th-century inventor of the sun-centered theory of planetary motion, it provided crucial support for Copernicus' notion that Earth doesn't sit still at the center of everything but instead moves in an orbit around the sun. Aristotle would have objected, asking, "What keeps Earth moving?" The new Newtonians would have answered, "It keeps moving because that is its natural tendency." In other words, because of the law of inertia.

Twentieth-century historian of science Herbert Butterfield provides this commentary:

> Of all the intellectual hurdles which the human mind has confronted and has overcome in the last fifteen hundred years, the one which seems to me to have been the most amazing in character and the most stupendous in the scope of its consequences is the one relating to the problem of motion, the one which ... every schoolboy learns to call the law of inertia.[3]

There's a lesson in the foregoing tale. The law of inertia seemed unbelievable when Galileo suggested it, and even more unbelievable when Copernicus, a century earlier and without entirely realizing it, suggested it indirectly in his sun-centered astronomy. It's quite counterintuitive. Think about it: How can our huge planet keep moving when there's nothing to push it? It's odd but true, just as quantum physics is odd but true. The quantum seems so counterintuitive that some scientists actually prefer to believe that quantum physics is merely a useful mathematical fiction. But to quote evolutionary biologist J. B. S. Haldane, "My own suspicion is that the universe is not only queerer than we suppose, but queerer than we *can* suppose."[4] The lesson of the law of inertia is that strange things can turn out to be true. As you will see, all the counterintuitive quantum phenomena are logically consistent and well supported by experimental evidence. Their apparent oddness is no reason to doubt their physical reality.

Isaac Newton (1642–1727, born the year Galileo died), took the law of inertia and ran with it. Once you conclude that, in the absence of external influences, objects keep moving in a straight line at unchanging speed, it follows that changes in an object's speed or direction of motion mean there are external influences on it. Physicists call such changes *accelerations*, a broadening of the common use of the word, which normally refers only to increases in speed. Physicists include slowing down and turning (changing the direction of motion) as accelerations, because these also are caused by external influences or, as they are called, external *forces*. Newton formulated a mathematically precise second law relating the external forces on any object to that object's acceleration, and a third law stating that the forces any pair of objects exerts on each other are always equal in magnitude and opposite in direction. I'll leave it to other books, including my nonmathematical textbook for non-science college students,[5] to present the countless wonderful implications of Newton's laws.

Most famously, Newton discovered his *law of gravity*, an algebraic formula saying that any two objects having mass must attract each other gravitationally, and that the strength of this gravitational force is larger for objects having greater mass and smaller for more widely separated objects. This was a historic breakthrough because, in addition simply to recognizing that there *is* a gravitational force, it expanded the notion of gravity from something that happens only on Earth to something that happens everywhere in the universe. Distant stars all exert gravitational forces on each other. Apples fall to Earth because Earth's gravity pulls them downward—a new way of looking at falling. Gravity holds planets, the solar system, stars, and galaxies together. The tides rise and fall because of the gravitational force on Earth exerted by the moon and sun. The law of gravity implies that astronomical bodies are, in a sense, falling through space. Because the only force acting on it is gravity, the moon is falling, but it's not falling downward. Its orbital motion, which was initiated long ago when Earth and the moon were formed, causes it to fall *around* Earth rather than downward. In the same way, Earth and other planets fall around the sun, held into their elliptical orbits by the sun's gravitational pull. As French poet and philosopher Paul Valery put it, "One has to be a Newton to see that the moon is falling, when everyone sees that it doesn't fall."[6]

Suddenly, everything fit together. Newton, a proficient mathematician, showed that, as logical consequences of the laws of gravity and motion, the planets must move in elliptical orbits around the sun. Similar reasoning explained the motions of all planets and moons, not to mention most of the motions then known. Newton's laws applied not only on Earth but also in the heavens where different rules had been thought to apply. This suggested, to many thinkers, new philosophies.

These pioneers ushered in our postmedieval scientific era. Newton's physics unveiled a rational consistency throughout the universe and inspired the 17th- and 18th-century Age of Enlightenment—a philosophical development that valued scientific thought, skepticism of conventional opinion, rejection of superstition, and opposition to intolerance, and that frequently challenged the Catholic Church.

Newtonian physics reigned for two and a half centuries. Its first problems didn't appear until late in the 19th century when, as we've seen, difficulties arose concerning the EM radiation emitted from heated objects—difficulties that were solved by Planck in 1900. Also, physicists encountered problems concerning motion that were solved ultimately in 1905 with Einstein's special theory of relativity. During 1900 to 1930, quantum physics along with Einstein's relativity largely replaced classical physics as our most fundamental understandings of how the universe works. But classical physics was by no means discarded. It remains a useful approximation for "normal" (to us) situations—sizes much larger than atoms, speeds much slower than light speed, gravitational forces much smaller than the forces near a black hole, and distances much smaller than a galaxy. In fact, some general classical principles, such as the law of inertia, conservation of energy, and the second law of thermodynamics (described later), are quite valid universally.

A Clockwork Universe

In a view similar to Democritus' statement about a-tomos and the void, Newton stated:

> It seems probable to me that God in the beginning formed matter in solid, massy, hard, impenetrable, movable particles ... and that these primitive particles being solids are incomparably harder than any porous bodies compounded of them, even so hard as never to wear or break in pieces.[7]

For Newton, "particles" meant small objects separated by empty space.

There's a machinelike quality to Newton's physics. For a classical physicist, a physical *system* (this word means a certain portion of the universe, such as an automobile, a set of pool balls, an electron, a molecule, Earth, or the solar system) is made of a finite number of particles, each particle located at a specific, but possibly changing, point in three-dimensional space. Certain specific forces act on each particle, perhaps exerted by other particles in the same system, or from outside the system. The game is then to use the principles of Newtonian physics to analyze and predict the behavior of the system by analyzing the motions of its particles. One applies Newton's laws to each particle, leading to a network of equations linking each particle's acceleration to the forces exerted by the other particles and also exerted by the "environment"—namely, the rest of the universe outside the system. Solving these equations yields predictions of the particles' positions at any later time, based on knowledge of their positions (and, as it turns out, their velocities) at some earlier time. In this analysis, precise causes (forces) act on precisely specified objects (particles) to produce precise effects (motions). It's all precise and predictable, like a well-oiled machine or a perfect clock.

The two-body gravitational problem is a simple but significant example. One imagines a system of two objects small enough that their spatial extension can be ignored. The only force is assumed to be gravity acting between the two objects. At some initial time, the two objects are at specified positions, moving with specified speeds in specified directions. This models the Earth–sun system, with the gravitational effects of other planets and moons ignored, with the spatial extensions of the sun and Earth ignored (because these objects are small in comparison with their separation), and with nongravitational effects such as solar radiation ignored. The general solution was worked out by Newton, along with every physics undergraduate today (it's easier since Newton invented calculus![8]). Both bodies move in elliptically shaped orbits (paths), provided that neither body is moving fast enough initially to eventually escape the gravitational pull of the other. Because the sun is so much more massive than Earth, it turns out the sun's ellipse must be far smaller than Earth's ellipse. Approximately, the sun stands still while Earth orbits it in an ellipse, in agreement with what was known at that time about Earth's motion around the sun.

A clock works in a similarly predictable way, and in fact the force of gravity is both the driver and the timer for a weight-driven pendulum-timed grandfather clock. Predictability is the essence of a clock. You start a grandfather clock by setting the hands and lifting the weights, and it runs itself until the weights have descended. Most educated people, for 250 years, believed the universe ran according to Newtonian physics. Although it's more complicated than a clock, a universe made of small particles obeying classical physics would run itself just like a clock. In such a universe, identical causes produce identical effects.

Like a grandfather clock, the mechanical universe works out its own future. Nothing you or I do can alter it in the least, for our bodies are made of particles that are part of the universe and the entire future of every particle was determined long ago by natural laws. French mathematician and astronomer Pierre-Simon Laplace put it this way in 1814:

> We may regard the present state of the universe as the effect of its past and the cause of its future. An intellect which at a certain moment would know all forces that set nature in motion, and all positions of all items of which nature is composed, if this intellect were also vast enough to submit these data to analysis, it would embrace in a single formula the movements of the greatest bodies of the universe and those of the tiniest atom; for such an intellect nothing would be uncertain and the future just like the past would be present before its eyes.[9]

The predictability of Newtonian physics presents philosophical difficulties such as the problem of free will. We like to think that we decide what we will do, but if everything, including each person's brain, is made of particles obeying classical physics, then we are not free to decide to scratch our nose or to compose a symphony. The universe decides for us, based on Newtonian physics. And if we decide, like Fyodor Dostoevsky's Underground Man, to assert ourselves by beating our head against a stone wall, it's really the universe that also decided on this supposed self-assertion.

René Descartes and others responded to the problem of free will with a dualistic philosophy according to which there are two interacting realities: an objective primary reality described by classical physics, and a subjective secondary reality of our thoughts and feelings. Determinism was thus a primary reality whereas free will became a mere secondary reality.

Today, there are reasons to question both classical physics and its philosophical implications. Quantum physics is one big reason. Classical physics is known to be only approximately correct within a limited range, and is far too limited to permit conclusions concerning free will. Quantum physics implies fundamental randomness, which contradicts the notion that identical causes produce identical effects, and arguably leaves room for some sort of free will (Chapter 6). It also implies space-filling quanta that are typically "entangled" with each other in such a way that their behavior contradicts a universe made of separate machine-like parts (Chapter 9).

Contrary to the classical universe, the quantum universe is anything but mechanical. Three key features of the classical worldview are, first, atomism: as Newton put it, "solid, massy, hard, impenetrable particles" that "never wear or break in pieces" form the basic reality. Second, predictability: once it got started, everything had to evolve precisely as it has evolved. Third, analyzability into parts: we can understand the universe by understanding the motions of its component parts. Quantum physics brings each of these mechanistic principles into question. First, the quantum universe is not made of permanent unmalleable particles, but instead is made of impermanent and highly malleable quanta the mere observation of which can change them radically. Second, identical causes no longer lead to identical effects. Future events, such as radioactive decay or chemical reactions, cannot be predicted, even in principle, regardless of how much we may know about the present. Third, the analytic process assumes it's possible to separate a phenomenon into component parts without changing it, but quantum nonlocality challenges this notion: many quantum systems are entangled so that it's not possible to think of them as independent objects, no matter how widely separated they may be.

Despite the breakdown of classical mechanistic physics, I hope to convince you in this book that the more general principles of the Age of Enlightenment do endure in the quantum age. Although the new rules are radically nonmechanical and the universe is not at all like a clock, the quantum universe nevertheless runs itself according to precise, rational, consistent rules.

Is there a quantum worldview? People's worldviews are mostly acquired thoughtlessly, as part of the cultural air. The mechanistic Newtonian worldview surely remains influential today, even (in fact, especially) among those who have never heard of Isaac Newton, and even though that worldview's founding principles are dubious today. No grand new metaphor, such as the clockwork metaphor for classical physics, has yet emerged to represent quantum reality.

The Four Fundamental Forces

Look around and spot some of the many forces—pushes and pulls— acting on various objects: Marie lifts a cup, a leaf flutters in the breeze, an apple falls to the ground, an airplane lifts into the air, a beef roast settles down against a spring scale. Surprisingly, all such common forces can be traced to just two: gravity and electromagnetism.

The roast pressing against the scale is a typical "contact force"; but at the microscopic level, what exactly is contact? Atoms have an outer halo of electrons. Electrons exert repellant electrical forces on each other, forces that become larger as the distance between the electrons becomes smaller. As two atoms come closer together, their electron halos begin to repel each other. The roast settles down onto the scale's metal surface until this repellent force between the scale's surface atoms and the roast's surface atoms becomes strong enough to hold (counterbalance) the roast's weight (the force of Earth's gravity pulling downward on the roast). At that point the settling stops, the electron halos in the two surfaces are distorted, the roast is at rest on the scale, and the overall or net force on the roast is zero because the upward contact force by the scale on the roast balances the downward gravitational force by Earth on the roast. All contact forces are traceable, in this manner, to EM forces acting between atoms.

Marie's fingers exert contact forces on the cup. Air molecules exert contact forces when they impact the fluttering leaf. Because of the way the airplane's wing is shaped, air molecules impact the bottom of the moving airplane's wing more strongly than the top, providing lift that counterbalances the force of gravity on the airplane. Falling apples are different, and this is why Newton was so interested in them. No contact forces act on a falling apple, so there is no counterbalance to the downward pull of gravity, which is why the apple falls.

All common forces are reducible either to EM or gravitational forces. Nineteenth-century physicists knew of no other *fundamental forces*, forces not reducible to other more fundamental forces. After Einstein's 1916 publication of his revolutionary general theory of relativity, he and others searched for a unification of that theory with electromagnetism. Because the general theory of relativity is basically a theory of gravity, this would have reduced all then-known forces to a single universal force.

Quantum physics was in a confused state in 1916, and Einstein worked entirely within a nonquantum framework. Einstein pursued his unified field theory until his death in 1955, unsuccessfully. His dream of unifying the fundamental forces remains alive and well today, but today's research is entirely within the quantum, not the classical, framework. There's a huge barrier to a quantum unification of forces: There is no acceptable quantum theory of gravity. Finding one is an enormous challenge that inspires many bright minds. The goal is a generalized gravitational force incorporating electromagnetism and all other forces. The best known of these programs, string theory, is described beautifully in Brian Greene's book *The Elegant Universe*.[10]

Here is the tale of science's discovery of two forces not reducible to gravity or electromagnetism. It begins in 1896, when French physicist Henri Becquerel stored a certain uranium compound away in a drawer for the weekend. By chance, an unexposed photographic plate was in the same drawer. On returning to his lab the following week, Becquerel was surprised to find the film exposed, despite being stored in a dark drawer. He found the effect reproducible by simply placing uranium near photographic film. Apparently, the uranium sent out invisible rays, a process now known as *radioactivity*. Becquerel subjected the uranium to various chemical treatments, but this didn't alter the effect, implying that radioactivity had little to do with chemistry, little to do with the electron halos around each atom. It was science's first brush with the nucleus.

Today, we recognize two additional fundamental forces, beyond gravity and electromagnetism. Called the *strong force* and the *weak force*, both act only at short, subnuclear distances, and both play important roles within the nucleus. The *strong force* binds *nucleons* (protons and neutrons in the nucleus) together. The simple understanding that nuclei are made of protons and neutrons demands some such force. After all, protons repel each other electrically and neutrons are not electrified, so the EM force acting alone would instantly blow apart any nucleus having more than a single proton. There must be some kind of glue. Gravity is such a binding force, but when acting between protons in the nucleus it's a trillion trillion trillion times weaker than the electric force. It would be a boring universe if electromagnetism and gravity were nature's only fundamental forces. The only existing atoms would be hydrogen. There would be no ice cream.

The strong force is an interaction between quarks, the fundamental quanta of which nucleons are made. This force binds each nucleon's three quarks together, but it doesn't act only within each nucleon. Some of the strong force leaks out of the nucleon to bind neighboring nucleons together, furnishing the needed nuclear glue. The strong force reaches just far enough to bind nucleons with their closest neighbors. This means larger nuclei are held together only loosely, because the EM force is quite long range and reaches far beyond any individual proton to repel all other protons within the same nucleus, whereas the strong force binds only close neighbors. So as nuclei increase in size, there must come a point beyond which the long-range repulsion overcomes the short-range attraction and the nucleus falls apart. This is why you can't have a nucleus as big as a house.[11]

The largest stable nucleus is lead, with 82 protons. Every nucleus with an *atomic number* (number of protons) greater than 82 is unstable or *radioactive*—meaning, it eventually falls apart, or *decays*, spontaneously.[12] Many nuclei with 82 or fewer protons are also unstable, depending on the balance between their proton number and neutron number. Neutrons are bound mutually by the strong force without being repelled mutually by the EM force, and hence act as pure nuclear glue. This is why neutron stars can exist but not proton stars.

Meanwhile, back at Becquerel's lab, the uranium nuclei in the drawer, with 92 protons in their nuclei, were unstable. A uranium nucleus decays spontaneously by emitting an *alpha quantum* (commonly called an *alpha particle* but I'm sticking with "quantum") made of two protons and two neutrons bound together by the strong force. As an alpha quantum separates from a nucleus, it receives a huge electrical push from the remaining protons. This hurls the alpha quantum away from the nucleus with enough energy to alter radically any molecules it slams into—molecules such as those comprising the photographic plate in Becquerel's drawer.

Alpha decay is one of the two most prominent forms of radioactivity. As we've seen, it can be understood in terms of the interplay between short-range strong nuclear forces and long-range EM forces. *Beta decay*, the other prominent form of radioactivity, is more subtle. It provides evidence for a second subatomic fundamental force, the *weak force*. In ordinary matter, this force acts primarily within individual nucleons.

The weak force causes free neutrons—neutrons that are outside of any nucleus—to be unstable. Here's how this happens. Each neutron is made of two different kinds of quarks: one *up quark* and two *down quarks*. In a free neutron, there's a chance that one of the down quarks will transform spontaneously into an up quark, turning the neutron into a proton (made of two ups and one down) while creating an electron in the process. This transformation is caused by a force acting at the microscopic level, but it is neither the EM force nor the strong force. Physicists call it the *weak force*. The electron is created with a high energy, causing it to exit the neutron at 5% of light speed. Because of this spontaneous transformation of free neutrons into protons, there are few free neutrons. Unlike free protons, which travel easily throughout the universe with no sign to date of ever transforming into anything else, free neutrons are inherently unstable. A typical free neutron transforms spontaneously into a proton in about 15 minutes, although quantum uncertainties make this time highly variable from one neutron to the next. The newly created electron is called a *beta quantum*, to indicate its origin in a high-energy interaction between quarks, although it's just like any other electron.

The weak force is far weaker than the strong force and also weaker than the EM force. But it's considerably stronger than the gravitational force acting between individual nucleons. In fact, gravity is so weak compared with the other three forces that it begs the question: Why is gravity so important in the universe? The answer is that, unlike the strong and weak forces, gravity acts over large i.e., macroscopic, distances. The EM force is also long range; but unlike the EM force, *gravity is always attractive*. Mass always attracts other mass, whereas electrified objects can either attract or repel each other. So you can pile up mass without limit, making the gravitational force stronger and stronger, without the whole thing falling apart, whereas if you pile up three or more electrified objects, some of them are going to repel each other and things will fall apart. Thus, gravity shapes the universe and it shapes the big things such as stars and galaxies within the universe whereas the EM, strong, and weak forces act over much smaller distances to shape molecules, atoms, nuclei, and other relatively small things.

In stable nuclei, the interaction between nucleons tames the weak force and prevents neutrons from decaying. This is good, because otherwise the only stable nucleus would be hydrogen and, once again, there could be no ice cream. But in certain nuclei that have an excess of neutrons, beta decay is favored. One of the neutrons in the nucleus transforms spontaneously into a proton by emitting a beta quantum that then flies out of the nucleus—a second form of radioactivity.

To date, all known forces are reducible to just four: gravity, electromagnetism, strong, and weak forces. Some further unification has occurred; as Chapter 5 explains, the EM and weak forces are now understood as different aspects of a single force, much as electricity and magnetism are understood as aspects of the EM force (Chapter 4).

This discussion hasn't mentioned 95% of the universe: dark matter and dark energy. Here, all bets are off. There's plenty of evidence for the existence of dark matter and dark energy, but little evidence of what they are. They might be describable within the framework of quantized fields and the four known forces, or they might not. One thing you learn from studying quantum physics is that nature has a large imagination.

We thus end the tale, begun in Becquerel's laboratory in 1896, of the strong and the weak sub-atomic forces, two fundamental forces in addition to the gravitational and EM forces we experience directly in our macroscopic world. It seems remarkable that all the forces—all the pushes and pulls—discovered so far can be ascribed to only these four.

Energy Is Forever

We all have some idea of what it means to say "Alice is full of energy," or "the rocket exploded with great energy." But what exactly do we mean by *energy*? It's arguably the most important concept in physics. Physicists spent lots of time thinking about energy before it's meaning was finally clarified in the mid 19th century, 150 years after Newton's work. Its presently-known significance was revealed only in the 20th-century theories of relativity and quantum physics.

Many physics textbooks, for vague and convoluted reasons, never define this all-important word clearly[13]; but today we know quite clearly what *energy* means, and it's simple. *Energy is the ability to do work.* OK, so what is *work*? In physics, *work is done whenever a force (a push or pull) acts on an object while the object moves through some distance.* When you lift a rock, you do work on it. When you wheel a suitcase through an airport, you do work on it (the suitcase, not the airport). But when you push against a rigid wall, you do no work on the wall regardless of how fatigued your muscles become, because the wall doesn't move. When you drop a dish, you do no work on it during its fall because you aren't touching it so you aren't exerting a force on it; however, the force of gravity (Earth's pull) does work on the falling dish. Note from this last example that work is not necessarily connected with humans: Earth does work on any falling object. A moving locomotive does work on the rail cars it pulls.

Energy is more abstract than work, because it's an ability: it refers to something that could happen, not necessarily something that does happen. A moving baseball, for example, has energy because it could (although it might not) smash into a catcher's mitt and push it backward, doing work on the mitt. All moving objects have this energy as a result of motion, called *kinetic* (motional) *energy*. There are several other important types of energy. A rock that's been raised above the ground has energy (relative to when it was on the ground) although it's just sitting still, because it *could* fall and impact a nail, pushing it into the ground, which requires work. This is *gravitational energy*, energy that arises from the gravitational force. Other major forms of energy are *elastic energy* (the ability of a deformed body to do work by snapping back), *thermal energy* arising from temperature (i.e., from thermal motion), *EM energy* arising from EM forces, *radiant energy* carried by a light beam or other EM radiation, *chemical energy* resulting from a system's molecular structure, and *nuclear energy* resulting from a system's nuclear structure.

I will occasionally discuss energy in quantitative terms. For this, you need to know that, in metric units, both energy and work are measured in *joules*. A joule is not a lot of energy; for example, when you lift a stick of butter by 1 m, you do about 1 joule of work on the butter and you increase the butter's gravitational energy by 1 joule.

Another needed quantitative detail is "powers of ten." For example, 10^4 means 10 to the fourth power, or $10 \times 10 \times 10 \times 10$, or 10,000. So you just start from 1.0 and move the decimal point four places. Similarly, 10^{-4} means 1/10,000, or 0.0001: you start from 1.0 and move the decimal point four places to the left. As another example, light speed is 3×10^8 m per second, or $3 \times 100,000,000$ m per second, or 300,000,000 m per second. So 3×10^8 is just 3.0 with the decimal point moved eight places.

So what does energy really mean? Well, it's the ability to do work, and doing work means exerting a force to move stuff around, so *energy is the ability to move stuff around* or, to put it more generally, *the ability to change things physically.*

We need to home in more closely on thermal energy, the understanding of which during the 19th century was the key to finally grasping the principles of energy.[14] In fact, the study of the general principles of energy is called *thermodynamics* for precisely this reason.

A pot of hot water has a greater ability to do work, greater energy, than a pot of cold water. Here's evidence. If the hot water is boiling, it can cause its pot lid to rattle, and work must be done (by steam exerting a force on the lid) to cause the rattling. Cold water can't do this. If the hot pot is not hot enough to boil, one way to get work out of it would be to find a second liquid, such as alcohol, that boils at a lower temperature and use the hot water to warm it so that the second liquid's pot lid rattles. *Thermal energy*, sometimes called by the misleading term *heat*,[15] is defined as this ability that warm objects have to do work.[16]

It took a while for scientists to learn that warmth is a form of energy. It was thought that warmth is a material substance that causes objects containing more of it to be warmer, and objects containing less of it to be colder. This sounds plausible, but if you think in terms of atoms and molecules, there's a better way to visualize warmth—one that is more simple because it doesn't invoke a new substance. This proper understanding of warmth arises only at the unseen microscopic level, which is why it took scientists so long to figure it out.

You can learn about the microscopic nature of warmth from experimentation. Try this: Blow a balloon partly full of air and tie it shut. Float it for 30 seconds in a large closed pot of slowly boiling water and quickly (before it cools) put it between bookends without squeezing it. Remove the balloon without moving the bookends, cool it in a freezer for 30 seconds, and quickly replace it between the same bookends. It should have shrunk significantly. Why?

Please ponder. . . .

When you blew into the balloon, you filled it with about 10^{22}—ten billion trillion—air molecules. The enclosed molecules move in various directions, impacting the wall of the balloon occasionally, pressing it outward (which is called *pressure*). After you blow air into the balloon, there will then be more molecules (per square centimeter of balloon surface area) hitting the inside than are hitting the outside, so the net effect is outward pressure on the wall of the balloon. This is why balloons expand when you blow into them. But how do we explain the contraction observed during cooling? The number of molecules inside the balloon was fixed after you tied the balloon shut. Contraction must be caused by decreased pressure against the inner wall, implying that the fixed number of molecules on the inside are now impacting the wall less strongly. A natural conclusion is that these molecules are moving more slowly. Apparently, cooling causes the molecules to move slower, and warming causes them to move faster.

This hypothesis is confirmed by numerous experiments and is certainly entitled to be ranked as a theory rather than simply a hypothesis—a theory sometimes called the *kinetic* (motional) *theory of thermal energy: Warmth is the random, or disorganized, motion of a substance's molecules.* This is true not only for gases, with molecules separated widely from each other; but also for liquids, with molecules packed together while allowing sufficient space to slide past each other; and for solids, with molecules packed so tightly they can only vibrate around fixed points. This important idea, that warmth is nothing but disorganized molecular motion, is called the *mechanical interpretation of thermal energy* because 19th-century physicists assumed Newtonian mechanical principles were fully applicable to molecular motion (but quantum physics tells us they were wrong). Even in the light of quantum physics, the insight that warmth is disorganized molecular motion is still quite correct. When you think about it, it's astonishing. When your skin warms, your skin's molecules are not individually warmer, they are simply moving faster.

Drop this book to the floor. Does it speed up as it falls?

Please ponder. . . .

Observe the book-drop carefully. It clearly speeds up, because it started from rest. But does it (a) speed up all the way down or (b) speed up quickly just after release and then fall the rest of the way at unchanging speed?

Keep pondering. . . .

To answer, try this: Hold the book a few centimeters above a hard surface and drop it. Now try the same thing from greater and greater heights. What do your ears tell you? The book's impact gets louder as the height increases, so it must move faster when it falls farther, so it must speed up all the way down.

You're seeing one of nature's grand organizing principles in action.

As the book falls, it steadily loses gravitational energy because it's losing height. But you just demonstrated that it's also steadily gaining kinetic energy. With the help of Newton's laws of motion, and with some simplifying assumptions such as an absence of air resistance, physicists can prove quantitatively that, at any point during the fall, the amount of kinetic energy gained equals the amount of gravitational energy lost.

More remarkably, even including air resistance and so forth, experiments show that the total energy of the entire system doesn't change throughout the process. In the real experiment, air resistance is significant. Air is set into large-scale motion (called *wind*), and air is also warmed, by the falling book, implying that thermal energy must be included. Nevertheless, the experimental result is that the total energy—the overall ability of the system (including the air) to do work—does not change throughout the fall. And even in situations involving quantum physics or the effects of Einstein's relativity, experiments show that the total amount of energy still remains unchanged. We say that *energy is conserved*.

As far as we know, it's an entirely general principle. *The total energy of all the participants in any process remains unchanged.* More briefly, *the amount of energy in the universe never changes.*[17] This is called the principle of *conservation of energy*.

Question: In this case, what happens to the energy when your dropped book hits the floor?

There's only one plausible option: It turns into thermal energy.

One expects, then, that driving a nail into wood with hammer blows should warm the nail. Try it. You can feel the warmth.

Equivalence of Energy and Mass

It was 1905, and a fabulous year for Albert Einstein.[18] He published the foundations of his special theory of relativity along with several other papers. One of these (but not his work on relativity, which was considered controversial) would eventually be the basis for his Nobel Prize. Another predicted a startling new principle of energy.[19] Although this principle was not motivated by quantum physics and would be true even if we lived in a classical universe, it proved crucial to understanding quantum physics.

In 1915, Einstein announced his general theory of relativity, a further development of his ideas about space and time that were introduced in the 1905 paper. The general theory explains gravity as a consequence of the properties of space and time. Although many physicists are searching fervently for a theory that incorporates Einstein's general relativity into quantum physics, general relativity does not enter into the quantum principles discussed in this book.

Part of Einstein's genius was his ability to imagine simple but fundamental thought experiments that could be analyzed easily but might be impractical to perform in a real lab. To develop his new energy principle, he imagined an object at rest that emits two identical, brief pulses of light in opposite directions. The point of imagining two pulses, rather than just one, was that the object would then remain at rest without recoiling as it would if there were only one pulse, simplifying the analysis. He then used the equations of his special theory of relativity to show that *emission of the light pulses would necessarily reduce the object's mass*. This is astonishing: an object emits two light pulses, which themselves have no mass at all, but nevertheless have radiant energy, and this reduces the mass of the object! One can then imagine that the object remains at rest while continuing to emit light pulses until its mass is entirely gone so that the object vanishes!

How much energy is emitted as light before the object is consumed entirely? Einstein's prediction, based on his special theory of relativity, was mc^2 joules, where m represents the object's mass before emitting any light, c represents light speed, and c^2 is the square of light speed (light speed times light speed). Because mc^2 joules of energy must be emitted to reduce the object's mass to zero, and because energy is conserved, this amount of energy must have been contained within the object before the emission process began. In other words, *any object at rest that has a mass of* m *kilograms must have an energy* $E = mc^2$ *joules*. Einstein argued that the converse is also true: *any object at rest that has an energy of* E *joules must have a mass of* m $= E/c^2$ *kilograms* (an equation that is logically equivalent to $E = mc^2$). In other words, *for objects at rest, mass and energy are entirely equivalent: any object having energy has mass, and any object having mass has energy, and the relation between the two is* $E = mc^2$. This is Einstein's principle of *mass and energy equivalence*.

Einstein considered this principle to be special relativity's most important prediction. It's startling for two reasons. First, Einstein's argument implies that, *although energy is conserved, mass is not*. The emission of light reduces the emitting object's mass, while creating light that itself has no mass, reducing the total mass of the universe. Previously, it had been thought that matter, as measured by its mass, is conserved in every physical process. So matter is ephemeral, but energy is forever. In this sense, it's not a material world, but rather an energetic one.

Second, Einstein's equation predicts that the amount of energy in ordinary macroscopic objects is enormous. Every 1-kilogram (2.2-pound) object contains c^2 joules of energy. Because c is 3×10^8 m per second, c^2 is 9×10^{16} in metric units. So 1 kilogram contains 9×10^{16} joules, the total energy output of a huge, 1000-megawatt power plant during 3 years of operation!

Modern physics is full of startling principles such as $E = mc^2$ that are, however, well verified by observations. Although early-20th-century scientists argued that Einstein's argument was faulty and his conclusion false, $E = mc^2$ has withstood the test of time and, more to the point, experiment. For one famous example, in nuclear reactions (changes, such as radioactive decay, in the structures of nuclei), the energy changes are so large the resulting mass changes can be measured easily. The experimental result is just as Einstein predicted; the mass of all the matter involved in any energy-emitting nuclear reaction (such as the operation of a nuclear reactor) drops by exactly $m = E/c^2$, where E represents the energy emitted by the nucleus during the reaction. The same happens in chemical reactions, and every other energy transformation, but the mass changes are usually too small to measure easily.

Electron–positron annihilation is a striking case. *Positrons*, also called *antielectrons*, are quanta with the same mass as electrons but with opposite EM properties. So they have a positive, rather than (like electrons) negative, electric charge, and are called *antiparticles* of the electron. Quantum field theory (the high-energy version of quantum physics) predicts, and high-energy experiments confirm, that particle–antiparticle pairs can annihilate each other entirely by simply being near each other. But energy must be conserved, so some form of energy must be created during particle–antiparticle annihilation. Typically, electron–positron annihilation creates two photons, the total energy of which always turns out, experimentally, to be mc^2 (plus any kinetic energy the pair might have had), where m represents the total mass of the pair, just as Einstein predicted.[20] Matter is not durable, and can be destroyed entirely, but energy is forever.

Matter can also be created. Under the right conditions,[21] a single photon can vanish spontaneously and be replaced by an electron–positron pair, provided the energy of the photon is greater than mc^2, where m is the total mass of the pair. Not only matter, but also radiation, is ephemeral.

The Glorious and Tragic Tale of Entropy: You Can't Go Home Again

You can break an egg, but you can't put it back together. When a bullet slams into a wall and buries itself inside, nearly all the bullet's kinetic energy converts to thermal energy, but you can't use this thermal energy to launch the bullet back toward the gun. There are many such *irreversible processes*, which can proceed easily from state A (the bullet exiting the muzzle of the gun) to state B (a warmer bullet at rest in a warmer wall), but can proceed from state B to state A only with assistance from outside energy sources. Other examples include an object falling to the floor or a child's swing "dying down."

Why is that? Here's a thought experiment to help us ponder.

Imagine two boxes of a gas made of "inert" atoms—atoms such as helium that cannot react chemically; one box is hot and the other is cold. Let's call the boxes H (hot) and C (cold). Because warmth is disorganized microscopic motion, H's atoms are moving faster, on average, than C's atoms. Now imagine joining the two boxes, separated by only a thin metal wall that allows the passage of thermal energy without allowing the passage of gas molecules. As you probably know, thermal energy then flows through the metal from H to C. This is often called the *law of heating*: given the opportunity, thermal energy flows spontaneously from hot to cold, cooling H and warming C. Thermal energy won't flow spontaneously the other way, from cold to hot.

So an energy flow from hot to cold is an irreversible process. Again, why is this? Let's look at it microscopically. H's fast-moving atoms impact the metal wall strongly, shaking the wall and transmitting these vibrations to C's atoms, whereas C's atoms impact the wall weakly. So on average, more microscopic kinetic energy flows from H to C than from C to H, so H cools while C warms. This continues until the boxes are at the same temperature, when net thermal energy flow stops because equal amounts are now flowing in both directions. H and C have then come to "equilibrium" at some fixed intermediate temperature.

This is a typical irreversible process. Could it possibly reverse, transferring thermal energy from C to H, bringing both boxes back to their previous hot and cold temperatures? Strictly speaking, the answer is a surprising yes. Here's how: At equilibrium, atoms on both sides move at a variety of speeds, some faster and some slower. It's possible that during, say, 1 microsecond (one millionth of a second), most of the impacts on H's wall would happen to be by H's slower molecules, and most of the impacts on C's wall would happen to be by C's faster molecules. This could transfer net thermal energy from C to H during that microsecond. If this continued for enough microseconds, H and C would eventually return to their original temperatures!

But the odds against this occurring are enormous. If the gas is at normal atmospheric pressure and temperature, there are some billion billion (10^{18}) individual atomic impacts against every square centimeter of the metal wall of each box every microsecond. Suppose each box is a cube measuring 1 centimeter along each edge, so the metal interface between H and C has an area of 1 square centimeter, and suppose the two boxes have come to equilibrium (same temperature). After 1 microsecond of the improbable reversed process, H is a little warmer than C. To make thermal energy flow from C to H for even one additional microsecond, the net effect of the 10^{18} impacts within box C must be larger than the net effect of the 10^{18} impacts within box H. But C's slower atoms typically impact the wall less strongly than H's faster atoms. Thus, even if there were only, say, 10 impacts on each side instead of 10^{18}, the odds would be against an energy transfer from C to H. These odds get worse and worse as the number of impacts increases, for the same reason that, in a fair coin toss, the odds of throwing 20 heads in a row are far less (1024 times less, in fact[22]) than the odds of throwing 10 heads in a row. So if a little thermal energy should happen, by chance, to flow from C to H, the chances are overwhelming that, during the next microsecond, much of it would flow back to C.

As you can see, so-called irreversible processes are reversible in principle, but the reverse process is wildly improbable. This is true even for broken eggs. To dramatize this, imagine running a Hollywood movie backward. Everything in the backward movie (waterfalls flowing upward, bullets moving backward into gun barrels, people growing younger, and so on) is possible physically, but highly unlikely. For example, a waterfall could flow upward if enough of the randomly moving water molecules in the pool at the bottom of the fall just happened to all be moving (as a result of their thermal energy) rapidly upward at the same time.

To summarize: When we say a certain process is irreversible, we really mean the reverse process is wildly improbable. The principles of thermodynamics are "statistical" in this sense. The explanation of irreversibility has everything to do with probabilities.

Here's an even simpler thought experiment that illustrates the same connection between irreversibility and probabilities, although it has nothing to do with thermal energy. Imagine a partially organized deck of 52 cards, with all the spades put to order at the top of the deck, but with the rest of the cards well shuffled. Shuffle the entire deck a few times. Although it's possible for this mixing to produce a deck that is better organized than before, it's overwhelmingly likely that the final state will be less organized than the initial state. In this experiment, it's easy to see the reason for the irreversibility. There are simply many more ways of further disorganizing a partially organized deck than there are ways of further organizing it. A similar principle applies to your living room, which, as you may have discovered, is easier to disorganize than to organize. This principle of disorganization lies behind all cases of irreversibility.

For physical systems made of atoms and molecules, organization at the molecular level—microscopic organization—is a crucial consideration. Returning to the boxes of inert gas, the initial state (one box hot, the other cold) is partially organized in that most of the faster molecules are in H and most of the slower molecules are in C. When the boxes are placed into contact, they are able to proceed to other states, and these are overwhelmingly likely to be less organized, having a less strict separation of fast molecules in H and slow molecules in C, simply because there are more of these disorganized states and because thermal molecular motion is random. Detailed analysis shows that, like the deck of cards, each of the more disorganized states is more probable in the sense that there are a much larger number of them. So it's highly probable that the system will proceed toward less organized states. This is why thermal energy will, if given the opportunity, flow from hot to cold.

The general principle is: *A system that is partly organized at the molecular level and that is given the opportunity to reorganize is highly likely to proceed to a less organized state.* This is called the *second law of thermodynamics*.[23]

Scientists have managed to quantify the notion of microscopic organization, or rather its opposite, disorganization. The quantitative measure is called *entropy*. We don't need to define this word precisely here. Suffice it to say that entropy is a macroscopically measurable property of typical macroscopic systems, and it's a measure of the system's disorganization at the molecular level. With the help of this notion, we can state the second law quantitatively: *The total entropy of all the participants in any macroscopic physical process is overwhelmingly likely to increase or remain unchanged; decreases in total entropy occur only randomly and can be neither sustained nor controlled.* I've stated the preceding sentence carefully, specifying that all participants must be included and that entropy decreases are possible but improbable and unpredictable, and hence uncontrollable.

The growth of a leaf is a significant and interesting example. A photosynthesizing leaf incorporates simple carbon dioxide (CO_2) and water (H_2O) molecules to make sugars such as glucose ($C_6H_{12}O_6$). The glucose molecule is more organized, more structured, than the randomly moving CO_2 and H_2O molecules that went into it, so the process of photosynthesis proceeds from less organized (higher entropy) to more organized (lower entropy), seemingly violating the second law. The answer to this dilemma is fairly obvious: The leaf, CO_2, and H_2O are not the only participants in this process. They had help from the sun. Radiation has the temperature of the body that emitted it. Radiation from the sun's 6000-°C surface has a temperature of 6000°C, although it doesn't feel that hot to our skin because the sun is far enough away that its energy is widely dispersed by the time it arrives at our skin, making the radiation's energy density (its energy per cubic centimeter) so low there is insufficient energy to immediately burn your skin. When this energy arrives at a leaf, the high-temperature sunlight is absorbed and reemitted as lower temperature radiation. During this process, the radiation comes to equilibrium with its surroundings and, just like the two boxes of gas, this coming to equilibrium at a common temperature represents a great increase in the entropy of the participants (radiation plus surroundings). This entropy increase drives the decreased entropy of the leaf's molecules. In agreement with the second law, the overall entropy of the universe increases despite the entropy decrease involved in photosynthesis. Sunlight provides not only energy, but also organization, to Earth.

As far as we know, the second law applies to the universe itself, implying that the universe is becoming more disorganized. Compared with intergalactic space at a temperature of 3° absolute (−270°C, also called 3 *kelvin*), stars and galaxies are gathered together in relatively small volumes and have enormous temperatures, amounting to a high degree of organization (separation of hot stuff from cold stuff) and thus a low entropy for the universe as a whole. This organization allows the sun to provide structure to Earth by driving biological growth and biological evolution. As the universe comes to equilibrium over many billions of future years, there are opportunities for organized structures such as stars, planets, and life. It is thanks to the energy flow and changes caused by disequilibrium that life, liberty, and the pursuit of happiness are possible.

Our universe has been increasing its entropy for 14 billion years, and must have started from an all-time low-entropy well-organized state. In fact, the Big Bang creation of the universe appears to have been a high-energy, submicroscopic quantum physics event starting within a volume some 10^{-35} m across, about a trillionth of a trillionth of a trillionth of a meter![24] Atoms are 10^{-10} m across and nuclei are 10^{-15} m across, so this is small—probably as small as is physically possible—because space itself might be quantized at this level. At the time of creation, the universe was perhaps as well organized as it's possible for anything to be.[25] The immediate and continuing expansion of space then represents a huge entropy increase, offering creative possibilities for additional entropy-producing reorganizations.

The Big Bang's low-entropy initial state is the reason there is a second law of thermodynamics, the reason you remember the past and not the future, the reason the universe is not in an eternal thermal equilibrium state with nothing more exciting than thermal vibrations, and the reason stars, life, and ice cream are possible. It's also the reason you can't go home again, and the reason all things must eventually end. The Big Bang, and the law of entropy that it entails, drives the universe's glorious possibilities as well as its inevitable tragedies.

4

Fields and Classical Electromagnetism

Everybody should have a chance to play with magnets. Toss a paper clip near one pole of a magnet; it's immediately swept onto the pole. Something is surely out there, invisible, in the "empty" space around the pole. Your intuition is correct: something *is* there—namely, a magnetic field. Imagine putting on a suit of iron armor and strolling between two huge magnetic poles. You'd be hurled toward the nearest pole. Something's definitely out there, at every nearby point.

Any physical entity that exists at every point throughout a region of space is called a *field*. Water in a pool is a field of water; a smoke-filled room is a smoke field. A field of wheat is a wheat field. Many fields are quantifiable: a pool of water typically has a particular density (mass per cubic centimeter), temperature, and speed at every point. Fields are the opposite of particles. An ideal particle would occupy only a single point in space, although the term *particle* also applies to entities occupying only a limited region of space. Most of the fields discussed in this book extend throughout all space, although they are often concentrated primarily within some limited (perhaps microscopic) region.

Beginning around 1830, Michael Faraday and James Clerk Maxwell created a historic shift in our physical worldview, from particles to fields. As one consequence of their work, the magnetic field is now viewed as just one aspect of a more general EM field that fills the universe and is the key actor in all electric and magnetic phenomena.

Albert Einstein, a convinced advocate of the field view of reality, recommended that it replace the old particle view:

> Before Maxwell, Physical Reality ... was thought of as consisting in material particles. ... Since Maxwell's time, Physical Reality has been thought of as represented by continuous fields ... and not capable of any mechanical interpretation. This change in the conception of Reality is the most profound and the most fruitful that physics has experienced since the time of Newton.[1]

Not surprisingly, fields are central to Einstein's general theory of relativity. Throughout most of his life, Einstein assumed the universe to comprise just two universe-filling *classical fields* (fields that don't exhibit quantum effects): the gravitational field and the EM field.

As we'll see, quantum physics also assumes the universe to be made entirely of fields (Chapter 5). These *quantum fields* differ from classical fields in two major ways. First, they are quantized—meaning, they are made of highly unified, spatially extended bundles of field energy called *quanta*. Each quantum is itself simply a disturbance in a universal field, analogous to a ripple that disturbs the smooth surface of a pond. This field-bundling feature obviously lends a particle-like aspect to quantum fields, because the bundles are somewhat like particles. Bundles can, for example, be counted, whereas the points filling a spatial volume cannot be counted. This countable, digital aspect of quantum fields is not found in classical fields.

Second, the various types of quantum fields include not only "force fields" for three of the four known forces (a quantum theory of the fourth force, gravity, is still a work in progress), they also include another quite unexpected category called *matter fields*. With matter fields, material objects such as electrons, protons, atoms, and molecules are brought within the field framework. In other words, *electrons, protons, atoms, molecules, and other material objects are made entirely of fields*, fields similar to the magnetic field and other force fields. As a prelude to quantum fields in Chapter 5, this chapter focuses on one nonquantum field: the classical EM field.

The Field Idea

Fields are one of physics' most plausible notions, arguably more intuitively credible than tiny particles drifting through empty space. Even Isaac Newton, father of the postmedieval particle-based mechanical view of the universe, was intuitively attracted to the field view of reality. Although Newton's own theory assumed the gravitational force to act directly and instantaneously between objects such as the moon and Earth, without mediation from any physical processes that might occur in the space between the two, he privately felt the underlying mechanism for gravity lay in a space-filling nonmaterial medium. In fact, "Newton worked obsessively to try to discover evidence for a medium filling space."[2] In an exchange of letters with Reverend Richard Bentley, Newton wrote:

> It is inconceivable that inanimate brute matter should, without the mediation of something else which is not material, operate upon and affect other matter without mutual contact.... That one body may act upon another at a distance through a vacuum, without the mediation

of anything else, by and through which their actions and force may be conveyed from one to another, is to me so great an absurdity that I believe no man who has in philosophical matters a competent faculty of thinking can ever fall into it.[3]

But in his published writing and lecturing, Newton was not willing to guess at something for which he had no evidence, maintaining firmly, "I do not feign hypotheses."[4] Because Newton never spoke or wrote of fields professionally, and because his authority was so great, it's not surprising that classical physics first developed as a particles-only theory. During the 18th and 19th centuries, scientists generally assumed the universe to be made of indestructible particulate atoms moving through empty space, interacting via direct contact or direct action at a distance involving no intermediate fields, as seemed to be implied by Newton's physics.

So it's not surprising that, from 1900 to 1930, the quantum fathers tried to place the new phenomena within the framework of Newton's particles-only physics, despite earlier and contemporary work on fields by Faraday, Maxwell, and Einstein.[5] However, Chapter 5 shows that *a particle framework cannot explain quantum phenomena*. One goal of this book is to show that a field framework can explain quantum phenomena.[6]

Because he conceived the modern field concept, Michael Faraday (1791–1867) is arguably the father of modern physics. He was born into a rather poor English family and received only the most basic schooling, but he read avidly while working as an apprentice to a bookseller for 7 years, beginning at age 14. An independent and eager learner, he benefited especially from the principles of study recommended in contemporary science education books, and from scientific textbooks. At age 20 he attended a series of lectures by eminent chemist Humphry Davy, and subsequently sent Davy a 300-page book based on notes he had taken during these lectures. Davy was flattered and impressed, and eventually employed Faraday as an assistant. Faraday soon proved himself as a successful chemist and even more successful physicist. His experiments led to his discovery of "Faraday's law of electromagnetic induction," the basis for the electric generator (which supplies electric current when you feed rotary motion into it), and to understanding more fully the forces between magnets and electric currents, the basis for the electric motor (which supplies rotary motion when you feed electric current into it—the opposite of an electric generator). He was offered a knighthood in recognition of his services to science, but turned it down on religious grounds, believing it to be against the word of God to accumulate riches and pursue worldly reward, stating he preferred to remain "plain Mr. Faraday to the end." When the British government asked him for advice on the production of chemical weapons for use in the Crimean War of 1853 to 1856, he refused, for ethical reasons, to participate.

Faraday was impressed, just as I am and you should be if you're not already, with the ability of magnets to reach out across what he called "mere space" to pull on iron objects, and to push or pull on other magnets. If you've never played with magnets, please do so. By "mere space," Faraday meant space with no matter in it, also known as *vacuum*. Mere space was altered, he thought, whenever a magnet was nearby. This alteration, this "magnetic field," was a property of space itself. Magnetic field was not some additional ingredient added to space, it was instead a "condition of" or "state of" space itself. The presence of a magnet changed the very nature of the nearby space.

Faraday was the first to conceive of fields the way Einstein did and the way we do today. *Fields are properties of space itself.* Space has properties. For example, space can have the magnetic field property. Faraday argued that magnetic forces, and also electric forces (he knew the two were related), never act via direct action at a distance, but instead act via fields that fill the space between the interacting objects. Although we can't see these fields directly, we'll find they are physically real and not just something scientists invent to concoct theories and make calculations.[7]

James Clerk Maxwell (1831–1879), born 40 years after Faraday, was more mathematical but less visionary than Faraday. He wrote that Faraday's uses of so-called EM force lines (a way of picturing fields geometrically) showed him "to have been in reality a mathematician of a very high order," although Faraday's formal mathematical abilities included only the simplest algebra and did not extend as far as trigonometry.[8] Maxwell found the correct mathematical relationships, known as *differential equations*, that describe how the EM field varies in space and time. Maxwell's equations describe how electric and magnetic fields behave to produce all known EM effects. These equations tie the electric and magnetic fields together as two aspects of a single EM field that has its own independent existence as a property of space.[9]

The EM Field

On a nonmetallic tabletop, rub two pieces of plastic wrap or two sheets cut from a plastic bag, with a piece of tissue paper. Bring one plastic sheet and the tissue close together, holding each by a corner, without touching. They should attract each other. Bring the two plastic sheets together without touching. They should repel each other. You've probably seen similar phenomena when taking clothes out of an electric dryer. They demonstrate one of nature's four fundamental forces: electromagnetism.

Such electrified objects are said to possess *electric charge*, or simply *charge*. Experiments like this demonstrate there are two types of electric charge, and that two objects possessing the same type repel each other whereas two objects possessing different types attract each other. Rubbing apparently causes the sheets to become charged, with the plastic and the tissue acquiring different types of charge.

This all jibes very nicely with what we know of atoms. The standard picture of an atom has permanently charged protons and uncharged neutrons in the nucleus, with permanently charged electrons orbiting relatively far outside the nucleus. All protons are charged one way, called *positive* whereas all electrons are charged the other way, called *negative*. It's reasonable to suppose that electrons, because they orbit far outside the nucleus, can be removed easily from their atoms. So rubbing a plastic sheet with tissue should rub some electrons off of one object and onto the other. The chemical properties of plastic and tissue determine in which direction the electrons move. It happens that the chemistry of the two substances (i.e., the degree of attachment of electrons to their atoms) favors electron transfer from tissue to plastic, so the plastic becomes negatively charged and the tissue positively charged. Both substances are "insulating" materials—meaning, electrons cannot flow easily within them or on their surface, so both tend to retain whatever excess positive or negative charge they may have. Metals, on the other hand, are "conducting" materials that allow the easy flow of electrons. Notice that the EM force acts at a distance: the two sheets are attracted or repelled even though they don't touch.

An electric current, such as the one that flows through light bulbs in your house, is simply a flow of electrons along a wire. All electric currents are motions, flows, of charged objects, usually electrons but possibly protons or electrically charged atoms or molecules (*ions*). Experiments show that electric currents exert forces on magnets, demonstrating a connection between electricity and magnetism. Such experiments demonstrate that *all magnetic forces are traceable to the motion of charged objects such as electrons and protons*.[10] At the microscopic level, there aren't any magnets; there are only electrically charged objects. Two charged objects exert electric forces on each other and two moving charged objects also exert magnetic forces on each other.

In this case, what explains permanent magnets, which don't appear to contain any charged objects at all? A permanent magnet's effects are caused by the spinning and orbiting motions of some of the electrons orbiting the magnet's atomic nuclei. All these microscopic electric currents are coordinated (lined up) to create the macroscopic effects observed among magnets, paper clips, and suits of iron armor. Experiments demonstrate that *all electric and magnetic forces have a common source—namely, electric charge.*

Maxwell's theory consummated electricity and magnetism's unification into a single EM force. Maxwell's equations describe the dynamical (this word means "over time" or "as time passes") behavior of the combined electric and magnetic fields and of electrically charged objects. In Maxwell's theory, an *EM field* surrounds every charged object. This field exists wherever any other charged object *would* (if it were present) feel electric or magnetic forces. It fills the space between separate charged objects, causing them to exert forces on each other. This field is a condition of the space around each charged object. Electric charges change the surrounding space, causing the space to exert a force on any other charged object placed within that space. A "stronger" EM field is one that would exert a stronger force on electric charges placed in its vicinity. As you might expect, larger amounts of electric charge create stronger EM fields in the surrounding space, and the EM field created by a charged object is stronger at points closer to the object and weaker at points farther from the object.

We don't see EM fields directly, but we see their effects. Because we don't see them directly, it's a common misconception to conclude they are not physically real, but merely useful mathematical abstractions. As we'll see, this conclusion is wrong. The EM field is a physically real condition of space.

Tales of the Mighty Kingdom: EM Waves

Evidence is close at hand showing light to be a wave phenomenon; in fact, evidence is as close *as* your hand. Try this: Make a long thin slit by holding the ends of your thumb and forefinger about a millimeter apart, several centimeters in front of one eye. Close one eye and look through this slit, focusing on a well-lit white wall in the background. Your fingers should look blurred. Where the blurs from the two sides of the slit overlap, you should see narrow bright and dark lines, similar to Figure 2.3, running parallel to your fingers.

 . . . *This is a break for experimenting.* . . .

These are lines of constructive and destructive interference, like those seen in Figure 2.3, except that they are formed not by the interference between light from two thin-slit sources, but rather by interference from different portions of the single wider slit—much wider than the wavelength of light—formed by your fingers. This millimeter-wide single slit acts like hundreds of thin-slit sources, all of them interfering with each other, with the retina of your eye acting as the observation screen, furnishing direct evidence that light is a wave.

But what is light a wave in? What's doing the waving? For decades after Thomas Young's 1801 double-slit experiment, nobody knew. Then, in 1864, Maxwell showed his equations implied EM fields could spread through empty space in a rapid interplay between electric and magnetic effects, and the speed of this spreading was roughly 3×10^8 m per second (300,000 kilometers per second), equivalent to about seven times around Earth in 1 second. This is the speed of light, as had been known since 1676, when Olaf Roemer made an ingenuous argument, based on observations of one of Jupiter's moons, showing light speed to be about 3×10^8 m per second.[11] Maxwell wrote, "[t]he agreement of the results [with the experimentally measured speed of light] seems to show that . . . light is an electromagnetic disturbance propagated through the field according to electromagnetic laws."[12] The EM field was doing the waving.

It was unprecedented. Before the work of Faraday and Maxwell, Newton's physics led to the view that all waves are disturbances of some material medium, with wave motion described by Newtonian mechanics. There are familiar examples of such mechanical waves: water waves are mechanical disturbances in water (disturbances of water molecules), and sound waves are mechanical disturbances in air (disturbances of air molecules). Water and air are material substances. Newtonian physics seemed to imply light must also be a wave in some as-yet-unknown material medium, but this turned out to be wrong. There is, in fact, no *material* medium for light waves. The medium is, instead, the universal EM field, which is a fundamental property of mere space (as Faraday had put it), and is not made of matter.[13] Maxwell's prediction that light is an EM wave was specific, quantitative evidence that *so-called empty space has properties; mere space is not really empty of physical activity*. Einstein would make much more out of this insight during 1900 to 1916, as he developed his special and general theories of relativity. This idea has been central to quantum physics since at least 1973. In hindsight, fields are central to everything quantum, but this was not fully understood until 1973.[14]

German physicist Heinrich Hertz, in 1887 at the Karlsruhe Institute of Technology, verified Maxwell's theory of EM waves experimentally. The theory predicted that, if electrically charged objects such as electrons or protons vibrate back and forth, they will radiate EM waves in all directions, with the wave vibrations perpendicular to the direction of motion of the waves. You can produce such waves yourself, for example, by rubbing a plastic comb with tissue and then shaking the comb (Figure 4.1). But you won't be able to detect these waves because you can shake only at a few vibrations per second, and the resulting EM waves carry too little energy to measure. Hertz constructed an electrical device that caused electric charge to collect on two small metal balls separated by a small air gap. The intense EM field between the balls caused the air to spark (become electrically conducting), causing electrons to vibrate back and forth across the gap 500 million (5×10^8) times per second. These dancing electrons sent out an *EM wave*: EM fields moving outward at light speed, vibrating 5×10^8 times per second.

To describe wave motion quantitatively, we need two new words. The number of complete back-and-forth vibrations sent out in each second is called the *frequency* of the wave, measured in vibrations per second or *hertz*, and abbreviated as Hz. The length of one complete spatial repetition of the wave's shape in space is the *wavelength*. For example, if you shake a charged comb three times per second, you'll send out three complete wavelengths every second and this wave's frequency will be 3 Hz. It turns out that the wavelength of this wave will be 100,000 kilometers. Clearly, Figure 4.1 is not drawn to scale.

Hertz's waves, with a frequency of 5×10^8 Hz, carried sufficient energy to cause electrons to vibrate at the same frequency, back and forth across an air gap in a simple metal ring *placed 12 m away*. This verified Maxwell's unprecedented

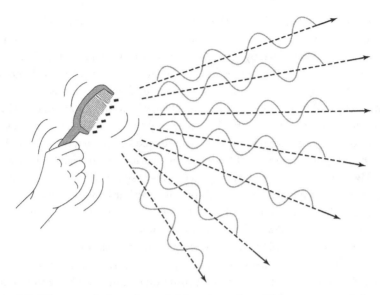

Figure 4.1 When you shake a charged object, it sends an EM wave outward through the universal EM field. The figure is not drawn to scale. Your hand can shake only a few times per second, creating wavelengths of some 100,000 km—a quarter of the distance to the moon!

prediction of EM waves. Please picture the experiment: Hertz's apparatus shakes electrons back and forth between two metal balls, and other electrons located *12 m away* across essentially empty space (the air in the room had little effect on the transmission) then shake also! It seems miraculous. What connects the electrons in the sender with the electrons in the receiver? The answer: the universe-filling EM field. This might seem unremarkable in our age of incessant electronic signaling by a plethora of media, but it was stunning in Hertz's day. Similarly, quantum physics offers stunning, unbelievable phenomena that will be regarded someday as normal.

The metal ring that received Hertz's EM waves also contained a device that could measure the wave's wavelength, which turned out to be 0.6 m, about 2 feet. This information, together with the known frequency, enabled Hertz to calculate the speed of the waves, which turned out to be light speed, just as Maxwell had predicted.[15] This was one of history's most significant experiments, verifying Maxwell's theory and ushering in modern communications technology.

Like all EM waves, Hertz's waves traveled at light speed: 3×10^8 m per second. But the wavelength of visible light, at around 600 nanometers, is much shorter than Hertz's 0.6-m waves. A nanometer (nm) is a billionth (10^{-9}) of 1 m, so Hertz's wavelengths are a million times longer than light wavelengths. As you may have guessed, Hertz's waves are *radio waves*, generated and detected by radio, television, radar, and other electronic equipment. Just as Hertz's experiment required a sender and a receiver, every EM wave travels through space from a sender that causes electrons or other charged objects to vibrate, to a receiver containing other electrons that vibrate in response. The sender for most light waves is electrons moving within atoms. The visible spectrum ranges over wavelengths of 400 to 700 nm, corresponding to the different colors of the rainbow from violet to red.

Because all EM waves travel at the same speed, 3×10^8 m per second, it must be the case that *waves with shorter wavelengths have higher (larger) frequencies; waves with longer wavelengths have lower (smaller) frequencies.* For example, if we compare waves from a comb (Figure 4.1) shaken at a frequency of two shakes per second (2 Hz), with waves from another comb shaken at a lower frequency of 1 Hz, the second comb must send waves of longer wavelength—twice as long, in fact. There's a second useful general relationship of this sort concerning energy: *Shorter EM wavelengths (higher frequency) transmit more energy; longer EM wavelengths (lower frequency) transmit less energy*, provided other things are equal. This stands to reason, because there's more shakin' goin' on—more motion—at higher frequency; hence, more energy is transmitted. For example, an EM wave with a frequency of 2 Hz (two waves arriving per second) delivers twice as much energy per second as a wave with a frequency of 1 Hz—provided the two waves have equal widths of vibration.

Figure 4.2 displays the entire EM spectrum, described by Hertz as "a mighty kingdom" and "the great domain of electricity."[16] Let's run through this kingdom from right to left.

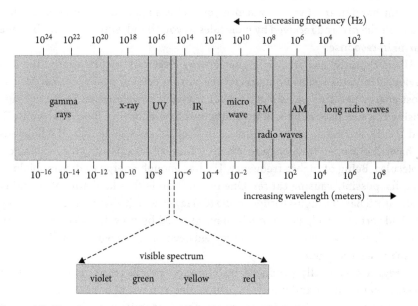

Figure 4.2 The electromagnetic spectrum.

Radio waves, the longest (meaning longest wavelength) EM waves, can be sent and received by human-made electronic equipment such as antennae and electric circuits. In a transmitting antenna, electrons create an EM wave by vibrating through short distances; in a receiving antenna, electrons are caused to vibrate in response to an incoming EM wave. Radio waves are also radiated ubiquitously in nature, from stars, galaxies, and the Big Bang. Radio waves range from very low frequencies, around 100 Hz with Earth-size wavelengths; through 10^3 Hz (kilohertz) with 100-kilometer wavelengths; 10^6 Hz (megahertz) with 100-m wavelengths; and up to 10^9 Hz (gigahertz) with 10-centimeter wavelengths. Above this, at higher frequencies of 10^9 to 10^{11} Hz, with wavelengths ranging from centimeters to millimeters, lies the *microwave* range, which can be radiated by microwave ovens and by natural processes. Jumping over the "IR" in the figure momentarily, the *visible spectrum* (recall that "spectrum" means "range"), ranging from red (700 nm, 4×10^{14} Hz) to violet (400 nm, 8×10^{14} Hz). What's special about these frequencies is simply that they are the ones to which the radiation-sensitive cells in the human eye respond. Most of this visible light comes from electrons moving in atoms.

The visible spectrum is a tiny region sandwiched between *infrared* (IR: "below red" in energy and frequency) on the long-wave side and *ultraviolet* (UV: "above violet") on the short-wave side. IR has wavelengths ranging from 1 millimeter down to 1/1000 millimeter—the size of a grain of baby powder. Molecules moving randomly in natural thermal motion typically radiate IR, implying that all objects radiate IR simply because of the irrepressible thermal energy of their molecules. Hotter objects, such as the sun or a burning log, emit more IR; cooler objects, such as Earth or an ice cube, emit less IR (per unit of surface area). If you put your hand near a fire, the warmth you feel is a result of IR radiated by molecules in the fire and received by molecules in your hand, which vibrate (become warm) in response.

UV is typically radiated by higher energy electrons within atoms. The sun obviously radiates visible light, in fact the sun's spectrum peaks (i.e., carries the most energy per unit wavelength range) at visible wavelengths. So it's not surprising the sun also radiates abundantly in the neighboring spectral regions: IR and UV. IR warms us and, of course, our eyes detect visible rays. UV happens to have the proper frequency to excite (cause vibrations in) many biological molecules. Being of higher frequency, UV has enough energy to disrupt biological cells, possibly causing cancer. One implication is that land animals couldn't evolve until Earth's high-altitude ozone (O_3) layer, which absorbs most incoming UV, had formed. And, as you might have guessed by now, UV also causes sunburns, which is why you can get sunburned even on a day when visible sunlight doesn't reach the ground.

X rays are typically produced by high-energy electrons or protons that strike a metal target and knock an orbital electron out of a metal atom's inner "electron shell" containing inner orbital electrons. Within each such atom, an outer electron then "falls" inward to replace the missing inner electron, emitting a high-energy X-ray photon with wavelength that is typically the size of an atom.

Gamma rays come from high-energy nuclear processes and have wavelengths about the size of individual nuclei. X rays and gamma rays have very high frequencies, around one billion billion (10^{18}) Hz for x rays, and one trillion trillion (10^{24}) Hz for gamma rays, and high energies that reflect high-energy processes occurring within all nuclei and certain atoms. X rays and gamma rays have sufficient energy to ionize (strip electrons from) biological molecules. Although this can be dangerous to us animals, it has a bright side: By penetrating deeply into biological matter, x rays can examine our insides without surgery and gamma rays can cure some cancers by killing diseased cells.

Fields Are Real

All EM waves resemble light. The only distinguishing feature of light is that the human retina happens, for evolutionary reasons, to be able to detect it.[17] Most of us regard light as something physically real. If it is real then it follows that, unless one makes the narrowly solipsistic argument that a thing is only real if humans can sense it directly without assistance from electronic gadgets, all EM waves are physically real.

The visible spectrum is comparatively tiny (Figure 4.2). Think of how much more surrounds you. Just among the EM waves (and there's much more out there),[18] you are always enveloped in radio waves from space and from every radio and TV station within your receiving range (as you can demonstrate by turning on your radio or TV), by microwaves from the Big Bang, and by IR and UV from the sun.[19] What's more, all the visible frequencies coming from the sun, light bulbs, visible reflections, and so on, are zipping all around you; you capture only a minute fraction on your retina. Democritus' and Newton's concept, that space contains only tiny particles separated by vast stretches of emptiness, is wildly false. On the contrary, even if we restrict ourselves to just the EM realm, every cubic millimeter of space is packed with an overlapping glut of physical phenomena—a mighty kingdom, indeed.

Heinrich Hertz's experiment and all similar EM transmissions across space are powerfully direct evidence for the reality of fields. As Hertz demonstrated, if you shake electrons *here*, then, after a predictable time interval, electrons *there* shake in response. This magic (I of course use this word only metaphorically) occurs every time you tune into a radio station. Zillions of such demonstrations testify daily to the physical reality of the EM field.

There is further evidence of the physical reality of EM waves in the well-substantiated principle of conservation of energy, which implies that *EM waves carry energy*. Here's why. Suppose Hertz's radio transmitter sends an EM wave to Mars, and that the travel time, at light speed, is 12 minutes. Energy must travel from sender to receiver because work must be done to cause the receiver's electrons to vibrate, and work requires energy. Where is this energy during the 12-minute travel time? It's not in the sender (which could be turned off after sending the message), and it's not yet in the receiver. And energy never vanishes. So it must be in the space between sender and receiver, in the EM field. QED (for the Latin "quod erat demonstrandum" or "which was to be proved"—a useful term for impressing others with your wise argument).

Philosophers might disagree about what is real, but for most physicists nothing is more real than energy. A similar energy-based argument can be made for the reality of the gravitational field and for other physical fields. This was Maxwell's and Einstein's argument for the reality of the EM field.[20]

A principle called *subjective idealism* was proposed during the 18th century by Anglo-Irish philosopher Bishop George Berkeley. It denies the existence of matter and instead holds that objects such as rocks and tables are only ideas in human minds and cannot exist without being "perceived." This seems to me, and perhaps to you, to be a wrong-headed fantasy. But it has actually been resurrected at the most fundamental physical level by some physicists who reject the physical reality of some of the quantum phenomena discussed in this book. It's a view that begs many questions: What qualifies as perception? Can a baby perceive rocks and tables and thus pop them into existence? Can a very low-IQ human? A chimpanzee? A flea? And are we to believe that rocks and tables vanish as soon as nobody is perceiving them? What if only a camera is perceiving them? You get the idea: subjective idealism has its problems. English writer and all-'round intellectual Samuel Johnson once had a famous conversation with Bishop Berkeley. Johnson's biographer, James Boswell, accompanied Johnson and talked with him immediately after the Johnson–Berkeley discussion. Boswell wrote:

> After we came out of the church, we stood talking for some time together of Bishop Berkeley's ingenious sophistry to prove the nonexistence of matter, and that every thing in the universe is merely ideal. I observed, that though we are satisfied his doctrine is not true, it is impossible to refute it. I never shall forget the alacrity with which Johnson answered, striking his foot with mighty force against a large stone, till he rebounded from it, "I refute it *thus*."[21]

Judging from the rebound of Johnson's foot, the stone was made from a rather large mass of matter. Like Johnson, most of us regard such solid and massive objects as the epitome of reality, sufficiently real to refute the idealist fantasy. So let's ask whether fields have mass and whether rocks might be made of fields.

Fields and $E = mc^2$

Recall Einstein's principle of mass and energy equivalence: For objects at rest, any object having mass has energy $E = mc^2$ and any object having energy has mass $m = E/c^2$. You've just seen that EM fields have energy. This is true even for fields that are at rest rather than radiating outward at light speed. An electric field at rest has the ability to exert forces on electrically charged objects and thus to do work on them; such ability to do work is precisely what we mean by energy. Mass–energy equivalence then implies that *EM fields at rest have mass*, implying that, like Johnson's stone, they resist acceleration and they have weight. If we needed a down-to-Earth reason to consider energy and fields to be physically real, this is surely it. Like Johnson's stone, fields have weight. If you could accumulate sufficient EM field within an enclosed region, you could kick it.

Try, or at least imagine, this: Put two magnets together so that they stick, and put them on a table. Call this *configuration A*. Now pull them apart and place them, separated, back on the table. Call this *configuration B*. Question: In which configuration does this system (the magnets with their fields) have the most energy, or is their energy the same in the two cases? (Hint: Because energy is the ability to do work, it must be true that, when you do work on some system, you expend some of your energy and you increase the energy of that system.)

Ponder now....

If you answered configuration B, you're correct. You had to do work to separate the magnets, so you added energy to the magnetic field.

Here's a follow-up question: In which configuration does the system have the most *mass*, or is their mass the same in the two configurations?

Ponder again....

The answer is, again, B. The configuration with the most energy has the most mass because $m = E/c^2$.

It seems miraculous. Not only energy, but also mass resides in the so-called empty space between the two separated magnets. This space isn't really empty after all! It has weight, just like a rock.

But EM field energy does not have a significant effect on the mass of ordinary matter, because ordinary matter is made of atoms and the EM field energy of an atom is millions of times smaller than the total energy (mc^2, where m is the atom's mass) of the atom. The dominant force acting within atoms is not the EM force but rather the strong force that holds its quarks together and holds the nucleus together. Just as the EM force is described by the EM field, the strong force is described in terms of a *strong force field*.

Does the energy of the strong force field have a significant effect on the mass of ordinary matter? Here, the answer is yes, definitely. If we consult a chart illustrating the "standard model of fundamental particles and interactions,"[22] we'll find that the three quarks comprising one proton have masses adding up to 0.012 mass unit whereas the total mass of one proton is 0.938 mass unit (the mass units in which these numbers are reported are specialized for high-energy physics experiments and needn't concern us). And the three quarks comprising one neutron have masses adding up to 0.015 unit whereas the total mass of one neutron is 0.940 unit. So the masses of the proton and neutron are 78 times and 63 times larger, respectively, than the sum of the masses of their constituent quarks! This doesn't add up! The reason is that the strong force field that holds the quarks together contributes almost all the energy, and therefore almost all the mass, to protons and neutrons. The individual quark masses contribute only a small fraction.

These numbers for the masses of the up and down quarks are inferred only indirectly from high-energy physics experiments and are highly inaccurate, because we cannot isolate individual quarks to examine them directly. A much better estimate of the masses of individual quarks comes from theoretical physicist Frank Wilczek's work. In a calculational *tour de force*, he used a rather complex facet of quantum field theory[23] to calculate the mass of the strong force field acting between the three quarks within protons and neutrons. The result of this theoretical calculation was that fully 95% of the mass of protons and neutrons comes from the energy of their strong force fields![24] This is our best estimate of the mass of these fields.

Let's ponder this. All ordinary matter is made of atoms, and some 99.95% of the mass of typical atoms resides in their protons and neutrons. We conclude, from Wilczek's calculation, that *95% of the mass of ordinary matter resides in the mass of its strong force fields.* The mass of the universe's ordinary matter comes not from so-called "solid" objects, but almost entirely from force fields! At the micro level, even rocks are not so solid; they are made of "nonmaterial" fields. Nevertheless, they are massive (i.e., they possess mass).

What about the remaining 5% of the mass of ordinary matter—the mass of quarks and electrons themselves? In 1964, British physicist Peter Higgs, along with several other physicists, developed a hypothesis now known as the *Higgs mechanism*. This hypothesis explains how quarks, electrons, and other mass-bearing fundamental quanta acquired their individual masses. Higgs' hypothesis received strong experimental confirmation when physicists at the Large Hadron Collider high-energy accelerator in Geneva, Switzerland, discovered a new type of quantum in 2012 and identified it in 2013 as a *Higgs boson*. This discovery means that the Higgs hypothesis is now an experimentally confirmed theory.

According to the Higgs theory, a previously unknown fundamental field—the *Higgs field*—exists throughout the universe. Some fundamental quanta, such as quarks and electrons, interact with the Higgs field whereas others, such as photons, do not. The nature of this interaction is best described as molasseslike. Interaction with the Higgs field gives quanta inertia—resistance to acceleration—and this resistance to acceleration is what we mean by mass

(Chapter 3). So *the Higgs interaction gives quanta mass*. Hence, matter differs from radiation only in that matter interacts with the Higgs field whereas radiation does not, and this confers mass on matter but not on radiation. Some material quanta have larger masses than others because they interact more strongly with the Higgs field. I'll discuss the nature of this field further in Chapter 5.

Like all known fundamental fields, the Higgs field is quantized—made of bundles of field energy (Chapter 5). The name given to the quantum—the energy bundle—of the Higgs field is *Higgs boson*. As stated, it was identified in recent experiments at the Large Hadron Collider. As the theory had predicted, the Higgs boson itself has mass—in fact, a rather sizeable mass equal to 133 proton masses.

Summarizing, rocks are made of quarks and electrons, whose interaction with the Higgs field provides 5% of the rock's mass. The remaining 95% comes primarily from the strong force field that binds quarks together into protons and neutrons, and binds protons and neutrons together into nuclei. Ultimately, a rock is nothing but fields—fields you can kick.

Four Early Tales of the Quantum

By 1900, most physicists concluded that matter is made of small material particles as described by Newton's theory, and that forces between particles are caused by direct contact and by the forces of gravity and electromagnetism acting at a distance, with at least the EM force acting by means of space-filling fields that radiate forces over great distances. These conclusions soon began to change.

In retrospect, we can see that the first flaw in this picture occurred toward the end of the 19th century when physicists found it impossible to explain the radiation from heated objects using the obvious tool for such an explanation—namely, Faraday's and Maxwell's EM field theory. We saw in Chapter 2 that Max Planck resolved this difficulty in 1900, but only with the help of a small assumption that had no basis in previous theories: The energy emitted by the atoms of a heated object is quantized (i.e., restricted to only certain allowed amounts). It was the quantum's first venture onto the scientific stage—a barely noticed straw in the winds of time.

The second phenomenon, historically, requiring a quantum explanation was the *photoelectric effect*. When light falls on a metal, electrons within the metal are often either ejected or caused to flow along the surface. This is the basis of "photovoltaic" electricity from sunlight. Classical EM theory can explain this effect, at least qualitatively, as follows: Light is a vibrating EM field, and when light hits a metal surface, this vibrating field spreads over the surface, shaking electrons until some break loose. This is plausible, but there's a quantitative problem. The classical EM field spreads out uniformly over the metal so that, at any particular point on the surface, considerable time should be required before an atom at that point absorbs enough radiant energy for an electron to be shaken loose. Calculations showed that several seconds were needed for sufficient energy to build up at any individual atom, whereas experiment showed that the first electrons were ejected as soon as light struck the surface.[25]

So there was a problem. In 1905, Einstein pointed out that the solution lay in a natural extension of Planck's idea. Planck had assumed the energy radiated by atoms is restricted to certain allowed values. As in the example of the quantized swing (Chapter 2), this implies energy is emitted in short bursts. Einstein followed this logic one step further and assumed these bursts created tiny lumps of radiation that traveled through space, and that light is made of such lumps. In this case, each lump (today we call it a *photon*) could deliver all its energy to one tiny portion of the metal surface. If this was sufficient to dislodge one electron, then each photon could dislodge one electron *with no time delay*. These lumps solved the problem, and eventually won Einstein his Nobel Prize.[26]

A third early sighting of the quantum occurred in connection with the EM spectra emitted and absorbed by various chemical elements. Chapter 1 discussed the emission spectrum of hydrogen; hydrogen gas emits radiation only at certain precise wavelengths. Why these wavelengths and no others? This phenomenon was first noticed as long ago as 1802, when scientists used prisms and other devices to detect the entire spectrum of wavelengths from the sun and found that certain specific wavelengths were absent from the sun's complete rainbow of wavelengths. We now know these missing wavelengths are absorbed by atoms in the gaseous atmosphere that surrounds the sun.

Niels Bohr, in 1913, used these emission and absorption data to figure out the internal structure of atoms. He combined the already discovered notion of the planetary atom (electrons orbiting a tiny nucleus) with Planck's idea about energy quantization to explain how atoms emit and absorb energy. Working mostly with the simplest atom, hydrogen, Bohr used Newtonian physics to obtain equations for the way the energy of the atom's single orbiting electron varied as the electron's speed and distance from the nucleus varied. He then assumed an additional rule that I needn't spell out here, a quantization rule similar to Planck's assumption and similar to our quantized swing analogy (recall that we quantized the height to which the swing could rise), according to which the orbiting electron could move only in certain allowed orbits. He also made the plausible assumption (based on conservation of energy) that, when an atom emits radiation, the energy of its orbiting electron drops from one allowed energy value to a lower allowed energy value. Similarly, when an atom absorbs radiation, the energy of its electron rises from a lower allowed energy to a higher one. Bohr's use of Newtonian physics along with the quantization assumption led to conceptual paradoxes that would be resolved only in 1925 by an entirely new, non-Newtonian theory of the quantum. In 1913, Bohr's theory gave quantitatively correct values for the EM spectrum of hydrogen. Despite the paradoxes, physicists recognized there was something right about this work; the detailed quantitative agreement between theory and experiment couldn't be just a coincidence. Bohr's theory was a bridge from the Newtonian clockwork universe to the world of the quantum. [27]

As mentioned, the conventional view in 1900 was that matter comes in tiny particles whereas radiation comes in spatially extended fields. The foregoing three breakthroughs challenge the field theory of radiation, concluding that radiation has a lumpy, particlelike nature, but none challenge the Newtonian notion that matter is made of tiny particles. For Louis de Broglie, it seemed odd that the new ideas affected so radically our view of radiation while leaving our view of matter more or less unchanged. His discomfort, which was as much aesthetic as scientific, led him to hypothesize a fourth early sign of the quantum.

de Broglie (pronounced "de Broy"), a PhD student at the University of Paris in 1923, felt there should be symmetry—balance—between radiation and matter. He considered it ugly that radiation should exhibit both wave properties, as implied by Maxwell's classical theory, and particle properties, as implied by the three preceding breakthroughs, whereas matter behaves always as particles. He suggested matter should also have both wave and particle properties, providing supporting technical details in his dissertation. His PhD committee didn't know what to make of this odd hypothesis and sent de Broglie's dissertation to Einstein for his opinion. Einstein commented that "it is a first feeble ray of light on this worst of our physics enigmas." The committee approved de Broglie's dissertation. For some scientists, including me, de Broglie's innovation is one of history's most beautiful scientific ideas.

For Newtonian-trained physicists, it was impossible to imagine how a tiny material particle such as one electron could be a spatially extended wave. A wave in what medium? What was waving? But de Broglie persisted. Based on the symmetry he envisioned, and working from Planck's paper on the radiation from heated objects, he deduced a formula predicting the wavelength of the wave that he thought was associated with every material particle.[28]

Four years later, Clinton Davisson and Lester Germer fired electrons at a crystalline (i.e., its atoms were arranged in a regular, or crystal, pattern) nickel target in such a way that the electrons bounced from the nickel surface and then impacted a detection device. Each electron impacted a specific point on the detector, in agreement with the notion that an electron is a tiny particle. But the *overall pattern* made on the detector by zillions of reflected electrons was astonishing. It was the kind of pattern that would be predicted if an EM wave had reflected from the surface, although material electrons, not nonmaterial EM waves, were doing the reflecting! The overall pattern made by zillions of pointlike impacts was a wave pattern, whereas the individual impacts appeared to be caused by tiny particles. Davisson and Germer were even able to measure the wavelength of the observed pattern. It agreed with de Broglie's predicted wavelength of the waves associated with electrons.[29]

This was puzzling, to say the least. In fact, there is still confusion about this wave-versus-particle puzzle (Chapter 5). Davisson and Germer's result confirmed de Broglie's hypothesis that so-called "particles" (which electrons were supposed to be) can have wave properties. The wave-versus-particle puzzle is easy to state: Matter and radiation both seem to have wave properties and also particle properties. Why is that? What is matter? What is radiation? Particles? Or waves? You'll soon see that both matter and radiation are made of quanta and that *each quantum* is extended spatially and must be classified as a wave in a field. Yet, each is a highly unified single object best described as a bundle of field energy, giving quanta a strong, but ultimately misleading, particle flavor.

From 1925 to 1930, Planck, Einstein, Bohr, de Broglie, Erwin Schrödinger, Werner Heisenberg, Wolfgang Pauli, Paul Dirac, Max Born, and many others proceeded from the breakthroughs described above to fashion a new theory that replaced both the classical particle theory of Newton and the classical EM field theory of Faraday and Maxwell as the fundamental theory of the behavior of matter and energy. Scientists and thoughtful people all over the world are still coming to terms with this theory of the quantum. I hope this book furthers this process.

The remaining chapters present the major principles of this distinctly nonmechanical "quantum mechanics," as it was called in 1930. Although this theory is barely noticeable (i.e., barely distinguishable from classical physics) in the world of our macroscopic perceptions, quantum physics looms large at the microscopic level and has yielded an extraordinary new understanding of that world, leading to technologies that have affected every aspect of our lives.

But more important in the long run might be the effect of quantum physics on humankind's habits of thought. The universe is deeply quantum: it was born in a quantum event; the layout of its galaxies reflect deep quantum realities; its very fabric is woven according to quantum, not classical, principles. The conceptual space between the classical and quantum descriptions is enormous, yet our intuitions remain in the grip of science's long-gone Newtonian era rather than its freshly astonishing quantum era. In my opinion, we are only beginning to explore these differences, only beginning to understand the philosophical, historical, and cultural significance of quantum physics. It's been difficult to explore the culture of the quantum seriously because there has been so little consensus about the physics of wave–particle duality, fundamental randomness, quantum states, superpositions, nonlocality, measurement, quantum jumps, and other quantum quandaries. Hopefully, this book is a step in the consilience—the enriching unification—of the world of the quantum with the broader culture.[30]

5

What Is a Quantum?

Democritus had an excellent question: Is the stuff of the universe spatially extended and infinitely divisible or is it made of indivisible particles? He suggested particles. Twenty centuries later, Newton said the same. Faraday and Maxwell suggested a space-filling EM field that provided an extended medium for EM forces acting between Newton's material particles. Einstein's gravitational theory sided with force fields. By the early 20th century, the universe appeared to be filled with fields of force connecting particles of matter.

We'll see in this chapter that quantum physics supports fields, with a vengeance. *Particles cannot explain the observed quantum phenomena, whereas fields can.* And quantum physics takes fields far beyond their classical status. *According to quantum physics, not only force, but also matter arises entirely from fields.* These fields are similar to classical EM fields, with one big difference: They are quantized. This gives them some particlelike aspects. For example, field energy can be increased or decreased only in indivisible steps. Democritus was half right. The universe is made of indivisible entities that you can count—1, 2, 3, and so on, but these quanta are not much like Democritus' particles. Each quantum is an extended bundle of field energy, quanta are not necessarily small, they overlap in space, they are flexible not rigid, and they are delicate not indestructible.

Quanta turn out to be indivisible, highly unified entities, despite being spatially extended. So just as he supposed, Democritus' chunk of gold *cannot* be subdivided endlessly. One eventually arrives at individual quanta, and you can't subdivide a quantum.[1] It's hard to imagine how all this can be. This chapter explains, with evidence, how it can be.[2]

Great art and great science have much in common. One definition of great art is that it's unexpected yet, once experienced, perfect: a symphony's surprising but perfect change of key, a story's surprising but perfect turn of plot. It's the same in science. Einstein's "field of spatial curvature" astonished scientists, yet turned out to be the perfect framework for gravity. In just this way, quanta are not what anyone could have expected, yet they turn out to be perfect. de Broglie, with his marvelously aesthetic intuition that matter must have wave properties, was the first to sense this. The universe, it seems, is an incomparably inventive artist.

What Is a Quantized Field?

Why is there something rather than nothing? It's perhaps the ultimate philosophical question, and the ultimate scientific question. In my opinion, science will get closer and closer to, but never arrive at, the answer.

Nevertheless, quantum physics comes surprisingly close to answering it. But of course you've got to accept a few quantum principles to get to the answer, and then you can always ask: Why are there these quantum principles rather than nothing? Here is the quantum argument.

Imagine a region of space in which there are absolutely no physical phenomena of any kind—no normal matter, no radiation, no dark matter or dark energy, no other physical entity, and no energy at all. Imagine, in other words, absolute nothingness throughout some region of space.

It is a fundamental principle of quantum physics that such a region of space cannot exist. Here's why.

Let's just consider, momentarily, EM phenomena. What does quantum physics tell us about the possibility of "EM nothingness," in which there are no EM phenomena of any kind—no photons, no electrically charged objects, no EM energy or EM fields of any kind. In this case, the EM field throughout this region of space would be precisely zero, with no uncertainty. Such a situation would violate Heisenberg's principle, according to which all fields must maintain at least a minimal degree of randomness. So the EM field must exist, and it must have at least a minimal, randomly varying energy everywhere. To put it another way, if a region of space somehow found itself with absolutely no EM field at all, that region would instantly vanish from the universe. As you'll see in this chapter, there are several different kinds of quantum fields—although most physicists think they are all simply different aspects of a single yet-undiscovered unified field. All these fundamental fields must exist everywhere in at least a minimal form. If space exists, there must be something in it, namely quantum fields.

This minimal field is called the *quantum vacuum* or *vacuum field*. It fills the universe. It's definitely not nothing. It's the starting point for quantum physics. That is, quantum physics begins from the notion that quantum fields (or perhaps a single unified quantum field) exist throughout the universe in at least their minimal vacuum form. These fields are "quantized"—in other words, they obey the principles of quantum physics.

The simplest example is the quantized EM field. The EM vacuum field extends throughout the universe, even in regions between the galaxies where "nothing"—no thing—exists. In such vacuum regions, there are no photons, no charged objects, no EM forces, but there is, a minimal, randomly fluctuating EM field that satisfies the indeterminacy principle.

Everything said about the classical EM field applies to the quantized EM field. It is a property of space itself. The quantized EM field fills space the way smoke can fill a room. It's the medium for transmitting EM forces between electrically charged objects, including transmitting all the EM radiations.

Sticking with only EM phenomena, *everything that happens electromagnetically can be described as disturbances—waves—in the EM vacuum field.* This is not a difficult concept. All EM forces and radiations are simply ripples in the vacuum EM field, analogous to ripples in a tub of water except that water waves happen on a two-dimensional surface whereas EM waves happen in three-dimensional space. Even in a tub of water with no ripples, we still have a tub of water—analogous to the EM vacuum field. Now imagine a single ripple carrying energy across the water's surface. This ripple is analogous to a single photon carrying EM radiation through the EM field. *Photons are ripples—waves, or disturbances—in the universal EM field.*

This sounds pretty much like classical EM waves (Chapter 4), but with the additional notion of the vacuum field. Here's what makes this a quantum field: In the real world, the EM field is *quantized*. This means that *all disturbances— all photons—come in the form of highly unified EM energy bundles called* quanta. So every EM interaction occurs via the instantaneous creation, destruction, or exchange of one or more entire quanta of energy. Each bundle carries a prescribed amount of energy, as explained in more detail later. Thus, a quantum of EM energy (a photon) has an all-or-nothing character: it's entirely present in the field or entirely absent. You can't have just part of a photon.

This all-or-nothing character entails that, like the quantized swing of Chapter 2, you must add or remove an entire quantum of energy instantaneously, giving energy exchanges a jerky, jumpy character. Furthermore, although each quantum is spatially extended (because it's part of the extended EM field), it always behaves as a single unit. Any alterations in a quantum extend instantaneously to the entire quantum, even if it's spread out over many kilometers. You can't alter part of a quantum because it doesn't have "parts"; it's a single thing.

In this chapter we'll examine several other kinds of quantized fields. Each obeys the previous principles: each has a vacuum field, all processes occur as waves or ripples in the vacuum field, and all such waves occur as highly unified but spatially extended quanta carrying a specified quantity of energy.

Max Planck, in 1900, was the first to notice quantization. It was not easy to notice. To see why, consider a typical quantum of visible light, such as a photon of orange light with a frequency of 5×10^{14} Hz. In his paper in 1900, suggesting—for the first time—the quantum idea, Planck stated the following formula for finding the energy of one photon[3]: Using standard metric units, multiply the radiation's frequency (measured in Hertz) by the number 6.6×10^{-34}. So the energy of our orange photon is

$$\text{energy} = (6.6 \times 10^{-34}) \times \text{frequency} = (6.6 \times 10^{-34}) \times (5 \times 10^{14})$$
$$= 3.3 \times 10^{-19} \text{ joules.}$$

The number 6.6 × 10⁻³⁴ is now called *Planck's constant*. As you know (Chapter 3), 1 joule is a rather modest amount of energy. So a typical quantum of light—a typical photon—has a seriously small energy: 0.000,000,000,000,000,000,33 joule! This is why you've never noticed that light is made of photons.

Planck's idea made all the difference. Quantum physics exists because the quantum exists (i.e., because quantized fields exist). Although we never notice quanta in our daily lives, were the universe not made of these small bundles of energy, our world would be classical. One can imagine a universe in which Newtonian physics rules. Our universe would be Newtonian if Planck's constant were exactly zero rather than 0.000,000,000,000,000,000,000,000,000,000,000,66. It would be boring, but that's OK because we wouldn't be here to endure it. Atoms could not exist if Planck's constant were zero, because electrons within atoms would have no state of lowest possible energy (Chapter 7) and would spiral down into the nucleus. The world would consist only of tiny electrically neutral nuclei, so there would be no chemistry. Things would be much less interesting, and there would certainly be no ice cream. Luckily, Planck's constant is not quite zero. It's plausible to claim that this (namely, Planck's constant being not equal to zero) is the reason there is something rather than nothing.

An Experiment with Radiation

Recall the double-slit experiment with light (Figure 2.2). The experimental result (Figure 2.3) demonstrates wave interference, showing light is a wave in an extended medium—an EM field. This section tells of a more detailed version of this experiment. It provides evidence that the EM field must be quantized—it's energy must be bundled in the form of photons—*but the experiment does not alter our conclusion that light is a wave in an EM field.* This experiment provides evidence for, and an example of, the theory of quantized fields presented earlier.

Imagine performing the double-slit experiment with extremely dim light. One expects the result to look like Figure 2.3 only dimmer, and this would be correct if the EM field were classical (not quantized). But this is not what happens. Figure 5.1 shows what does happen.[4] In sufficiently dim light and with a short photographic exposure time, *light impacts at only a few points on the screen*, with no discernible trace of interference (Figure 5.1a)! Longer exposure simply yields more impacts (Figure 5.1b). But with sufficiently long exposures (Figure 5.1c–e), *an interference pattern emerges in the distribution of impacts.* The same phenomenon shows up in ordinary photographs (Figure 5.2).[5] It's similar to the pointillist paintings of some French impressionists.

It's natural to conclude from the tiny bright spots that light is, after all, made of tiny particles. This would be hasty, to say the least. One reason it's hasty is that the spots are not all that small. The small points of light in Figure 5.2 represent the exposure of a single silver photographic grain that is 2 millionths of

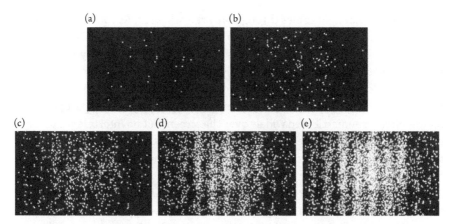

Figure 5.1 (a–e) Double-slit experiment with dim light and varying exposure times. The interference pattern builds up from particlelike impacts.

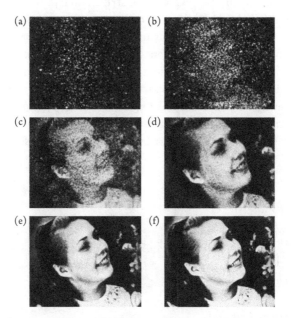

Figure 5.2 As the exposure time gets longer and longer, a photo emerges from individual particlelike impacts

a meter across and is made of a billion atoms. When viewed at the atomic level, this is not tiny. So let's just say these small spots are made by "photons," without committing ourselves to any picture of what a photon might be—particle, wave, or something else.

This issue of what light is made of has been debated since Newton's day, and we'll spend some time on it. First, consider the double-slit experiment with one slit covered so that light comes through only the other slit. As you know (Chapter 2), waves coming through a narrow opening will spread out or "diffract" on the far side of the opening. Bright light coming through a single, very narrow slit (having a width comparable with the light's wavelength) does the same thing: the light widely illuminates the viewing screen, as shown in Figure 5.3. If we then use dim light and short exposure times, we again see small impact points of light—photons—distributed widely over the screen, but no interference pattern emerges as we increase the intensity. Instead, the entire screen fills with small spots of light, and, like a pointillist painting, the smooth distribution of Figure 5.3 emerges as the number of small spots increases.

Suppose we assume that photons are small particles. It's easy to imagine how the wide single-slit pattern of Figure 5.3 might then be explained. Perhaps each photon interacts with the edges of the slit, causing the photons to veer sideways and impact over the entire width of the screen with each photon impacting at only one point but collectively filling the entire screen.

But how can small particles explain the striped double-slit interference pattern of Figures 2.3 and 2.4? By the very definition of the word *particle*, one particle cannot go through both slits. It's unreasonable to suppose that a particle coming through one slit could experience forces resulting from the other slit, or could "know" in any other way whether the other slit was open or closed; this is because, from the viewpoint of a tiny particle passing through one slit, the other slit is very far away. As you'll see, this experiment has been done not only with photons, but also with electrons, neutrons, atoms, and many types of molecules, and we always get an interference pattern similar to Figures 2.3 and 2.4. It's

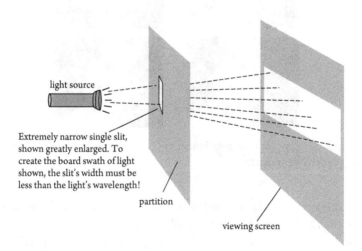

Figure 5.3 Single-slit experiment with light. Light arrives in a wide swath across the screen, with no interference pattern. Contrast this with Figure 2.4 which shows what happens when two slits are open.

difficult to imagine that any known force—gravity, electromagnetism, strong, or weak—could cause a tiny particle of light coming through one slit to act differently depending on whether the other slit was closed or open.[6]

Perhaps the interference pattern is caused by photons coming through one slit interacting with photons coming through the other slit. We can test this hypothesis by allowing only one impact to occur at a time. Even when the beam is so dim that impacts occur at, say, one per minute, the interference pattern still forms after a sufficiently large number of impacts. So interactions between different photons cannot explain the pattern.

Because the stripes in Figure 5.1 emerge even when photons pass through the slits one at a time, and do not form when only one slit is open, we are forced to conclude that *each photon comes through both slits*. Whatever the objects are that make the small spots in Figure 5.1, each one must extend over, and pass through, both slits. As one of quantum physics' founding fathers, Paul Dirac, put it, "Each photon ... interferes only with itself."[7] But the slits could be, say, a millimeter apart, over a million times bigger than an atom. We are thus led to an important conclusion: *A photon can't be a small, atom-size particle.*

Here's an entirely different argument supporting this crucial conclusion. It happens that, when we increase the separation between the two slits, the interference pattern squeezes up (i.e., the spacing between bright lines gets smaller).[8] This squeezing implies that *all* the photon impacts on the screen shifted in position. But we can increase the separation by moving only the left-hand slit farther to the left. The particle hypothesis implies that half the photons come through the right-hand slit. If photons are particles, how could the impact points of these right-slit photons be altered when only the left-hand slit was moved? Apparently, *each impact was made by a single photon that came through both slits.*

This is odd, but for Richard Feynman it was not simply odd, it was paradoxical—logically contradictory. Although this great and highly intuitive genius helped invent modern quantum field theory (which combines quantum physics with Einstein's special theory of relativity) during the 1940s, Feynman was strictly a particles guy. As quantum theorist Frank Wilczek put it, "uniquely ... among physicists of high stature, Feynman hoped to remove field–particle dualism by getting rid of the fields."[9]

As a preface to his own lecture about the double-slit experiment, Feynman advised his students:

> Do not keep saying to yourself, if you can possibly avoid it, "But how can it be like that?" because you will get "down the drain," into a blind alley from which nobody has yet escaped. Nobody knows how it can be like that.[10]

But "How can it be like that?" is always an excellent question. The particles guy is really saying that the interference pattern cannot be explained by his own preconceptions—namely, that the universe is made of particles.

But if photons are not particles, how can we explain the particlelike impact points? Here's how: Each photon is a spread-out wave, an extended bundle of field energy that comes through both slits and then fills the screen just before interacting with it. *Each photon then interacts with the entire screen.* But the screen is made of zillions of individual atoms. Because a photon cannot be subdivided, it must deposit its entire energy into just one of these atoms "chosen" randomly (Chapter 6). This atomic-level interaction is then amplified so that a tiny visible flash appears on the screen.

The small, limited extent of each flash is a feature of the photon–screen interaction, not of the photon itself. Imagine that the photon is a large balloon settling down on a bed of nails. The interaction—the bang—is going to occur at only a single nail. This does not imply that, before the bang, the balloon was present only at that single nail.

As each photon approaches the screen, it incorporates the entire interference pattern; it contains information favoring an interaction at the bright lines rather than the dim lines. In a typical double-slit experiment with light, the pattern (and hence the pre-impact photon) is a few centimeters wide. Each photon contacts all the screen's atoms within the pattern, filling the entire region, yet it must give up its energy to only one atom because it's a unified quantum. But the photon only carries information about the overall striped pattern, not about any particular atom where it should interact. Hence, the selection of a particular atom is made randomly but within the confines of the overall pattern. The photon–atom interaction causes the photon to vanish.

Summarizing: When both slits are open, each photon comes through both slits and approaches the screen as a single object spread out over the entire interference pattern. Because it is a single unified object, the quantum must then interact with a single atom, selected randomly, at the screen. The quantum then gives up all its energy to this atom, i.e. it "collapses" by vanishing into this atom. It acts like a wave as it passes through the slits and approaches the screen, but then it hits like a particle. It is, at all times, a wave in a field.

How big is a photon? According to Louis de Broglie, a founding father of quantum physics, an individual quantum "fills all space" (see his statement quoted later). He was talking about electrons, but the same principle applies to photons. Quantum theory implies that every free (unconstrained) quantum is extended indefinitely in space and is not contained within any region of finite volume. So any talk of the size of a photon or of any other quantum must refer to some finite region within which the photon has a large probability (say, 99%)

of interacting with a detector such as a viewing screen. How large, then, is such a high-probability region for a single photon?

I've argued that each photon in a double-slit experiment comes through both slits and extends over the entire interference pattern. The visible double-slit pattern (Figure 2.3) can easily be a few centimeters wide, so the high-probability region of each of these photons must be this wide. But photons can be far larger than this. The Very Large Telescope on Cerro Paranal mountain in northern Chile verifies visible and infrared photons that are at least 100 m wide (transverse to their motion) by detecting interference in the light received by four separate, movable telescopes separated by a total of 100 m. Because "each photon interferes only with itself" (Dirac), the photons detected by this array must have had lateral extensions of at least 100 m. Radio photons, having longer wavelengths, can be far larger than 100 m. The Very Large Array near Socorro, New Mexico, in the United States, is the world's largest radio telescope array. It stretches along a 36-kilometer line, and demonstrates the interference of radio photons over this distance, verifying that photons can be 36 kilometers wide.

Using a technique known as *very-long-baseline interferometry*, a team of radio astronomers is linking up several radio telescope arrays located around the world into a single Event Horizon Telescope. Each individual array will record photon arrival data very accurately in space and time, and computers will, at a later time, combine the separate observations to obtain results that are equivalent to observations that could be made by a single Earth-size radio receiver. The Event Horizon Telescope will focus on the giant black hole, with a mass of four million suns squeezed into a region the size of Mercury's orbit, at the center of our Milky Way galaxy. It will observe processes taking place just outside the edge or "event horizon" of this black hole—the spherical surface from which nothing can escape. This will be by far the most extreme gravitational environment ever observed and will provide an ultimate testing ground for Einstein's general theory of relativity. The photons detected by this combined system will be as wide as our planet, yet, when detected by a receiver, they will collapse instantaneously into a small region like the points of light seen in Figure 5.1.[11]

An Experiment with Matter

You've seen evidence that at least one fundamental force, the EM force, is a universe-filling field, just as Faraday and Maxwell had thought, but that it's a *quantized* field, which is something new.

Now let's turn from force to matter, which has been thought, traditionally, to be made of particles. Recall de Broglie's suggestion that, like EM radiation, matter should also exhibit both wave and particle aspects; and recall Davisson and Germer's demonstration that electrons, a form of matter, exhibit wave aspects. Today, experiments show quite directly that each electron is indeed a wave in a field. But how can an electron, which was supposed to be a particle, be a wave? We'll find the answer to this question by studying a variation on the preceding double-slit experiment. As Feynman put it, the double-slit experiment "has in it the heart of quantum mechanics. In reality, it contains the *only* mystery."[12] This time, we'll assume this experiment uses electrons rather than photons.

There are many ways to create a steady stream of electrons—for example, by heating a wire electrically until electrons boil off and fly away from the wire. Suppose we send a stream of electrons toward a double-slit setup, just as in Figure 2.2, but with an electron source instead of a photon source.[13] Such a double-slit experiment with electrons was done for the first time by Claus Jonsson in 1961.[14] The result is surprising. The image on the viewing screen is an interference pattern that looks exactly like Figure 2.3! The only difference is that the pattern is much smaller (thinner bright and dark lines) because typical electron wavelengths are far smaller than light wavelengths.

What's going on? How can a stream of electrons form an interference pattern? It's supposed to be a stream of tiny particles, after all. What's waving?

A revealing experiment was performed by Akira Tonomura and colleagues in 1989.[15] They lowered the intensity of the electron stream so that it carried only about one electron per millisecond (a thousandth of a second), guaranteeing that it was highly unlikely for more than one electron at a time to be moving from slits to screen. This rules out the possibility that interactions or cooperative effects between electrons (similar to a sound wave) could be responsible for the outcome. They also shortened the exposure time. One might guess this would produce a dimmer version of the interference pattern observed by Claus Jonsson in 1961, but this guess would be wrong. Figure 5.4 shows what this does produce.

Exactly as in Figure 5.1, the interference pattern builds up from the statistics of thousands of small individual impacts. The final impact pattern is a "pointillist" effect from many small impacts. Physicists had predicted this result since about 1930, but this difficult experiment couldn't be done until 1989. The interference pattern is quite clear in the final photograph.

Perhaps you've guessed what is causing the individual impacts: electrons! In Figure 5.4a, just a few electrons have impacted; in Figure 5.4e, zillions have impacted.

Tonomura's experiment reveals the heart of quantum physics. The experiment shows electrons play the same role in forming an interference pattern in the electron double-slit experiment (Figure 5.4) that photons play in forming an interference pattern in the photon double-slit experiment (Figure 5.1). So electrons are analogous to photons!

That is, electrons are to matter as photons are to EM radiation!

But photons are quanta of the EM field. So for all the reasons discussed in the preceding section, *electrons are the quanta—the energy bundles—of a space-filling*

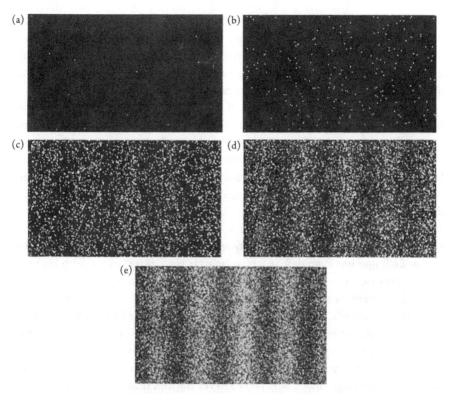

Figure 5.4 (a–e) Buildup of an interference pattern in the double-slit experiment using electrons. As in Figure 5.1, an interference pattern builds up from individual impacts over longer-time exposures.

field. But that field cannot be the EM field, because its quanta (its bundles of field) are already known to be photons. Photons are radically different from electrons. For one huge thing, photons are nonmaterial (zero mass, travel at light speed), whereas electrons are material (nonzero mass, travel slower than light speed). So electrons must be quanta of some field other than the EM field.

This new physical field is the central discovery of quantum physics.[16]

The new field goes by several names: *psi* (for the Greek letter Ψ), *electron–positron field*, or *matter field*. For obscure historical and technical reasons, physicists often call waves in this field *wave functions* or *probability waves*. These latter two mathematical terms are off-putting for most people and, far worse, they give the wrong impression that this new field is merely a mathematical abstraction when actually it's quite real, it's the heart of quantum physics, and it's at the heart of all matter. Be careful not to confuse this psi field with our other featured quantized field, the EM field.

This new result is quite far-reaching. Experiments with other forms of matter show that it applies not just to electrons, but also to protons and neutrons. In appropriate double-slit experiments, for instance, protons form an interference pattern similar to Figure 5.4, and so do neutrons. In fact, even atoms form a similar interference pattern, and so do molecules; this is surprising, because atoms and molecules are "composite" objects, made of many electrons, protons, and neutrons. This means all these material objects—protons, neutrons, atoms, and molecules—are quanta of a matter field.

According to high-energy physics, there are several kinds of these matter fields,[17] but in most situations on Earth they are all essentially identical and we use the same symbol Ψ for the matter field for electrons, protons, neutrons, atoms, and molecules. Scientists have performed interference experiments similar to the double-slit experiment not only with electrons but also with protons, neutrons, and various kinds of atoms and molecules; all these experiments produce interference patterns similar to Figure 5.4, demonstrating that each of these objects is the quantum of a field—a quantum that can, for example, come through both slits of a double-slit experiment. University of Vienna physicist Anton Zeilinger has even demonstrated matter wave interference with C_{70} molecules. Because each molecule contains 70 atoms, and each atom contains six protons, six neutrons, and six electrons, a C_{70} molecule is quite a hefty object by atomic standards. Zeilinger's interference pattern was like Figure 2.3, only very much smaller because the wavelength, and thus the pattern, is not only smaller for material quanta, it gets smaller still as the masses of the quanta get larger. This is what you'd expect, because quantum effects become negligible as masses become large. That's why you've never noticed that cabbages, kings, and playground swings follow the quantum rules.

The existence of matter fields, and the fact that they are quantized, is the elusive principle scientists were groping toward since 1900. It's one of the great unifying principles of physics that *not only radiation, but also matter are made of quanta*. It's not for nothing that the experiment that produced Figure 5.4 has been called "the most beautiful experiment in physics."[18]

Quantized matter fields explain the absolute uniformity of microscopic matter. As far as we can tell, every electron is precisely identical to every other electron, every proton is identical to every other proton, all hydrogen-1 atoms (i.e., atoms with a nucleus of just one proton) are identical, and so forth. Why is this? The explanation remained a profound mystery until the discovery of matter fields showed that electrons are identical because every electron is a ripple in a single matter field that fills the universe. This not only explains the absolute identity of all electrons, it also represents another unification. The zillions of electrons, for example, are instances of one thing: the universal electron–positron field.

So electrons are not tiny particles at all: they are ripples in a field that fills the universe. Here is how Louis de Broglie put it memorably in 1924:

> The energy of an electron is spread over all space with a strong concentration in a very small region.... That which makes an electron an atom of energy is not its small volume that it occupies in space,

I repeat it occupies all space, but the fact that it is undividable, that it constitutes a unit.[19]

So-called particles of matter are not particles. They are quanta of matter fields in exactly the same sense that photons are quanta of the EM field. All these things are merely ripples in unseen fields. Not only electrons and quarks, but also composites of electrons and quarks such as protons, neutrons, atoms, and molecules are quanta of a matter field. As quantum field theorist Steven Weinberg puts it, "The basic ingredients of nature are fields; particles are derivative phenomena."[20]

But Democritus' and Newton's notion that matter is made of tiny particles dies hard. Although the double-slit experiment with light, along with Faraday and Maxwell's theories of electromagnetism, convinced 19th-century physicists that light is a wave in a field, de Broglie's idea that matter, also, is a wave in a field remains difficult for many physicists to accept despite its repeated experimental confirmation. In the remainder of this chapter I discuss quantum fields further and describe additional evidence that the universe is made entirely of them.[21]

Quantum Field Theory: Quantum Physics Meets Special Relativity

Because quantum physics developed during 1900 through 1930, when Newton's physics still ruled, physicists inappropriately interpreted obvious field concepts in terms of Newtonian particles. As it turned out, a field view ultimately prevailed in "high-energy" or "relativistic" quantum physics, whereas an inconsistent particles view prevailed in lower energy nonrelativistic quantum physics. This has created untold confusion throughout the years.

In a burst of inspired guesswork in 1925, Erwin Schrödinger constructed an equation that de Broglie's hypothesized matter waves should obey. To this day, Schrödinger's equation is arguably quantum physics' most widely used mathematical formulation and the basis of most quantum physics instruction. It's a so-called "partial differential equation" that is clearly analogous to Maxwell's partial differential equations. Just as Maxwell's equations describe waves of radiation propagating through the EM field, Schrödinger's equation describes waves of matter propagating through a matter field that he represented by the Greek letter psi (Ψ).[22] Schrödinger showed that his equation had something strikingly right about it by using it to correctly predict the EM spectrum of the hydrogen atom. Today, in view of the parallel between the quantized EM field (Figure 5.1) and the quantized matter field (Figure 5.4), and for many other reasons, it's clear that the quantum physics of both radiation and matter is about fields, not particles, and that Schrödinger's equation describes matter fields in a manner parallel to the way Maxwell's equations describe EM fields. Unfortunately, most physicists have ignored these signs that quantum physics is all about fields, insisting on interpreting Ψ in terms of particles rather than fields.

During the late 1920s, Heisenberg, Born, Dirac, and others applied the new quantum principles to the classical EM field. Although Schrödinger's equation was "nonrelativistic"—meaning that it didn't take Einstein's special theory of relativity into account—any quantum theory of the EM field must take special relativity into account because EM radiation is the ultimate high-speed relativistic phenomenon. As you know (Chapter 3), the classical EM field exists at every point in space, so any quantum theory of this field must apply the quantum principles at every point in space. Thus, the quantum theory of electromagnetism was necessarily a theory about fields, not about particles. This and other theories that combined quantum ideas with relativistic ideas became known as *quantum field theories*.

Paul Dirac, in 1927, put together the first reasonably complete quantum field theory of the EM field and electrons, and showed that bundles of light (photons) emerged as a logical consequence of applying quantum physics to the EM field.[23] In physics jargon, Dirac "quantized Maxwell's equations." Two basic EM phenomena are the emission and absorption of radiation. In Dirac's theory, these phenomena occur via the creation and destruction of photons. Emission occurs when, for example, an electron in an atom "falls" into a state of lower energy and consequently creates a photon. Quantum field theory describes such processes in quantitative detail. This is new and revolutionary ground. Classic atomism—irreducible particles that can't be created or destroyed—is inconsistent with photon creation and destruction as predicted by quantum field theory's conjunction of quantum physics and relativity.[24]

Photons are quite unlike the indestructible particles Democritus and Newton had in mind.

It soon became apparent that not only photons, but also material quanta such as electrons and protons could be created and destroyed. For example, a sufficiently high-energy photon can transform into an electron–positron pair (Chapter 3), creating material quanta entirely out of EM field energy. Today, we know that quanta of every sort are created and destroyed in wild abundance in such high-energy phenomena as exploding stars and cosmic ray impacts on Earth's atmosphere, as well as in experiments at Geneva's Large Hadron Collider, where pairs of protons smash into each other to create a plethora of quanta from the collisional energy.

So when you picture the microworld, do not envision isolated particles moving in stately solitude through empty space; picture, instead, an ever-changing, energetic, unpredictable, crowded kaleidoscope of field quanta in spontaneous destruction and creation. The fully fledged tale of the quantum—namely, quantum field theory—describes and predicts all of this. It informs us that, through and through, this dynamic universe is both quantum and relativistic.

The universe's most characteristic phenomena, including the Big Bang, involve the creation and destruction predicted by quantum field theory. These processes arise spontaneously because the universe is made of energy, because energy assumes a multitude of forms randomly, and because mass and energy are equivalent. Reality is a set of space-filling fields with quanta rippling through and energy, like wind stirring a lake, generating the possibility of change, destruction, and creation.

Fields are all there is. This seems incredible when one considers, say, rocks and tables. How can these be simply fields, analogous to a magnetic field? *It's worth pondering. . . .* Stop reading and slam your hand down on a table. The table slams back, as you can tell from your stinging hand. It's a reality check, like the rock that Samuel Johnson kicked. Now look at the microscopic picture: The table and your hand are made of atoms, with electrons that create EM fields. Slamming brings your hand's electrons near the table's electrons, and their EM fields repel each other. This repulsion by force fields against other force fields stops your hand. It also distorts the molecules (more fields) in your hand, causing nerve cells (more fields) in your arm to transmit a signal to a pain center in your brain (more fields, surely). It all comes down to fields.

Although Schrödinger's equation provides a firm foundation for the quantum physics of slow-moving matter, it runs into problems at higher energies because it happens not to conform with the principles of special relativity. Paul Dirac fixed this in 1928 by inventing an equation that generalized Schrödinger's equation so that it incorporated the principles of special relativity and thus described high-energy relativistic matter correctly. Dirac's equation was even more accurate than Schrödinger's in predicting the hydrogen spectrum and also incorporating a new, distinctly quantum, aspect of the electron known as *spin*.[25] Like Schrödinger's equation, Dirac's equation was a partial differential equation similar to Maxwell's equations, and it described the same physical field Ψ that Schrödinger's equation described. In fact, when applied to slow-moving matter, Dirac's equation becomes identical to Schrödinger's equation.

Physicists initially regarded Dirac's equation in the same way as they regarded Schrödinger's equation—as a description of material particles. But Dirac's equation can also be viewed as a description of Ψ that is analogous to the way Maxwell's equations describe the classical EM field. This led Dirac and others to discover something new: Dirac's equation could be "quantized," in the same manner that Dirac had quantized Maxwell's equations. Just as the quantization of Maxwell's equations predicts the creation and destruction of photons, quantization of Dirac's equation predicts creation and destruction of material quanta. If you're guessing that these quanta are electrons, you're on the right track. But quantization of Dirac's equation predicts something unimagined: the creation and destruction of *pairs* of material quanta, with each pair comprising one electron *and one positron*—an object just like an electron but with positive, instead of negative, charge.

The predicted positron was discovered experimentally a few years later, by Carl Anderson in 1932. The prediction and experimental confirmation of a previously unknown quantum was one of the greatest achievements of quantum physics, and implied something new and profound in the quantum way of looking at fields such as the EM field and the Ψ field. Of equal significance was the understanding of the positron as the antiquantum of the electron. It was soon predicted, and verified experimentally, that all material quanta have corresponding antiquanta: there are antiprotons, antiquarks, and antineutrinos.[26] Antihydrogen atoms—a positron orbiting an antiproton—have been synthesized and stored for more than 15 minutes, which is a difficult task because antiquanta are annihilated quickly on encountering the ordinary (not anti-) quanta that dominate the universe.[27]

Previous theories whether Newtonian or relativistic or quantum, had described motion, but not creation or destruction. *Quantum field theory tells us not only how things move, but also what kinds of things can exist and how they are created.* And quantum field theory puts matter and radiation on the same footing. No longer bifurcated into tiny material particles and space-filling force fields, the universe is made of one kind of thing: quantized fields.

Note that, in this new theory, electrically charged objects such as electrons are still related to the EM field much as they were related in classical physics: electrons exert forces on each other by means of the EM field acting between the electrons. In classical (pre-quantum) physics, the electrons are particles while the EM field is a spatially extended classical (non-quantized) field. But in quantum field theory, both the electrons and photons are the quanta of fields. This is an enormous unification: There are no longer both fields and particles, there are only quantized fields. The electron is the quantum of the psi-field (or electron-positron field), while the photon is the quantum of the EM field; thus the electrons interact via processes occurring among photons. Photons are said to *carry* or *transmit* the EM force that acts between electrons.

Paul Dirac, Werner Heisenberg, Wolfgang Pauli, and others laid the foundations of quantum field theory during the 1920s and '30s, but despite its practical successes the theory was plagued by mathematical inconsistencies and unable to explain certain experimentally measured details. Along with the rest of the world, most physicists went to war during 1939 to 1945. Right afterward, in 1947/1948, a Japanese and two Americans—Sin-Itiro Tomonaga, Julian Schwinger, and Richard Feynman—each working independently, published quantum field theories of the EM force. Although the three theories looked different, they turned out to be logically equivalent, they solved the existing difficulties of previous quantum field theories, and they won a shared Nobel Prize for all three. Despite the equivalence of their theories, Schwinger interpreted his work in terms of fields whereas Feynman interpreted his in terms of particles.

Remarkably, the Dirac equation in 1928, its quantization shortly thereafter, and its epochal confirmation in 1932 with the discovery of the positron, failed to convince the physics community that quantum physics is entirely about fields. After all, Dirac's equation, especially its quantized version, obviously described an extended matter *field* Ψ that was analogous to the EM field of Maxwell's equations. Yet, most physicists continued to insist that this same entity Ψ, when it appeared in Schrödinger's equation, described low-energy material *particles*. So the same physical concept, Ψ, somehow represented particles in Schrödinger's equation but a field in Dirac's equation, even though Schrödinger's equation was simply Dirac's equation restricted to the case of slow-moving electrons. It was as though slow-moving particles, by simply moving faster, somehow transformed into space-filling fields.

This contradiction probably arose for pedagogical reasons. Every college physics department teaches nonrelativistic (low-energy) quantum physics at all levels of instruction, based on Schrödinger's equation. But relativistic quantum physics, based on quantized fields such as the Ψ of the quantized Dirac equation, is taught only in advanced graduate courses and is familiar only to specialists. This splits physicists into two noncommunicating camps: a nonspecialized low-energy particles camp and a specialized high-energy fields camp. The contradictory opinions about the meaning of Ψ probably arise from the lack of communication between these two camps.

It's important to use the right words. We've seen that the double-slit experiment must be understood in terms of quantized EM fields and quantized matter fields. This makes the term *quantum field theory* misleadingly superfluous. It's all just quantum physics, and quantum physics is entirely about fields. It also makes the word *particle* superfluous and hugely misleading. There's a simpler (fewer syllables) and far more accurate word: *quantum*.

There are no particles. Do not imagine, when visualizing a photon or electron as a ripple in a field, that there is a tiny particle floating within the ripple. The ripple *is* the photon or electron. Fields are all there is.

The Fundamental Quanta and the Standard Model

Unification is a recurring theme of science. Copernicus unified Earth with the other planets; Newton unified physics on Earth with physics in the heavens; Faraday and Maxwell unified electricity, magnetism, and light. As you can see from Figure 5.5, quantum physics continues this trend and seeks to extend it. This section summarizes the status of this progress.[28]

Many chapters are already written in the book of quantum unification. The zillions of quanta are disturbances in a few kinds of fields. Many quanta are composites of other more fundamental quanta. For instance, each proton and each neutron is made of three quarks. Thus, according to the Standard Model of

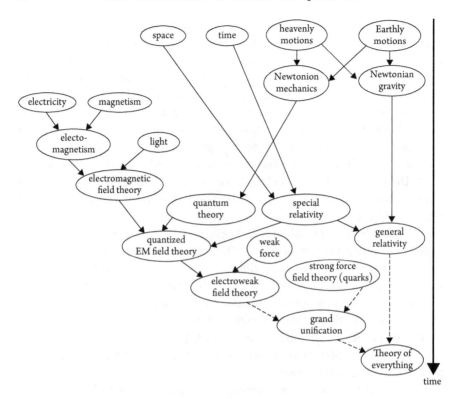

Figure 5.5 Unifications during the history of physics. Time runs from top to bottom. Arrows represent scientific advances and unifications. Dashed arrows represent possible future unifications.

Fundamental Fields and Quanta—a boring title for the world's most powerful scientific theory—there are only a few quantum fields and all quanta are either bundles of energy of these fields, or composites of such bundles. Table 5.1 lists the standard model's fields and their associated quanta. The standard model represents a consensus among physicists that has existed since the 1970s. I'll summarize a few highlights here; see the end notes for further references.[29]

The standard model describes normal matter and nongravitational forces; it doesn't describe gravity, dark matter, or dark energy. Contrary to what the word *model* might imply, it's really a well-confirmed theory, although it's known to be incomplete. According to this theory, there are two kinds of matter fields (quark and lepton) and two kinds of force fields (strong and electroweak) plus the Higgs field. The quark field has six kinds of material quanta called *quarks,* and the lepton field has six kinds of material quanta called *leptons.* The difference between quarks and leptons is that quarks experience both the strong and weak forces whereas leptons experience only the weak force. Protons and neutrons are not listed in the table because they are composites of quarks, and not fundamental.

Table 5.1 **Summary of the Standard Model of Fundamental Fields and Quanta, a theory of the fundamental fields. Each field has one or more kinds of fundamental quanta. The table omits the anti-quanta. The theory is known to be incomplete because it doesn't include dark matter, dark energy, or gravity.**

matter fields:	first-generation quanta	second-generation quanta	third-generation quanta
quark fields:	up quark	charm quark	top quark
	down quark	strange quark	bottom quark
lepton fields:	electron	muon	tau
	electron neutrino	muon neutrino	tau neutrino
force fields:			
strong field:	8 massless quanta called "gluons"		
electroweak field:	4 quanta: photon, W+ boson, W- boson, Z boson		
Higgs field:	1 confirmed quantum: Higgs boson		

For reasons nobody understands, the 12 material quanta come in three *generations* that exhibit parallel, or analogous, properties. Each generation contains two quarks and two leptons, as shown in Table 5.1.

First-generation quanta are stable, but the second and third generations are unstable—meaning that they decay quickly and spontaneously into lower generation quanta. All three played important roles for a fraction of a second during the Big Bang, after which the second and third generations decayed. Today, these higher generation quanta appear only briefly in isolated high-energy events such as cosmic ray impacts on Earth's atmosphere, explosions of stars, and high-energy physics experiments.

Nine of these material quanta carry electric charge and interact electrically, but each generation has a *neutrino* that is not electrically charged and interacts only via the weak force and gravity. Neutrinos are nearly undetectable and have almost zero influence on what goes on around us. Millions of them are zooming through your body right now; within a second they fly beyond the moon, many of them first passing undisturbed through the entire Earth.

Because the second and third generations are unstable, our normal surroundings are made of only the first-generation quanta. And the neutrino, despite its ubiquity, isn't noticeable in our normal surroundings. So *all ordinary matter comes down to just three kinds of quanta*: up and down quarks, and electrons. That's it! Protons are made of two ups and one down, neutrons are made of one up and two downs, and atoms are made only of protons, neutrons, and electrons. It's quite a unification.

Turning to the forces: Matter exerts forces via two force fields: the *strong field* and the *electroweak field* that transmits both the weak nuclear force and the EM force. In 1967, Pakistani physicist Abdus Salam and US physicist Steven Weinberg, working independently, uncovered a connection between the weak and EM forces. They proposed a new quantized field incorporating both—a unification comparable with the nineteenth-century unification of electricity and magnetism. The new *electroweak* theory predicted three distinct weak force quanta—called the W^+ *boson*, the W^- *boson*, and Z *boson*—so the electroweak force is "carried" by four different kinds of quanta (these three plus the photon). I'll clarify what *boson* means in Chapter 8. Quite unlike the massless photon, the W^+, W^-, and Z each have a mass of around 90 proton masses. In a major triumph for the new theory, all three were verified experimentally in 1983 following the construction of an accelerator sufficiently energetic to create these quanta's large masses.

The strong field has eight kinds of force-carrying quanta called *gluons* that transmit the strong force the way photons transmit the EM force. This field is the glue that binds quarks into protons and neutrons, and binds protons and neutrons into atomic nuclei (hence, "gluons"—cute, eh?).

Like photons, gluons travel at light speed and have zero mass. But there's a funny thing about W^+, W^-, and Z bosons that transmit the weak force. Because they are quanta of force fields, you might guess that, like photons and gluons, they would be massless and move at light speed. However, they have mass. Why is that? The answer is tied up with the following tale of the origin of mass.[30]

In 1961, Sheldon Glashow developed a theory that attempted to unify the EM and weak forces. His calculations implied that the weak part of the unified force should be transmitted by massless quanta, similar to photons. But photons move easily across the entire universe, giving the EM force its long range, and other massless force quanta should also carry long-range, not short-range, forces. The weak force has a tiny range—smaller than a nucleus—so its quanta must have mass, in disagreement with Glashow's theory.

Surprisingly, this puzzle's resolution turns out to be related to what happens inside a superconductor. Many metals, when cooled to near 0 kelvin (–273°C), become able to conduct electricity with no expenditure of energy— no *electrical resistance*. This *superconductivity* occurs because, when the temperature drops sufficiently, a "phase change" occurs that's a little analogous to what happens when the temperature of water drops below freezing. The superconductor's electrons then link up into two-electron pairs (note that this is a pair of *electrons* and not an electron–positron pair), the character of which is radically different from normal unpaired electrons.[31] This is best conceptualized in terms of the electron matter field, which transitions into a new state whose quantum is no longer a single electron but is instead an electron *pair*, that is, the basic "bundle" of this field is two paired electrons.[32] This transformed matter field causes photons to behave in a surprising manner. *Photons acquire inertia—mass—and move at less than light speed.* As you saw in the preceding paragraph, massive force carriers transmit forces only over short distances, so massive photons imply that the EM force becomes short ranged within a superconductor—so short ranged that the electron pairs can move through the metal freely, without experiencing the usual EM forces from their surroundings. This is why superconductors work: The paired-electron field causes photons to have mass and the EM force to have only a short range, and this neutralizes EM forces in the superconductor's interior, so electrons move freely.

Now let's return to the weak force. To explain its weakness, Peter Higgs and others suggested in 1964 that *an effect analogous to the superconducting phase transition exists throughout the universe.* According to Higgs' theory, there is a previously unknown universal quantum field that is analogous to the paired-electron field inside a superconductor. Photons do not interact with this *Higgs field*, but W and Z bosons do, and *this gives them mass*, just as the paired-electron field gives mass to photons inside a superconductor. The Higgs field affects W and Z bosons the way molasses affects a pebble moving through it. The W and Z become sluggish, in other words they acquire inertia or mass. As you might have already guessed, the quantum—the characteristic bundle of energy—of this new Higgs field is the famous *Higgs boson*.[33]

This explains why the weak force has such a short range. Just as the paired-electron field causes the EM force to be short ranged inside a superconductor by giving mass to photons, the Higgs field causes the weak force to be short ranged throughout the universe by giving mass to W and Z bosons.

This explanation comes with a huge bonus. It turns out that *all quarks and leptons must also interact with the Higgs field, giving these quanta an intrinsic mass.* This is definitely a good thing for you and me. Without an intrinsic mass, quarks and leptons would be zipping throughout the universe at light speed and could never be bound together to form atoms, or ice cream.

Electroweak unification illustrates the relationships between the fundamental fields. The theory connects EM and weak forces much as Faraday and Maxwell connected electric and magnetic forces. Such connections inspired Einstein's vision of a unified field that would incorporate all fundamental physical phenomena. Physicists now hope to find such a unification within quantum physics. In a burst of either enthusiasm or hubris, it's already been named: *the theory of everything*.

The Higgs boson remained the sole undiscovered prediction of the standard model until 2012, when it was created at the Large Hadron Collider. Its mass had been predicted to be large, and in fact turned out to be 133 proton masses.

As explained in Chapter 3, 95% of the mass of ordinary matter comes from the energy of the strong force binding quarks together, but the remaining 5%, resulting from quarks' intrinsic mass, which exists because of the Higgs field, is crucial. Without it, quarks would move at light speed and solid matter could not exist.

Beyond the Standard Model

There's more to the universe than the standard model. Although the standard model needed a Higgs field for logical consistency, additional Higgs–type fields could await discovery. A glance at Table 5.1 reveals similarities between quarks and leptons. There are six of each. They come in three generations, each comprising two quarks and two leptons. Such patterns inspire the search for a *grand unified theory* of the electroweak and strong fields. Dark matter and dark energy are not included in the standard model. They are almost certainly quantized fields. Dark energy is perhaps related to the Higgs field.[34]

Takaaki Kajita in 1998 and Arthur McDonald in 2001 made new discoveries lying beyond the standard model. In monumental experiments using enormous neutrino detectors buried far underground, they showed that at least one of the three types of neutrinos must possess mass, contradicting the standard model, which predicts that neutrinos have no mass and move at light speed. Their work also shows that neutrino masses are tiny—a million times smaller than an electron's mass. Why should neutrino masses be tiny rather than, simply, zero? Nobody knows; perhaps neutrinos interact with the Higgs field, or perhaps there is some mechanism other than the Higgs field that confers mass on them. Kajita and McDonald received the 2015 Nobel Prize in Physics for this discovery.

And there's a huge problem with gravity. The first fundamental force to be recognized, gravity will apparently be the last to be "quantized" (understood in quantum terms). Quantum physics describes correctly all microscopic phenomena encountered so far, whereas Einstein's general theory of relativity describes all large-scale cosmological phenomena correctly, including gravity.

Although the standard model and general relativity reign supreme today, they can't be all there is. General relativity is a classical theory whereas the standard model is a quantum theory, and there's no known way to make them get along together without altering one of them. According to general relativity, matter causes gravity and gravity in turn affects the structure of space and time.[35]

Microscopic quantum phenomena must alter general relativity's implications for space–time geometry at the microscopic level. But lacking a quantum theory of gravity, physicists don't know how to describe such alterations. Experimentally, the problem is complicated by the circumstance that quantum gravitational effects don't show up at the level of atoms, because gravity is practically negligible at this level. For example, the gravitational force between the proton and electron in a hydrogen atom is 1000 trillion trillion trillion times smaller than the EM force between them!

Surprisingly, it's at a level far smaller than atoms—a level so small that it has never been accessed by experiments—that the contradictions between general relativity and quantum physics really show up. Here's why. Imagine two protons smashing into each other as they do within the Large Hadron Collider, in slow motion, and at far higher energies than is possible at the Collider. We've seen that gravity is negligible at distances such as 0.1 nm (the distance across a hydrogen atom). But as our two protons get closer and closer, the forces between them increase, so the energy stored in their force fields increases. But *energy has mass, so the masses of the protons increase as they get closer.* This mass increase affects the gravitational force but not the EM force, causing gravitational attraction to increase enormously faster than the other forces as the protons get closer. Eventually, gravitational attraction exceeds all the repulsive forces, and the situation becomes unstable. Beyond this point, the distance between the protons continues to decrease spontaneously as their masses increase toward infinity. There must be something wrong with this picture, because an infinite mass would cause the entire universe to collapse!

The turnaround point, when repulsive forces and gravitational attraction become equal in magnitude, happens when the protons are about 10^{-35} m apart—a distance known as the *Planck distance*. In fact, theorists think all four known forces become about equal at this supersmall distance. At this distance, gravitational effects must become important in quantum physics, implying that general relativity must play a major role. A theory that incorporates the quantum principles into general relativity is needed.

This problem is reminiscent of a similar problem that arises in the case of an electron orbiting the nucleus of a hydrogen atom. Classically, the circling electron is predicted to radiate energy and spiral down into the nucleus, with the attractive EM force becoming infinite as the electron gets closer to the nucleus. But this instability would mean that atoms as we know and love them couldn't exist, and by now you are well aware that this means no ice cream. Niels Bohr found the solution to this conundrum in 1913, by introducing quantum notions into atomic structure. These notions implied a lowest possible energy state, corresponding to a smallest possible electron orbit, in which the electron couldn't radiate and so didn't fall into the nucleus. The infinite gravitational force at microscopic distances is paradoxical in the same way, and the solution obviously requires introducing quantum notions into general relativity. But despite the efforts of many, this hasn't yet been accomplished.

The suspected equality of the fundamental forces at the Planck distance hints at a single underlying, unifying force. The main candidate for such a unified theory is the *string hypothesis*. I call this beautiful and promising idea a *hypothesis* rather than, like most authors, a *theory* because it currently has no direct supporting evidence. Extensive experimental verification, which doesn't yet exist for the string hypothesis, is a requirement for a scientific theory. Words are important, and the word for a promising guess, regardless of how wonderful and inventive and thrilling it might be, is *hypothesis*.

The string hypothesis arose from the problems of quantizing gravity. Before the string hypothesis, the plausible theories of quantum gravity predicted infinite forces and other absurdities. The source of this problem was that quantum field theories had always assumed an individual quantum could, at sufficiently large energies, be squeezed down to an arbitrarily small size. Because large energies imply large masses, this pointlike nature of quanta at high energies implies infinities that ruin the theory. *The string hypothesis avoids this high-energy pointlike structure by assuming that all quanta are strings.* A string is a quantum that, on being squeezed at the highest energies into its smallest configuration, reduces to a one-dimensional loop, like a rubber band, rather than to a point. A loop configuration spreads the quantum out in just the way needed to prevent the infinities.

This idea might lead to the perfect unification: *All fundamental quanta might be strings that are ultimately identical, with the difference between them lying only in their manner of vibration.* Physics would truly be the "music," the "vibes," of the universe. Photons and electrons would be identical strings vibrating in different modes, analogous to a bugle vibrating in one or another of its natural harmonics. The universe would be made of only one kind of thing: bugles.

No, seriously, it would be made of a single quantized field whose quanta are identical strings in various states of vibration.[36]

Poet T. S. Eliot wrote: "We shall not cease from exploration and the end of all our exploring will be to arrive where we started and know the place for the first time."[37] Since Democritus, physicists have dreamed of a theory that would explain everything in terms of one primitive idea. Perhaps strings will bring us to this place.

Particles, Fields, or Both?

The notion that quanta are particles is contradicted by the double-slit experiment. As you've seen, we must picture each quantum as an extended object—a wave in a field—that comes through both slits, spreads over the entire screen, and collapses to an atom-size disturbance on interacting with the screen's atoms. This section looks at a few of the other contexts in which particles-versus-fields arises, concluding in every case that an all-fields view is the only proper way to look at quantum physics.

Microscopic phenomena are generally both quantum and special-relativistic. For example, electrons in atoms move at 1% to 10% of light speed, speeds for which special-relativistic effects are far from negligible. Dirac, Heisenberg, Tomonaga, Schwinger, Feynman, and others recognized this by creating quantum theories that were inherently consistent with the rules of special relativity.

What Is a Quantum?

Most of these appeared to be theories about spatially extended fields, although Feynman's version looked more like a particle theory than a field theory. Debates ensued: Were relativistic quantum theories about fields or about particles?

It isn't easy to fit quantum physics into the framework of special relativity. Einstein created special relativity in 1905 as a purely classical theory, with no hint of the quantum. The 1926 Schrödinger equation, the fundamental equation describing quantum behavior, contradicted special relativity's so-called *covariance principle*—the notion that the laws of physics should take the same form in every nonaccelerating laboratory or "reference frame" no matter what the speed of that reference frame might be relative to other reference frames. Einstein's covariance idea says that, if you are in a reference frame such as a nonaccelerating passenger jet that's moving at unchanging speed in a straight line, everything should seem "normal" to you. For example, if you pour coffee into your cup, the coffee should flow directly downward (as observed by you) rather than along a diagonal line that somehow takes the airplane's forward velocity into account (making it extremely difficult to pour your coffee into your cup). The covariance principle has been verified innumerable times, including by you if you want to try pouring a cup of coffee during your next airplane flight—or you could drop a rock the next time you are a passenger in a nonaccelerating automobile moving straight ahead—the rock should drop from one hand directly downward into the other hand. Paul Dirac and others fixed this problem by modifying Schrödinger's equation to make it covariant; in fact, we saw earlier that the resulting "Dirac equation" is even more accurate experimentally than Schrödinger's equation. However, there are some quantum phenomena that seem, at first sight, to conflict with special relativity. I'll explore such phenomena in Chapter 9, where we'll find that the conflicts are only apparent, not real. The lesson is that, in thinking about a theory that is both quantum and relativistic, one must think quite carefully in order to avoid contradictions.

In particular, in 1974, Gerhard Hegerfeldt analyzed the consistency of the particle idea within any theory that is both quantum and relativistic.[38] He began from a clear definition of *particle*. He defined this word broadly as any entity that can be "localized"—meaning that we can track the thing down sufficiently to be able to say, with certainty, that it is entirely within some bounded region of space. A "bounded region" means a spatial region that doesn't extend to infinity in any direction. Your living room is bounded. So is the Pacific ocean. Hegerfeldt's particles could be as big as the sun, but there must be *some* bounded region within which they certainly lie. This doesn't seem like too much to ask of any entity that we would want to call a *particle*.

Hegerfeldt then showed that *particles cannot exist in a theory that obeys the standard principles of both quantum physics and special relativity*. The reason for this surprising conclusion lies in the standard prescription for the way any quantum entity must move, and in a key principle of special relativity.

Hegerfeldt showed that, if a presumed particle is localized at any single instant (such as, say, noon), and if it's a "free particle" that is not constrained by external forces to remain localized (as it would be if, for example, it were bound to an atom), it will be nonlocalized at *any* time after noon. For example, there will be some nonzero probability that, at a tenth of a second past noon, it will be detected a million kilometers away from its original bounded region. This certainly seems odd. If localized particles exist, said Hegerfeldt, they can remain localized only for an instant because they would expand immediately in such a way as to stretch across an infinite (i.e., unbounded) distance. Furthermore, as I now explain, this behavior would violate a key principle of special relativity.

You might have heard that "nothing can go faster than light," but this notion is actually false. For example, cosmologists think the universe expanded at speeds greatly exceeding light speed during a very short time interval after the Big Bang. This rapid expansion does not violate special relativity because space itself is presumed to have expanded, and special relativity implies only that nothing can move *through* space faster than light moves through space.[39]

Here is a correct statement of *relativity's universal speed limit: Objects and information cannot be transferred through space at faster than light speed.* Here's why. Because of special relativity's conclusions about space and time, if an object moved faster than light, then we could find some reference frame—some moving laboratory—within which that object would be observed to move backward in time. This would violate the normal "causal" order of things, because within such a laboratory effects would happen before causes: scrambled eggs would unscramble, for example, violating the second law of thermodynamics. And if information could be transferred through space faster than light, it would be possible, in principle, for you to send a message backward in time with instructions to a gang of bandits to kill your grandfather before he meets your grandmother—a logical contradiction.

So you can see the problem. If there's the slightest chance that our particle could be detected a million kilometers away at a tenth of a second past noon, the particle would have to move through space faster than light to get there, contradicting relativity's universal speed limit. So if we accept special relativity and the notion that causes precede effects, then *every free, or unconstrained, quantum extends to infinity.*

The quantum vacuum poses additional embarrassments for quantum particles. Recall that indeterminacy requires every quantum field to maintain at least minimal random fluctuations even when it contains no quanta at all. *If particles are the basic reality, then what is it that has these fluctuating values in vacuum where there are no particles?*

The quantum vacuum is not a figment of theorists' imaginations. It exerts real forces. These forces play measurable roles—for example, in the magnetic moment of the electron (Chapter 1), in determining a detail of the hydrogen atom's EM spectrum known as the *Lamb shift*, and in an effect called the *Casimir force*, whose tale I will now tell.

The Casimir force was predicted in 1948. If two electrically neutral flat metal plates are placed a short distance apart and parallel to each other, and if the two plates are within an evacuated region that is free of all outside forces, quantum physics calculations show the plates will attract each other with a force that increases rapidly as the distance between the plates decreases. Experiments confirm this result to within 1% of the expected theoretical value. [40]

What could cause such a force? The plates remain electrically neutral throughout. The force of gravity is far too small to cause the observed force. The theory assumes no external forces, and the experiments achieve this to a high degree. According to the theory, the force is caused by vacuum fluctuations of the EM field. Here's why. Although there are no quanta in a vacuum—no waves traveling through the vacuum field—there, nevertheless, are random nontraveling waves called *standing waves* similar to the waves that form along a violin string or in the air column within a bugle. These fluctuations of the vacuum EM field can be broken down[41] into many individual standing EM waves with various wavelengths. In an infinitely large region such as outside the parallel plates, there are no restrictions on the wavelengths of these waves. But in the region (which is box shaped if the plates are square) between the two plates, there are restrictions: Each standing wave must fit into this so-called *cavity*. Most wavelengths don't fit in such a way as to come out properly at the plate boundaries. Another way to say this is that most wavelengths are not "resonant" within the cavity, the way that most wavelengths are not resonant on a string of fixed length.

So there's a big difference between the character of the EM vacuum inside and outside the cavity. A much broader range of fluctuations is allowed outside the cavity, whereas only those fluctuations that "just fit" are allowed inside. These fluctuations exert pressure against the inner and outer surfaces of both plates, but the force on the outside of each plate is larger because more kinds of fluctuations are allowed outside than inside. So there's a net inward force on each plate. This is the Casimir force. It exerts forces on metal plates in vacuum where there are no quanta. If the universe is made of particles, then what is it that presses inward against these plates in vacuum, where there are no particles?

Unruh radiation is another fascinating puzzle that the vacuum poses for the notion of quantum particles. In 1976, William Unruh predicted, on the basis of special relativity and quantum physics, that an accelerating observer moving through a vacuum detects quanta that a nonaccelerating observer does not detect.[42] In addition, if both observers measure the temperature of the space around them, the nonaccelerating observer finds a temperature of 0 kelvin whereas the accelerating observer finds a temperature greater than 0 kelvin because of random thermal motion of the quanta she detects. In other words, a thermometer waved around (and thus accelerated) in empty space will record a nonzero temperature despite the fact that there are no quanta in empty space to *have* a temperature!

This prediction, which has not been observed conclusively but is thought by most experts to be correct,[43] is embarrassing if you regard quanta as particles. How can real particles be present for one observer but absent for another who observes the same region of space? But if the universe is made of fields, things fall into place: Both observers experience the same universe-filling vacuum field, but one observer's acceleration causes her to interpret this field differently.

As you might have guessed, the acceleration required to produce a measurable amount of Unruh radiation is superenormous. To observe a vacuum region to have a temperature of just 1 kelvin, an acceleration ten million trillion times larger than the acceleration of a freely falling object on Earth (larger than 10 million trillion "g's") is needed.

In what might be the most convincing possible demonstration of the field nature of reality, the predicted vacuum energy fluctuations were recently detected directly in a delicate experiment led by Alfred Leitenstorfer and Claudius Riek at the Center for Applied Photonics in Konstanz, Germany. The idea of the experiment was to use precise laser optical techniques to study extremely small regions of empty space over time intervals as short as 0.000,000,000,000,001 (10^{-15}) second. Any random EM disturbances detected in such a small region would necessarily have a very short wavelength, and thus a high frequency and large energy—large enough to be detectable directly. The experiment demonstrated convincingly the existence of these EM oscillations that exist even in absolute darkness where there are no photons.

Again one must ask: If reality is made of particles, then what is oscillating in this vacuum state where there are no presumed particles?

The particle misconception runs deep, in part because of the widespread use of the misleading informal word *particle*. Despite the evidence, and although leading quantum theorists argue explicitly for the field view, a survey of textbooks in my university's library shows most physics students are still learning that the universe is made of things called "particles."

Quanta are not much like sand grains or tiny marbles or other classically intuitive particles. As explained in this and succeeding chapters, quanta are highly unified, extended, flexible, changeable, unpredictable, fragile, gossamer entities that flit about or vibrate at mostly enormous speeds, that can extend over vast regions yet collapse in apparently a single instant of time to atomic or far smaller dimensions, that can vanish in an instant by transforming their energy into some other form, that can burst into being from the energy of a decaying atom or other quantum or from a random energy fluctuation, and that can entangle with each other to form highly unified yet delicately connected networks of composite correlated objects that appear to be in instant contact over perhaps cosmological distances.

Particle is exactly the wrong word to describe such an object. It promotes a serious misconception about our most fundamental scientific theory. For the world to become scientifically literate, as many of us think it must if it is to remain habitable, scientists must use terms that indicate their proper meaning, or that at the least do not flagrantly contradict their proper meaning. Why not use the right word: *quantum*?

The all-fields picture has interesting implications. Physicists regard fundamental fields to be conditions of space. So electrons, photons, atoms, and molecules are conditions—"states" or "shapes"—of space. So cabbages, kings, and your foot are made of states of space.

Fantastic? Yes, but experimentally verified in great detail, and far less fantastic than the tortured, unverified, and pseudoscientific notions (Chapter 10) some have imagined to try to explain how the universe could be filled with particles.[44]

PART 2

HOW QUANTA BEHAVE

6

Perfect Randomness

Richard Feynman put it well: "A philosopher once said 'It is necessary for the very existence of science that the same conditions always produce the same result.' Well, they do not.... The future is unpredictable."[1] *Quantum randomness* flies in the face of a fundamental implication of classical physics: predictability.

Albert Einstein frowned on quantum randomness. To his friend Niels Bohr, he once said, "Quantum mechanics is very impressive. But an inner voice tells me that it is not yet the real thing. The theory produces a good deal but hardly brings us closer to the secret of the Old One. I am at all events convinced that He does not play dice."

Bohr's response: "Einstein, stop telling God what to do."[2]

Classical physics worked like a charm from 1650 to 1900, describing fairly accurately nearly everything observed under ordinary conditions on Earth. As we saw in Chapter 3, given a precise description of the present, classical physics in principle predicts the entire future and past. You might object that a coin flip is classical but unpredictable, but it's predictable in principle. If we took into account the force exerted by the coin-flipper's thumb, air resistance, distance to the floor, elastic properties of the coin's bounce, and so on and so on, classical physics could predict the outcome.

Quantum randomness goes deeper than coin flips. Fundamental randomness usually accompanies interactions between quanta. Nothing in nature determines whether a particular quantum will interact with another quantum with which it could interact, and nothing in nature determines the particular outcome of such an interaction. Such things are usually random, indeterminate. For example, an alpha quantum (Chapter 3) within a radioactive nucleus might or might not interact with the strongly repulsive EM field lying just outside the nucleus' surface; and if it does interact, the direction in which it will fly out of the nucleus is also indeterminate.

As I've urged before, words are important. Quantum randomness is commonly referred to as *quantum uncertainty*, but it's a big misconception to think it has anything directly to do with *humans* being uncertain about this or that. Like all of quantum physics, quantum randomness is woven into the objective universe and has nothing specifically to do with humans. Throughout his groundbreaking 1927 paper on this topic, Heisenberg used the German word *ungenauigkeit* (indeterminacy), switching to *unsicherheit* (uncertainty) only in the endnotes. Unfortunately, this latter term has stayed with us. As you've guessed by now, this book will use less anthropogenic words: *randomness* and *indeterminacy*. These two terms are equivalent: an event is fundamentally random or indeterminate if it cannot be predicted even after every detail has been taken into account.

Evidence and Description of Fundamental Randomness

A beam splitter, such as the partially reflecting window of Chapter 1, splits light into transmitted and reflected beams. A 50–50 beam splitter can be made by depositing aluminum vapor on a small plate of glass until the aluminum is sufficiently thick to reflect 50% of an incoming beam.

Consider again the experiment of Figure 1.1, with beam splitter 2 absent. Suppose the incident beam is dimmed to one photon per second. We've seen in Chapter 1 that each photon is detected by either detector 1 or detector 2, verifying the wholeness of individual quanta: A photon never splits, never goes to both detectors. And if the experiment is performed over and over with many identical photons, the statistics of the impacts are entirely random, with 50% of the photons going to each detector. Every photon faces precisely the same situation, yet half the photons must go to one detector and half to the other. The wholeness of the quantum forces the choice to be random.

It's natural, but ultimately fruitless, to ask: What determines which detector is impacted? Is it, for example, irregularities in the glass plate? But when the glass surface is made more perfectly uniform, one finds that the splitting becomes more perfectly random, not less. Thus, the answer appears to be that *nothing* determines which path. Certainly, nobody has discovered any determining factor. Each photon faces a perfectly symmetric situation, yet each photon must remain whole. There is only one way nature can resolve this dilemma. The outcome must be perfectly random, with 50% probabilities for an interaction at either detector. In this manner, *the wholeness of the quantum forces randomness on nature*.

Quantum randomness has a perfection about it that is unequaled by apparently random macroscopic phenomena. Businesses, gambling casinos, and others frequently need sequences of reliably random numbers. Generating them isn't easy. Flipped coins, rolled dice, and the like, as well as certain mathematical formulas and computer programs, can generate lists of seemingly randomly chosen numbers. But, when subjected to thorough statistical tests, such lists deviate from perfect randomness. For example, in a supposedly random sequence of zeroes and ones, too many strings of six zeroes in a row might be found consistently. The problem is that the outcome of any classical process can be determined ultimately if one has sufficient information about its initial conditions, so it isn't "really" random. When quantum processes such as beam splitters and radioactive decay are used to create long sequences of zeroes and ones, statistical tests show these sequences to be easily distinguishable from, and more random than, any classical method of generating random numbers.[3] This perfection of quantum randomness provides substantial evidence for its fundamental nature.

For most of us, this demonstrates that God does indeed play dice, that quantized nature is perfectly random in a way that classical randomness cannot match. But science is never absolutely conclusive and some physicists argue that, when light passes through a beam splitter, unobserved "hidden variables" direct each photon either to reflect or transmit, implying that quantum processes only appear to be random but are ultimately predetermined, like classical physics. I'll discuss hidden variables later. For most physicists today, randomness is real and there are no hidden variables. A group of 17 quantum fundamentals experts was recently asked, "Does quantum mechanics imply irreducible randomness in nature?" Eleven responded with a fairly clear yes, two responded no, and four were noncommittal.[4] It's the consensus of working physicists (most of whom are not quantum fundamentals experts) and of standard textbooks that quantum phenomena are inherently random.

Quantum randomness intervenes deeply in the classical worldview. It crops up, for example, whenever light interacts with matter. Think of light passing through a single-slit experiment. If the slit is very narrow, the light beam spreads out or "diffracts" (Chapter 2) across the entire viewing screen, so individual photons impact at different points all over the screen even though they all came through the same narrow slit. Narrowing the slit only enhances this effect by further widening the beam. Perhaps we could collimate the beam by sending it through a second narrow slit immediately beyond the first slit—surely those photons that make it through both slits are collimated into a single narrow beam moving straight ahead. But all such attempts to control the impact point fail. *Each photon coming through the slit spreads out broadly beyond the slit and then collapses by interacting with a particular molecule in the screen.* What determines this molecule rather than some other molecule? Nothing. All the molecules of the screen are essentially identical, and the interaction point is fundamentally random.

We see the same indeterminacy in the two-slit pattern of Figure 5.1. What was true of photons in Figure 5.1 is also true of electrons in Figure 5.4. All photons and electrons, as well as atoms, molecules, and all other quanta, interact randomly. Physicist Edward Teller put it well: "In order to understand atomic structure, we must accept the idea that the future is uncertain. It is uncertain to the extent that the future is actually created in every part of the world by every atom and every living being. This point of view, which is the complete opposite of machinelike determinism, is something that I believe should be realized."[5]

The future is not encoded in the present and does not exist until it actually happens. According to quantum physics, there is an immense qualitative difference between past and future: The past exists, in the sense that it is in principle possible to know precisely what occurred at past times and places. The future does not exist in this sense, because of quantum randomness. Classical physics recognizes the distinction between past and future only via the second law of thermodynamics: total entropy will be greater in the future, and it was less in the past. But quantum physics appears to ascribe a unique role to the present instant, to "now," as the instant that separates the fundamentally determined from the fundamentally undetermined.

Probability is used in connection with processes having uncertain outcomes. The probability of an event is a numerical measure that has the value 0 if the event is impossible and the value 1 if the event is certain, and is greater when we are more certain the event will occur. It's easy to define quantitatively for such repeatable processes as coin flips. When we say the probability of a coin flip coming up heads is 50%, we mean that, in a long string of trials, the fraction of heads approaches 50% more and more closely as the number of flips increases. The *probability* of a particular outcome is the fractional number (or *relative frequency*) of occurrences of that outcome in a long series of trials. Relative frequencies make sense only for processes in which a long series of trials is possible. It's often possible to calculate probabilities based on inherent symmetries. For instance, in rolling a single die, the symmetry between the six outcomes implies the probability of each to be one sixth (assuming the process really is symmetric or fair). In a roll of two dice, we recognize 36 different symmetric outcomes, implying a probability of 1/36 for each; so, for example, the probability of 12 is 1/36, whereas the probability of 11 is 2/36, and so forth. It's fun to test these predictions experimentally, as my Dad once encouraged me to do by drawing a graph of the predicted statistics over 72 rolls. Try it. Dad would have been proud.

But what are the probabilities for each of six horses to win in the opening race this Saturday at the county fair? If you place bets based on some such easy answer as "one sixth on each horse," then I want to play poker with you. Or what is the probability that a particular Mars landing vehicle will get to Mars and deploy appropriately? It's not easy to apply relative frequencies to one-of-a-kind processes. Nevertheless, it can be useful to assign a number between zero and one that represents our personal degree of belief that a particular outcome will occur. Fortunately, we needn't worry about this issue because quantum probabilities are not subjective, not one-of-a-kind. Because all quanta of a given type are identical and quantum experiments are repeatable, at least in principle, the relative frequency definition is adequate and there is only one set of probabilities that describes correctly any particular quantum process. This set of probabilities can be checked by relative frequencies over repeated trials.[6]

Statistics is another word that crops up in connection with probability. Probability is a theorist's ideal, the relative frequency in a long run of precisely repeated trials, whereas *statistics* are the specific results of an actual finite set of trials. The probabilities in a fair coin flip are 50% heads and 50% tails, but the statistics in a run of 100 coin flips might be 57 heads, 42 tails, and once when your dog swallowed the coin.

Radioactive Decay: Perfectly Random

As you know (Chapter 3), certain types of nuclei are radioactive—meaning that they decay spontaneously by transforming into another nuclear type, usually via emission of either a small alpha or beta quantum. The tale of how this works is a fine example of quantum randomness.

Because an alpha quantum (two protons and two neutrons) is an exceptionally stable combination, it forms easily as a substructure within many nuclei. In a nucleus that's prone to alpha decay, nuclear forces favor the formation of alpha quanta near the outer surface of the nucleus. As a spatially extended matter field, each such alpha quantum sprawls entirely around the nuclear surface. In this outer region, the attractive strong nuclear force of the other protons and neutrons dominates only out to the nuclear surface;[7] beyond this surface, this short-range attractive force is reduced severely and the long-range repulsive EM force, by the other protons pushing outward on the alpha quantum, dominates.

If classical physics prevailed, there would be no possibility of an alpha quantum appearing outside the nuclear surface because the attractive nuclear force would prevent it from escaping in the same manner that a marble rolling around slowly inside a bowl is prevented by the bowl's sides from escaping. But if the marble and bowl were microscopic and dominated by quantum physics, a portion of the marble's matter field could extend through the sides of the "bowl" and beyond. In this way, our alpha quantum extends into the classically forbidden region outside its confining "bowl" (its nucleus), and there is some chance the alpha quantum will "decide," randomly, to interact at some point in this region; if it does, the alpha quantum is said to "tunnel" through the nuclear surface. The enormous force exerted by the other protons then hurl the alpha quantum outward at high energy. This is alpha decay. The questions of whether such an interaction occurs, and in which direction the alpha quantum will fly if the interaction does occur, are indeterminate.

It's a common misconception to think that we can, by studying a particular nucleus, determine whether it's about to decay. If we could peer into an individual nucleus, we would find it to be in a state of steady vibrations that show no significant change over time. The nucleus might then decay within the next microsecond or it might maintain its nondecayed state for millennia. It's like rolling dice: snake-eyes might come up on the first roll or not for the next 500 rolls.

Despite quantum randomness, there's an important element of deterministic order in alpha decay and other quantum processes. Whether we're talking about radioactive decay or beam splitters, the long-run statistics are predictable. Viewed one photon at a time, the beam splitter experiment is entirely indeterminate. Yet with 1000 photons, we can predict that about 500 will go to each detector. Fundamental randomness guarantees this statistical regularity. The same statistical determinism operates in the double-slit experiment, as you can see from the predictable patterns that eventually emerge in Figures 5.1 and 5.4.

Quantum interactions are individually random; but, like other random processes, the statistics of quantum processes over a large number of trials are approximately determinate. It is precisely this statistical determinism that leads to the appearance, at the macroscopic level, of classical determinism. For example, when zillions of photons pass through a double-slit experiment, the overall outcome is essentially predictable (Figure 2.3). However, when only 25 photons pass through, the results are far from predictable (Figure 5.1a).

It's ironic. The wholeness of the quantum entails the fundamental indeterminacy of nature, but determinism emerges at the classical level *because of* this perfect randomness at the quantum level.

Before delving further into quantum indeterminacy, I need to explain a point about nuclear structure. The properties of any nucleus are strongly affected not only by its number of protons, but also by its number of neutrons. So to discuss a particular nucleus we need to specify not only the element to which it belongs (its number of protons) but also the *isotope* to which it belongs within that element (its number of neutrons). For example, the element carbon (C) has six protons in its nucleus, but it can have six, seven, or eight neutrons. We abbreviate these three isotopes C-12, C-13, and C-14, where the number indicates the total number of nucleons (protons plus neutrons). These three isotopes of carbon have the same number of electrons and hence identical chemical properties, but quite distinct nuclear properties. For example, C-14, with 6 protons and $14 - 6 = 8$ neutrons, is radioactive, but C-12 and C-13 are not.

Consider a tiny sample of uranium-235 (U-235) atoms, perhaps half a microgram (half of a millionth of a gram), amounting to a thousand trillion atoms. The nucleus of this isotope has 92 protons and $235 - 92 = 143$ neutrons. Every U-235 nucleus has a certain number of alpha quanta floating near its surface, with their matter fields extended slightly beyond the surface. So every U-235 nucleus has a certain probability that, during any given period of time, such as 1 second, an interaction will occur that will hurl an alpha quantum outward. Because our sample contains a thousand trillion nuclei, these probabilities translate into predictable statistics.

The most useful such statistic is the *half-life*, the time during which half of a large sample of identical radioactive nuclei can be expected to decay. An equivalent definition is that an isotopes' half-life is the time during which a single nucleus has a 50% probability of decaying. For U-235, this half-life is about 700 million years. Half-lives vary all over the map: U-239 has a half-life of

24 minutes, U-238 has a half-life of 4.5 billion years, and polonium-214 has a half-life of 0.000,164 second.

For our sample of a thousand trillion U-235 atoms, the average number decaying in any given day is about 3000, or an average of 1 decay every 30 seconds. Each one that enters your body can do biological damage by dislodging electrons from, or "ionizing," biomolecules. However, 1 emission every 30 seconds is a low level of alpha radiation, carrying a very low risk to nearby humans. For example, it's far less than the natural level of alpha and beta radiation an average human receives naturally from the air, rocks, and soil around us. Small samples of U-238 are even less radioactive because of U-238's longer half-life. But you wouldn't want to hold a thousand trillion U-239 atoms in your hand, because about 300 billion would decay every second.

Both the indeterminacy of individual radioactive nuclei and the statistical regularities of large samples of radioactive nuclei testify to the perfect randomness of quantum phenomena.

Quantum Physics Is Statistical

As noted earlier, the beam splitter experiment demonstrates that quantization leads necessarily to indeterminacy. Each photon has two symmetric options, yet it cannot split. Indeterminacy is the way out of this puzzle. Yet, beam splitters and other quantum phenomena are statistically predictable: the accumulated statistics over many trials are predictable. Diffraction provides an instructive example that allows us, in this and the next section, to delve further into the fundamentals of quantum indeterminacy.

Figure 2.1 shows water waves spreading out—diffracting—after passing through a barrier containing two slits. Such diffraction always occurs whenever a wave passes through a small opening, simply because the opening provides a source for the disturbance to spread into the entire region on the far side of the barrier. Experimentally, significant diffraction occurs only when the opening is smaller than or comparable with the wavelength, and is just a small effect at the edges of the opening when the opening is much larger than the wavelength. Figure 6.1 illustrates three cases, as described in the caption.

Figure 6.2 presents photographic evidence of the diffraction of light that has passed through a single slit having a width a few times larger than the wavelength (Figure 6.1b). On the viewing screen, we see a broad central region of intense light surrounded by one or more narrow fringes caused by interference effects between the waves coming from different points within the slit. This figure is similar to the pattern you created in Chapter 4 and viewed directly on the retina of your eye using a single slit formed by your thumb and forefinger. That experiment, as well as Figure 6.2, provide additional evidence for the wave nature of light.

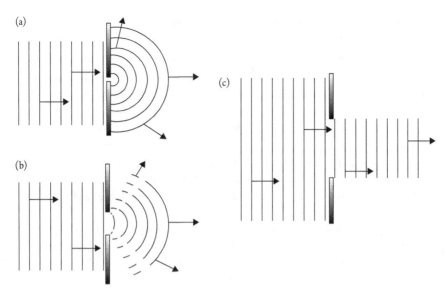

Figure 6.1 Diffraction when a wave passes through a single slit. The straight lines on the left of the opening represent wave crests. The wavelength is the distance between adjacent lines. (a) If the slit's width is smaller than the wavelength, the slit acts like a point source and waves diffract into all directions. (b) If the slit's width is a little larger than the wavelength, the wave partially diffracts, with interference (between different waves coming through different parts of the single slit) at the edges. (c) If the slit's width is much larger than the wavelength, waves come straight through except for slight spreading at the edge that is not shown in the diagram.

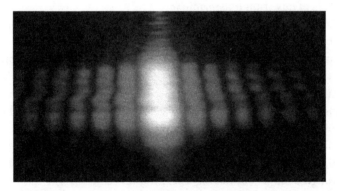

Figure 6.2 Photograph of light that has diffracted through a slit whose width is a few times larger than the wavelength. The pattern has a broad central region and narrow side-bands due to interference from different parts of the single slit.

Figures 6.1 and 6.2 show classical aspects of the single-slit diffraction of light. What happens if we dim the beam and amplify the detection at the screen to the point where individual quanta (photons) show up?

Figures 5.1 and 5.4 show us what to expect, although these figures show a double-slit rather than single-slit experiment. In the experiment of Figure 5.1, we found that each photon comes simultaneously through both slits as a spatially extended wave that then collapses upon interacting with the viewing screen. Although individual impacts appear to be scattered randomly, an interference pattern emerges as the statistical consequence of zillions of impacts.

With this background, it's easy to predict the result of the single-slit experiment at the individual photon level. The position of each impact is determined only probabilistically, in a manner consistent with the resulting pattern, which we will assume to be Figure 6.2. So as each photon passes through the experiment, it fills the entire slit (because it's an extended field), spreads as it approaches the screen, and interacts and collapses at the screen. The interaction occurs with high probability in the brighter regions of Figure 6.2, lower probability in the dimmer regions, and essentially zero probability in the dark regions.

The same reasoning extends to a beam of material quanta such as electrons or atoms. Electrons behave much like photons. Being waves in extended fields, they diffract when passing through a single narrow slit; their statistical pattern, formed from a zillion electron impacts, is similar to Figure 6.2 only smaller, because of electrons' smaller wavelength. The next section looks at this experiment in detail to explain one of the key quantum fundamentals.

Heisenberg's Principle

Every quantum is a field that, like a small cloud, has a range of positions and velocities, so its interactions with other quanta can occur at a variety of places and a variety of velocities, leading to a range of possible outcomes. *Heisenberg's principle* places general limits on this range of possibilities. I'll illustrate the general picture by considering a typical example: the diffraction of an electron passing through a single slit.

Figure 6.3 pictures this process.[8] An electron source S emits one electron that I will name "Tiny." As you know from Chapter 5, Tiny is a ripple in a universal matter field. I've pictured this ripple as four closely spaced wave crests moving across the page from left to right, which I'll call the y-direction. The x- and y-axes, shown at the right of the figure, are oriented in an unusual manner for reasons explained in the figure's caption. The five parts (a) through (e) of Figure 6.3 represent Tiny at 5 different times as it proceeds through this experiment. At (a), Tiny moves away from S. By the time it reaches (b), Tiny has expanded, as ripples do, to a lateral width W and is proceeding toward a single slit with a narrow dimension of width w (the slit's long dimension extends into the page). Tiny might

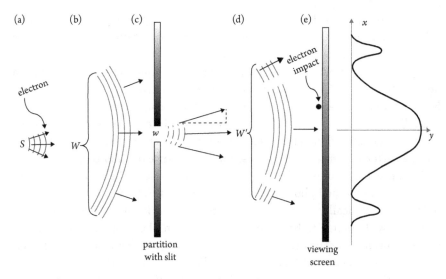

Figure 6.3 Five "snapshots" of one electron passing through a single slit. (a) The electron moves away from its source S. (b) It moves toward a partition containing a single slit having width w. (c) It diffracts (spreads) outward from the slit, with side-fringes. (d) It moves toward a viewing screen. (e) It impacts (interacts with) the screen and collapses into a small randomly-chosen atom-sized region. The diffraction pattern from zillions of impacts is graphed at the far right. To understand the graph, tilt your head to the right so the y axis appears upward with the x axis leftward. The number of impacts y is graphed, versus the impact position x. Quantum physics predicts this pattern, and predicts the electron's evolving shape as shown.

happen to interact randomly with the first barrier, in which case I'll neglect Tiny and consider instead an electron, also named Tiny (electrons are, after all, identical), that gets through the slit without impacting the barrier.

At (c), Tiny narrows to the slit's width w. Note that Tiny has now broken up into a central region and two narrow side "fringes" because of interference effects at the slit; nevertheless, Tiny remains a single electron—a single quantum. At (d), Tiny has again expanded to a lateral width W' (pronounced "W prime") and approaches the viewing screen. As you know (Chapter 5), each electron becomes as wide as the entire pattern that eventually appears on the screen—perhaps a few millimeters.[9] At (e), Tiny interacts at a random point on the screen and collapses into a particular atom of the screen; the screen amplifies this interaction to make the impact visible. The impact alters Tiny enormously, collapsing Tiny into something much smaller, of atomic dimensions. This collapse is subtle, controversial, and discussed in detail in Chapters 10 and 11.

Beyond the viewing screen, I've graphed (along the y-axis) the intensity of the macroscopic diffraction pattern resulting from zillions of electrons impacting the screen at different lateral positions x. The graph, which can be predicted in detail from Schrödinger's equation, represents an intensity pattern similar to

Figure 6.2, but it happens to have only one visible interference fringe on each side. The intensity y of the pattern at each point x on the screen is a measure of the probability that Tiny will impact at that point.

Let's analyze Tiny more carefully just after it passes through the narrow slit at (c). A classical point particle would have a definite position and emerge from the slit moving in only one direction; but Tiny is a field extending in all three spatial dimensions and emerging from the slit in a variety of directions. Let's consider only the lateral dimension, x. As Tiny exits the slit, its lateral extension is the slit's width w. I'll call this Tiny's "indeterminacy in position" at that particular time (corresponding to (c) in the figure).

But Tiny spreads as it exits the slit, spreading more widely when the slit is narrower (recall Figure 6.1). To factor this into our analysis, we must consider Tiny's velocity. By *velocity*, physicists mean not only speed, but also direction. For example, driving north at a speed of 80 kilometers per hour is quite a different velocity from driving east at this same speed. Velocities are typically represented in diagrams by an arrow whose length indicates the speed.

Tiny's spreading at (c) means Tiny has a range of velocities, with different directions as indicated by the three differently-directed arrows shown at (c). I've shown the upper velocity's horizontal (or y) and vertical (or x) "components" as dashed lines. The upper velocity points partly along the x-axis and thus has a positive x-component that I'll call v. The middle velocity has no x-component and the lower velocity's x-component is $-v$ (minus v), the same magnitude as the upper velocity's x-component but negative. Thus, the x-component of Tiny's velocity can have any value from $-v$ to $+v$, so the indeterminacy of this velocity component is $2v$.

Now you can see the result I'm driving at here: These two indeterminacies—Tiny's lateral position indeterminacy w and lateral velocity indeterminacy $2v$—are related. This is because, as Figure 6.1 illustrates, if you narrow the slit, the wave (i.e., the electron—namely, Tiny) spreads more widely so the lateral velocity develops a larger spread. If you widen the slit, the vertical component of tiny's velocity has a smaller spread—in fact Figure 6.1c shows that the vertical component of this spread is practically zero for a very wide slit. This is how waves behave: they spread widely on the far side of a very narrow slit, but they go straight through a wide slit with little spreading. Quanta such as electrons, being waves (and obeying Schrödinger's wave equation), do the same thing.

Working from the general principles of quantum physics, Heisenberg discovered this reciprocity between a quantum's spread in position and its spread in velocity to be a general principle. Specifically, *Heisenberg's principle* states that *a material quantum's indeterminacies in any component of position and the same component of velocity must change in inverse proportion to each other:* When one decreases, the other must increase. When one halves, the other doubles; when one triples, the other reduces to one third; and so forth. As another way of putting this, *when we alter either of these two indeterminacies, the other indeterminacy must change in such a way that their product remains unchanged.* As you've seen, this is simply because quanta are spatially extended waves and this is what waves do; for instance, they diffract.

Although this book is not the place for algebraic manipulations, here is the mathematical form of Heisenberg's principle as he derived it:

$$\Delta x \times \Delta v \geq h / 4\pi m.$$

Translation into English:

- Δx (pronounced "delta x") is the range or indeterminacy (in meters) of x-values (spatial positions) occupied by a single material quantum
- Δv is the indeterminacy (in meters per second) of the x-component of this quantum's velocity
- $\Delta x \times \Delta v$ is the product of Δx and Δv
- $\geq$ means greater than or equal to, implying that Heisenberg's principle establishes a lower limit on the product of the indeterminacies
- h is Planck's constant, 6.6×10^{-34} when standard metric units are used
- π is the Greek letter pi, the value of which is about 3.14
- m is the quantum's mass in kilograms

I need to clarify a point about the meaning of *indeterminacy*. Recall from Chapter 5 that de Broglie pointed out that each quantum "fills all space," and Hegerfeldt proved the same result more rigorously by showing that quanta cannot be localized within any finite-size region. In other words, the size of every quantum (provided it's not restricted by external forces) is infinite. Yet, Heisenberg's indeterminacy Δx, which represents the x-component of the extension of a quantum, is some noninfinite number. This is because Heisenberg's Δx represents only the high-probability region within which the quantum is highly likely to interact if given the opportunity. More precisely, Δx is the "standard deviation of x from its average value," the definition of which, if you really want to know, is in the endnotes.[10] A similar clarification applies to Δv.

An Indeterminate Universe

Heisenberg's principle says that indeterminacies must be at least large enough to satisfy the formula. It's not surprising they could be larger, perhaps far larger, because there are all sorts of reasons why a quantum's position or velocity could be indeterminate. For example, a quantum's indeterminacies usually grow with time, unless an interaction occurs that reduces it. Thus, Tiny has a small Δx at (c), but by the time Tiny gets to (d), its different velocities have caused it to spread so that it has a larger Δx although Δv has not changed. There are also all sorts of nonquantum reasons, often classified as experimental error, for a quantum's indeterminacy.

The idea of Heisenberg's principle is that a material quantum has a certain intrinsic "size" $\Delta x \times \Delta v$—a certain overall extension in its combined position and velocity—and this "size" is at least $h/4\pi m$ for each of the three spatial dimensions. Heisenberg's principle says there is a limit to how far you can squeeze a quantum into a small "size": beyond a certain point, further squeezing in either Δx or Δv must necessarily increase the other indeterminacy. As Figure 6.1 illustrates. a narrower slit implies a smaller Δx at the slit, but this means the quantum diffracts more widely beyond the slit (i.e., Δv is larger). It's like paste in a leaky tube: squeeze it in one place and it squirts out some other place.

Choosing an electron that manages to get through the narrow slit at (c) is a way of getting information about or "measuring" the electron's lateral position. As might be expected, the measurement reduces Tiny's lateral indeterminacy, from W to w. This is an example of a general principle that we consider in more detail later: Measurements often alter quantum states. Another measurement occurs when Tiny impacts the viewing screen, reducing its lateral dimension from W' to atomic dimensions.

To prevent misconceptions: Keep in mind that "measurements" have nothing necessarily to do with physics laboratories or humans (Chapter 2). The slit could be a gap between two pebbles on a distant planet, and the analysis would be essentially unchanged.

Tiny might wind up attached to a hydrogen atom in the screen, in which case its position indeterminacy Δx would be roughly 10^{-10} m, the diameter of the most common electron orbit in hydrogen. Note that Tiny retains some nonzero spatial extension Δx and does not transform into a point particle when it is absorbed into an atom, as is sometimes concluded from pointlike experimental results such as those shown in Figure 5.4. Electrons are fields, and they remain fields even when they make particle*like* (small) impacts on viewing screens.

Let's do a few numbers to see that Heisenberg's principle is still obeyed, even *after* the electron is absorbed into a hydrogen atom. Tiny's speed in this smallest orbit can be calculated (from Schrödinger's equation) to have an average value of about 2×10^6 m per second (about 1% of light speed), which means any particular velocity component could range up to this value, so the velocity indeterminacy is about 2×10^6 m per second. Thus, $\Delta x \times \Delta v$ is roughly $(10^{-10}) \times (2 \times 10^6) = 2 \times 10^{-4}$ in metric units. For electrons, the value of $h/4\pi m$ is about 6×10^{-5} in metric units. Because 2×10^{-4} is some three times larger than 6×10^{-5}, Tiny's indeterminacy is sufficiently large to satisfy Heisenberg's principle.

You might have heard that quantum indeterminacy arises from the unavoidable disturbance caused by measurements. This is a misconception. Quantum indeterminacy arises from each quantum's intrinsic extensions Δx and Δv, regardless of whether it's ever measured. Quanta are blobs, clouds, bundles that are spread out in x and v, and the indeterminacy in their interactions is a consequence of this natural spread. Δx and Δv are both highly variable, but their product $\Delta x \times \Delta v$ cannot drop below $h/4\pi m$. Any indeterminacy caused by measurement only adds to this intrinsic indeterminacy. For example, at (d) in Figure 6.3, Tiny's Δx extends over the entire diffraction pattern and has little to do with disturbances during the experiment. This electron is so spread out that it hardly has a position; it takes on a somewhat accurate position only when it collapses on impact with the screen.

This is a fundamental point: Quantum indeterminacy does not arise from human ignorance, measurement uncertainty, or disturbance. It arises because quanta are extended fields, and hence do not *possess* precise positions or velocities. Indeterminacy is a feature of the quantum universe.

It is true, however, that quantum physics studies properties of nature so fragile that it's nearly impossible to observe them without changing them. For example, we can determine an electron's approximate position by shining EM radiation on it and deducing the position from the behavior of the photons. Photon–electron interactions will necessarily disturb this electron's velocity, creating further indeterminacy in velocity and preserving Heisenberg's principle. The upshot is that experiments and theory show that every quantum obeys Heisenberg's principle, regardless of what nature or physicists might do to nail down both x and v.[11]

When Ernest Rutherford in 1911 discovered evidence suggesting the planetary atom, with its tiny nucleus and its orbiting particlelike electron, classical physics was unable to explain how this structure could maintain itself. Classical electromagnetism predicts that an orbiting electron must radiate energy quickly and spiral into the nucleus, collapsing the atom. Quantum physics resolves this dilemma. To spiral into the nucleus, the electron would have to violate Heisenberg's principle. It's a telling example of how the universe is ruled not by classical physics but by the quantum.

Heisenberg's principle applies to all material quanta: protons, molecules, and even macroscopic objects such as baseballs made of many quanta. Because the minimum $\Delta x \times \Delta v$ is $h/4\pi m$ and this gets smaller as the object's mass get larger, Heisenberg's principle poses less restriction on more massive objects, and poses practically no restriction on baseballs. For example, the minimum $\Delta x \times \Delta v$ of a 0.15-kilogram baseball is 3×10^{-34}, a million trillion trillion times smaller than the minimum $\Delta x \times \Delta v$ for an electron, and practically indistinguishable from zero. Heisenberg's principle is of no practical consequence for macroscopic objects. This is one of the reasons we can use classical physics in the macroscopic world.

Protons and neutrons are 2000 times more massive than an electron, so their minimum $\Delta x \times \Delta v$ is 2000 times smaller. This is a leading reason why nuclei are so much smaller than atoms. The other reason is the short range of the nuclear force that holds protons and neutrons together. This short range implies you can put only about 100 protons into a single nucleus before their long-range repulsive electric forces push the nucleus apart.

Concepts such as an electron's or proton's "radius" can be ambiguous. In experiments such as that shown in Figure 5.4, an electron just before impact can be spread out over 1 millimeter (10^{-3} m). An electron in an atom can range from as small as 10^{-10} m across to as large as 10^{-6} m.[12] These numbers refer to the Δx of electrons in various physical situations. You might have heard that the radius of the electron is so small that we've never been able to measure it, and it might even be zero. Such statements refer to the *minimum* Δx for an electron, the size of the smallest spatial volume into which an electron can be squeezed by allowing Δv to become correspondingly large. Today, it's known only that this minimum Δx is smaller than the smallest measurable distance (to date): about 10^{-19} m. Many physicists consider the electron to be a point object or a string with a size that is the Planck distance, but it could turn out that the electron has a substructure made of other quanta.[13]

Electrons in atoms have Δx values of about 10^{-10} m. Squeezing electrons into smaller sizes than this can only be done at the expense of a larger Δv, which implies the electron's speeds must range up to some high value, which in turn implies that the electron's average speed must be quite large. In other words, a spatially squeezed electron must have a high (large) energy. For similar reasons, all experiments to detect small things must be conducted at high energies at places such as the Large Hadron Collider. Heisenberg's principle is the reason we need high energies to detect small things.

It's likely that the universe had a strikingly quantum origin. The current theory of how the Big Bang got started involves random quantum energy fluctuations within the smallest conceivable region of space: the Planck distance. These high-energy fluctuations established a new quantum field that expanded or "inflated" rapidly to a much larger (but still tiny) size. Because of quantum randomness, the distribution of matter and energy within this tiny region was not uniform, and these variations remained imprinted on the larger scale universe as it inflated. Fourteen billion years later, we see these variations, enlarged by expansion and amplified by gravity, in the layout of giant galactic clusters of matter and energy—interspersed with huge near-vacuum voids—that fill the known universe. Thus do our magnificent telescopes observe, forever imprinted in the large-scale layout of the cosmos, quantum indeterminacy writ large.

Is There a Hidden Determinism?

When you flip a coin the outcome seems random, but it's not really random. Such unnoticed or "hidden" variables as the speed of your thumb, the height above the floor, the elasticity of the coin's bounce, and so forth, determine the flip's outcome but aren't normally taken into account. Could a similar hidden determinism stand behind quantum randomness so that nature only appears to be indeterminate but is actually determinate?

Einstein thought so. It turned out he was wrong about this, but wrong in a brilliant way. Here is the tale of why he thought so, why he was wrong, and why his analysis was nevertheless brilliant.

In 1935, Einstein and colleagues Boris Podolsky and Nathan Rosen (EPR) published an argument, based on quantum physics, that appeared to demonstrate the real and simultaneous existence of precise values of such variables as the position and velocity of an electron, contradicting Heisenberg's principle.[14] Their paper describes a procedure, based on accepted quantum principles, for measuring both the position and velocity of an electron with unlimited precision, in violation of Heisenberg's principle. If true, this would imply there is no true quantum randomness, but only an apparent randomness that arises from our failure to measure all the variables. This would mean quantum physics, despite being correct in what it does predict, is incomplete because it fails to predict some properties that exist in the real world. It was the most serious challenge ever made to the quantum principles. To tell you about it, I'll need to introduce *entanglement*, a far-reaching phenomenon that we will discuss more fully in Chapter 9.

Two random (unpredictable) outcomes, such as rolls of two dice, are said to be *correlated* if the outcome of one partially or entirely determines the outcome of the other. For example, if two dice were somehow rigged in such a way that the sum of their two outcomes is always odd, their outcomes would then be mutually correlated. If one comes up even, the other must be odd. It turns out to be possible to prepare pairs of quanta so they behave in a highly correlated manner. They are then said to be *entangled*. For an idealized example, suppose two electrons pass through separate but simultaneous single-slit experiments, using two separated setups such as that shown in Figure 6.3, and suppose that, before the experiment, the electrons are entangled in such a way that the difference between their two impact points on their own viewing screens is guaranteed to be $x_2 - x_1 = 0.3$ millimeter, where x_1 is the impact point of the first electron on its screen and x_2 is the impact point of the second electron *as observed on the other distant screen*. Then, the impact point of either electron could be predicted from the *other* electron's impact point. For example, if the first electron impacted at $x_1 = 12.2$ millimeters on its screen, we could predict that the second electron impacts its own screen at $x_2 = 12.5$ millimeters. As you'll see in Chapter 9, a similar situation occurs in the real world.

Experiments in recent decades show that quantum entanglement is unaffected by the distance between the entangled quanta, and it acts instantaneously. For instance, our two electrons could be on different continents, yet the impact point of one electron would still determine the other electron's impact point, even though the two impacts occur simultaneously and even though both positions were highly indeterminate before measurement.

EPR noticed this odd phenomenon is predicted by quantum theory and argued that it implies a real and precise position for both electrons. Here's why, according to EPR. It seems inconceivable that a measurement of one electron could in any way *instantaneously change* the real physical situation of the other electron. After all, neither energy nor information can be transmitted through space faster than light (Chapter 5), so it appears impossible that a measurement of one electron could *instantaneously* alter the real situation of the other. Thus, EPR argued the second electron must itself *have* a precise position, regardless of whether the first electron's position is actually measured.

This is a plausible argument. For example, suppose I'm in New York and I mail envelopes, one containing a gold coin and the other containing a silver coin, to Alice in Amsterdam and to Bob in Beijing. I inform Alice and Bob of this in advance, without telling them who will receive which coin. If Alice, on receiving her envelope, finds a silver coin, she knows instantly that Bob in Beijing got the gold coin. But Alice's observation of her coin obviously does not change Bob's coin into gold. It's simply Alice's knowledge, rather than the real situation of either coin, that has changed.

In the same way, EPR argued that measurement of one electron cannot have changed the other electron's real situation, but merely changed the experimenter's knowledge concerning the preexisting situation. Just like the gold and silver coins, they argued that the entangled electrons must both *have* real and precise positions, and that the human experimenters simply don't know them. A complete theory should go beyond quantum theory by providing these precise positions. Putting further icing on this already-convincing cake, EPR also presented an argument demonstrating the existence of precise velocities for two entangled quanta. They concluded that electrons must have preexisting, precise, positions and velocities, violating Heisenberg's principle and the entire notion of quantum randomness.

The plausible assumption leading to EPR's conclusion was that *physical processes occurring at one location can have no instantaneous (immediate) effect on the real physical situation at another location*. There can be, as Einstein phrased it, "no spooky action at a distance."[15]

This assumption seems to be part and parcel of the theory of relativity's limitation on superluminal transport. A measurement of the position of one electron, said Einstein, cannot instantaneously cause the other distant electron to have a precise position; instead, the other electron must have already had that position. It's analogous to saying that Alice's discovery that her coin is gold cannot instantly turn Bob's coin into silver. I'll call this plausible notion the *locality principle*.

Despite this significant challenge, quantum physicists continued their successful progress without much concern for these theoretical and philosophical objections. The prevalent attitude was that everybody should "shut up and calculate."[16]

That was 1935. It wasn't until 1964 that John Bell shined new light on the EPR argument. He showed (Chapter 9) that *quantum theory actually does violate the locality principle.* The kind of situation to which Bell's result applies is precisely the one envisioned in EPR: two entangled quanta, some distance apart. Bell showed that, according to quantum physics, a measurement made on one quantum can indeed instantaneously alter the real physical situation of the entangled distant quantum! Any such instantaneous physical action across a distance is said to be *nonlocal.* Bell's theoretical prediction was later confirmed experimentally. Even more surprising, later experiments, together with Bell's analysis, show that *nature herself is nonlocal, regardless of the truth or falsehood of quantum physics.*

Thus, contrary to Einstein's opinion, there is in fact spooky action at a distance. The spookiness is predicted by quantum physics, and it's found in nature.

But in this case, what about relativity's prohibition against superluminal transport? Recall (Chapter 5) what relativity says about motion faster than light: Neither *energy* nor *information* can travel faster than a light beam. It turns out that (1) there is no energy transfer in the nonlocal processes analyzed by Bell and tested by experiment, and (2) the inherent randomness of quantum physics camouflages these processes in such a way that there can be no information transfer. Thus, the intuitive basis of EPR's analysis was wrong. Although quantum physics predicts action at a distance that some might consider spooky, it does not violate relativity's prohibition against transfer of energy or information faster than light speed.

Nonlocality seemed spooky to Einstein, but there is, in fact, an intuitive quantum basis for it. According to Chapter 5, a quantum is a single unified thing. As we'll see in Chapter 9, entangled quanta behave, in many respects, as a single unified quantum. So perhaps it is not so surprising that, although they are separated, one electron's real situation *can* change instantly when a second entangled, but distant, electron changes—contrary to what Einstein thought. The instantaneous nature of the change in both members of an entangled pair is a consequence of the unity of the quantum: when a quantum (or an entnagled pair of quanta) changes in any way, the entire extended quantum undergoes the change, all at a single instant.

This work by EPR marks the discovery of the nonlocal implications of entanglement, and was brilliant in that respect although its underlying assumption—that the universe obeys the locality principle—was wrong. This was no mean accomplishment and a testament to Einstein's genius.

David Bohm's Deterministic Model

Because quantum physics is odd and thought by some to be inherently inconsistent, there have been proposals to resolve the supposed paradoxes by altering quantum physics. EPR questioned one of these odd features—namely, quantum randomness—and called for a more complete and deterministic theory. Here's the tale of one prominent deterministic alternative: David Bohm's 1952 *pilot–wave model*.

Quantum physics describes a universe in which everything is a consequence of ripples in matter fields and radiation fields. Bohm's notion is that, in addition to these fields, for every quantum (every ripple) there is a precisely located point particle that is guided, or piloted, by the ripples, and these particles are the things we observe directly in experimental outcomes such as those shown Figures 5.1 and 5.4. According to the pilot–wave model, quantum fields also really exist, so the universe is made of both extended quantum fields and isolated point particles. The entire system—fields plus particles—is deterministic. Given the initial configuration of all the fields, and the initial positions and velocities of all the particles, the entire future and past of the universe is determined. Such a deterministic universe was Bohm's reason for devising this model.

Here's how Bohm's model works for a single electron passing through the double-slit experiment. Two physically real entities are present:

1. The *electron* is assumed to be a point particle that is always located at a specific position, that starts from a single point within the electron source, that moves along a single path through one or the other slit (not both), and that impacts the screen at a precise point.
2. The *quantum field* is the matter field described in Chapter 5. It obeys the quantum rules, including Schrödinger's equation. It passes through both slits and interferes at the screen in the double-slit pattern of Figure 2.3. This field can be observed only through its action on the electron. The field guides the electron, in a precisely predictable manner, to a specific impact point.

Suppose now that a number of electrons pass through the slits one at a time. According to Bohm's model, the experimental result of Figure 5.4 shows the impacts of these many electrons, each one following one precise trajectory through one slit, guided by an extended field that goes through both slits and forms an interference pattern all over the viewing screen. The reason different electrons impact at different places is simply that each one started from a different place. These different starting places are the hidden variables of Bohm's model, analogous to the additional variables in a coin flip that would, if measured precisely and incorporated into the calculation, enable one to predict the outcome. Although the experimental procedure for preparing these electrons appears, from our macroscopic viewpoint, identical for all electrons, Bohm's model assigns each one a different precise initial position and velocity and this, rather than quantum indeterminacy, accounts for their different paths and different impact points. According to Bohm, the various electrons in this experiment all actually *have* precise and predictable positions, although our experimental procedures prevent us from *knowing* them.

Bohm arranged his model so that the statistics of large numbers of electrons would agree with the statistical predictions of quantum physics *if* the initial (at the beginning of the experiment) spatial distribution of these electrons agreed with quantum physics. Thus, to get the distribution of electron impacts shown in Figure 5.4e, Bohm must assume the electrons have an initial distribution (at the source) that agrees with the initial state of the matter field. Because each of Bohm's electrons is not a field, but rather a point particle with a specific position and velocity, there would seem to be no reason why we couldn't (for example) start all the electrons with identical positions and velocities (but at different times so they wouldn't interact with each other), in which case they would all impact the screen at the same point. Nobody has ever observed anything like this. But by assuming that the hidden variables are, for unknown reasons, distributed in a manner consistent with conventional quantum physics, Bohm created a model that agrees with Heisenberg's principle and all of quantum physics.[17]

Figure 6.4[18] shows the theoretical trajectories of some 100 particles passing independently through a double-slit experiment, as calculated from Bohm's model using initial conditions that reflect the double-slit quantum state at the slits. Each electron starts from one slit at a specific location and velocity. As the electrons move toward the right, their paths become bunched in such a way that they impact the screen (not shown) in the standard interference pattern of Figure 2.3. In Bohm's theory, the guidance of the matter field Ψ causes this bunching: The electrons are guided by the matter field.

Things get more interesting when we apply Bohm's model to a system of interacting electrons—for example, the eight electrons in an oxygen atom. These electrons interact with each other as well as with the nucleus (with a large mass that keeps it nearly fixed in place) via the EM force. In such cases, the motions

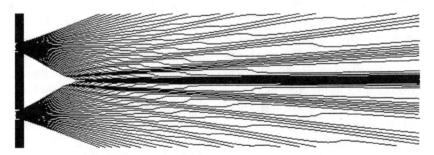

Figure 6.4 Double-slit trajectories of 100 point particles, calculated from Bohm's model. In this model, there is no quantum uncertainty. Each particle has a precise starting point within one or the other slit at the left (only the slits' narrow dimensions are shown), a precise and predictable trajectory, and a precise impact point. Electrons are guided to this impact point by the matter field Ψ.

predicted by Bohm's model are quite erratic and highly nonlocal: the motion of each electron depends strongly and instantly on the positions of all eight electrons! Bohm's model is nonlocal with a vengeance. You've seen that nature actually *is* nonlocal, and that conventional quantum physics also predicts nonlocal phenomena, so Bohm's nonlocality is not a deal-breaker. However, we'll see in Chapter 9 that the nonlocality inherent in standard quantum physics does not violate special relativity, whereas Bohm's more extreme nonlocality has proved difficult to reconcile with special relativity. Current theories of the Higgs boson, neutrinos, dark matter, and other microscopic phenomena are relativistic quantum field theories that combine an all-fields view of quantum physics with the principles of special relativity. A striking and experimentally verified prediction of every quantum field theory, for example, is the creation and destruction of quanta, but Bohm's model hasn't been able to describe such highly relativistic phenomena.[19]

"Although Bohm's model has its enthusiasts, it has never been broadly accepted by physicists."[20] One good reason is the problem with special relativity, but another less substantial reason is simply the vagaries of history. Although de Broglie proposed a Bohm–like model in 1927, the formulation of quantum physics that gained consensus during the theory's developmental period (1925–1930) was nondeterminate, so Bohm's deterministic and highly nonlocal model appeared quite radical in 1952. Because conventional quantum physics agrees with all known observations, physicists were reluctant to devote much time to thinking about an alternative.[21]

Bohm's model is important because it demonstrates that new variables *can* be added to quantum physics to obtain a deterministic theory that agrees with conventional nonrelativistic quantum physics. Before Bohm's model, physicists thought this was impossible.

One of the most telling arguments against Bohm's model is stylistic: it's ad hoc and cumbersome. Why should the universe be made of two such different kinds of things, extended quantum fields *plus* isolated point particles, when standard quantum physics makes the same experimental predictions using fields alone? Occam's razor advises us to choose the simplest explanation that fits the facts.

Randomness and Nothingness

Imagine a coil spring hanging from a ceiling with an iron ball attached to the bottom coil. The ball will stretch the spring to some equilibrium length at which the ball remains at rest with the spring's upward pull on the ball just balanced by gravity's downward pull on the ball. If you then pull the ball downward by a few centimeters, stretching the spring slightly, and release it, the ball oscillates up and down around the equilibrium position. This oscillating system is one of the simplest examples of motion under the influence of a force.

Classically, this system can vibrate with any amplitude—any width of vibration. But like our quantized playground swing in Chapter 2, microscopic oscillator systems are quantized—they are permitted to vibrate only with certain specific amounts of energy and thus only at certain specific amplitudes (widths of vibration). Most interestingly, Heisenberg's principle implies that *an oscillator cannot be at rest at its equilibrium position; it must instead always oscillate around this position.* This is because, if it were at rest, its precise velocity (zero) and precise position (the equilibrium position) would violate Heisenberg's principle.

Quantum systems can never rest. This simple, thoroughly quantum, notion has deep implications for quantum fields. As you know, floating a small cork while a water ripple passes by shows that the cork, and thus the water, simply oscillates up and down as the ripple passes. It's the same with quantum fields: at each point in the field, as a quantum (a ripple) passes by, the field simply oscillates around a fixed value. Heisenberg's principle applies to these oscillations, so that *quantum fields must oscillate at every spatial point. Even when the field is in its lowest energy or "vacuum" state having no quanta, it must vibrate randomly everywhere.*

In fact, if a particular quantum field such as the EM field exists at all, then it must exist everywhere. Any region from which the EM field is absent would be a region within which the EM field is precisely zero, and Heisenberg's principle doesn't allow this. So a region from which any particular quantum field is absent cannot exist. To put this another way, if quantum fields suddenly vanished entirely from some region, that region would itself vanish from the universe.

An empty universe is inconsistent with the quantum rules. A spatially extended universe must, at a minimum, be entirely filled with vacuum quantum fields, because true emptiness would violate Heisenberg's principle. This obviously doesn't entirely answer the philosopher's question as to why there is something rather than nothing, but perhaps it's part of the answer. [22]

7

Quantum States and How They Change

Just as you can be in a situation of running, falling down the stairs, or uncontrollable giggling, electrons, photons, atoms, and other quanta can be in various *quantum states*. A quantum can be in numerous different states of motion or even in a superposition (Chapter 8) of several states of motion simultaneously. A quantum can spread or contract, move this way or that way, have high or low energy, be located around this point or that point, be submicroscopic or enormous, or be in a superposition of any of these.

Physicists during the late 1920s puzzled out the mathematics of quanta, describing their possible states and how these states can change over time. In this chapter I describe some of the possible states of fundamental quanta such as electrons, and of compound quanta such as atoms, and the manner in which these states can change over time. We will ask a deep question: Are quanta and their states physically real (i.e., *ontological*) or are they merely useful predictors of experiments (i.e., *epistemological*)?

States of Fundamental Quanta

In connection with the double-slit experiment, I stated in Chapter 5, "As each photon approaches the screen, it . . . incorporates the entire interference pattern in the sense that it 'knows' it must interact at the bright lines rather than the dim lines"; yet, "the photon has no instructions about which atom should absorb it." All of this is an expression of the photon's quantum state. A field, after all, is a condition or state of space, and a photon or other quantum is a bundle (or portion, or package, or piece) of a field, so different states of a photon are simply different conditions of space. The notion of a quantum's state is merely an extension of the field concept.

I argued in Chapter 5 that the double-slit experiment, when conducted one electron at a time (Figure 5.4), demonstrates that each electron embodies the entire interference pattern. The electron approaching the screen is "in the double-slit state". Similarly, an electron passing through a single-slit experiment "in the single-slit state" (Figure 6.3).

An electron in the two-slit state behaves differently from an electron in the one-slit state. The evidence for the existence of these two states is that a two-slit electron's impact on the screen is always consistent with the two-slit pattern, whereas a one-slit electron's impact is always consistent with the one-slit pattern. Figure 6.3 shows the one-slit state as it develops over time. Figure 6.3c and d show the electron just after being "prepared" in the one-slit state by passing through the single slit.

Two electrons or other identical quanta are in different states if the statistical predictions that can be made about them are different. As you know, the word *statistical* is important here, because of quantum randomness. For example, you can't always tell from a single impact whether an electron is (or rather was, just before impact) in the one-slit or two-slit state; but as Figure 5.4 shows, you can tell from a zillion impacts of electrons that have been prepared identically in the same double-slit quantum state. However, there are occasions when a single impact does distinguish between different quantum states. For example, a single impact can distinguish between the one-slit and two-slit states if it happens to show up in a dark region between two-slit interference lines, for in this case the two-slit pattern is ruled out.[1]

The state of a quantum is represented in the theory by a mathematical object,[2] conventionally symbolized by the Greek letter Ψ and often called a *wave function*—a boringly abstract mathematical phrase that I will avoid. Because an electron is a disturbance in a matter field, Ψ describes the shape of a three-dimensional *matter wave*. Figure 6.3, for example, pictures such a state Ψ of a single electron at five different times as the electron passes through a single-slit experiment.

Figure 7.1a graphs a typical state of a single quantum, such as the electron in Figure 6.3 at a single instant, in a more detailed way. A short, compact wave of this sort is called a *wave packet*. This wave packet is the closest we can come to picturing a single electron. Do not imagine that this wave represents some tiny pointlike particle that is not actually shown in the figure. The wave is all there is.

Figure 7.1a shows only a single dimension, labeled x, of the three-dimensional electron. This wave packet is a series of ripples extending over some distance (usually microscopic) and is moving in the x-direction. The other axis, labeled Ψ, represents the strength, or value, of the matter field (the disturbance) at each point x. The figure also indicates the indeterminacy Δx in the electron's position x along the x-axis. Electrons in atoms usually have position indeterminacies Δx in the range 10^{-10}m to 10^{-6} m, but electrons in careful experiments can be much larger.

What exactly is the relation between this graph of a matter field and experiments? There's always been debate about this. Erwin Schrödinger and others who sorted out the general theory of quanta during the 1920s were strongly

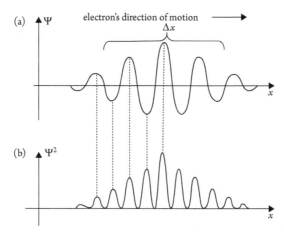

Figure 7.1 (a) Graph of a typical quantum state of one electron at one instant. The numerical value of Ψ represents the strength of the quantum field, or matter wave, at various locations x. The graph shows only one of the electron's three spatial dimensions. (b) The probability of this electron interacting with another quantum at different locations x is represented by Ψ^2 (psi squared), shown. The dashed lines show how Ψ and Ψ^2 are related at several locations.

influenced by Newtonian physics, which assumed a universe made of tiny, separated, structureless particles, perhaps point particles. These physicists discovered how to predict wave patterns such as that shown in Figure 5.4 that appear in quantum experiments. Because quanta are individually unpredictable but statistically predictable, it was natural to assume the wave pattern Ψ must represent the experimental likelihood of finding a pointlike quantum to be *at* the spatial point in question.

Some physicists still support this particle viewpoint, but the arguments of Chapter 5, and of those who study relativistic quantum physics, inform us that the universe is made of spatially extended fields, not point particles. In this case, results such as those depicted in Figure 5.4 show that the value of Ψ at some point x represents not the likelihood of finding the electron to *be* at that particular point, but rather the likelihood that *an interaction will occur* at that point. In Figure 5.4, this interaction is between the electron and the viewing screen. The older terminology, about the electron itself "being at" a particular point, suggests the incorrect notion that each quantum is at a single spatial point but we don't know which point so we must represent our lack of knowledge by a statistical distribution, just as we would assign a 50% probability of heads to a coin that has been flipped but not yet looked at. This notion is wrong. Quanta are inherently extended in space and cannot be said to be at any single point. Quantum probabilities are not just a matter of a lack of knowledge, but rather a tally of nature's assigned likelihood of the electron interacting at any particular location within the electron's extended matter field. As mentioned before, you should imagine the electron as a large balloon and the detection screen as a bed of nails: the electron extends over many nails but the interaction is going to occur at only one of them. The quantum state assigns a probability to each possible point of interaction.

The graph in Figure 7.1a shows a wave packet with crests where Ψ is positive and valleys where Ψ is negative.[3] Experimentally, a quantum field must have both positive and negative values because of the phenomenon of interference, which requires crests and valleys to interfere both destructively (when crest meets valley) and constructively (when crests meet or valleys meet). Experiments show Ψ behaves in this wavelike way. However, probabilities are relative frequencies and these are always positive. What's wrong?

What's wrong is that the physical impact or "intensity" of any wave—the amount of energy it carries—turns out to be proportional to the *square* of the wave's strength or value. Because the square of a negative number is positive, these intensities are never negative. Therefore, although Ψ represents the state of a quantum, Ψ^2 (psi squared, Figure 7.1b) represents the intensity of a quantum's impact on the world as measured by the probability that the quantum will interact with some other system at the point x.

For example, the number of electron impacts in any small region of the screen in Figure 5.4 is proportional to the value of Ψ^2 within that region, so when you look at Figure 5.4e, you're looking at the intensity Ψ^2 with which electrons impact the screen at various points or, equivalently, the probabilities for a single electron to interact with the screen at various points. Keep in mind that when you look at a record of impacts of many identical quanta (such as seen in Figure 5.4e), you're looking at an image of the quantum field, just before impact, of a *single* quantum in a *single* trial.

For another example, the graph at the right side of Figure 6.3 represents Ψ^2 for various lateral positions x within the electron, as one electron approaches the viewing screen—the probability that any single electron will interact with the screen at x.

The Tale of the Origin of Light

Where does light come from? The answer emerges from the related tale of atoms and their quantum states. It is best told by beginning with the simplest atom, H-1, an electron attached to a proton.

If you isolate one proton and one electron from other objects, and if they aren't moving too fast, their mutual electrical attraction will attach them. Both are matter waves, but because the proton's mass is some 2000 times larger, Heisenberg's principle tells us its indeterminacies are 2000 times smaller than the electron's. So the proton is relatively classical compared with the electron, and it's a good approximation to think of it as a small particlelike object around which the electron's matter field (i.e., the electron) is spread out like a cloud. Schrödinger's equation then describes the configuration and motion of this electron field.

A natural first question that physicists ask in dynamical (involving motion) situations like this is: Are there any states of the electron that persist without changing over time? In physics, such "stationary states" of dynamical systems are often key to finding the system's motion in more general situations. In the case of wave motions, the stationary states are called *standing waves*.

A purely classical example is the standing wave established on a violin string of fixed length when a bow is drawn across it to excite the string into vibrational motion. Figure 7.2 pictures four different standing waves on a string stretched between two fixed points, A and B. The entire string has a regular wave shape at any single instant, and each point on the string oscillates back and forth between the shape shown by the solid line and the shape shown by the dashed line. The figure shows four possible standing waves. All are waves that "fit" properly onto the string. For example, in the second state from the top, one complete wavelength fits onto the string; in the third state, 1.5 wavelengths fit.

In the first stationary state, at the top, the string vibrates in a single segment, with only the ends fixed in place. In the second state, the string vibrates in two

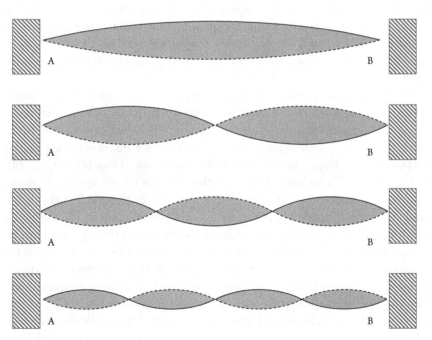

Figure 7.2 A string between two fixed points, A and B, in several states of vibration. Each vibrational state is a "standing wave" in which the string vibrates back and forth between the wave shapes shown by the solid and dashed lines. The four standing waves shown have different wavelengths and different vibrational frequencies.

segments; in the third and fourth states, it vibrates in three and four segments, respectively. States having five segments, six segments, and so on, are also possible. To achieve these standing waves, one must choose specific frequencies (rates of vibration) for the string, frequencies that will result in wavelengths that fit properly on the string. Let's assume, for example, that a frequency of 5 Hz (5 vibrations per second) causes vibrations for which a half wavelength just fits on the string, as shown in the first of the four states in Figure 7.2. The frequency that causes the second state turns out to be twice the frequency of the first state, or 10 Hz in our example.[4] The frequencies for the third and fourth states turn out to be 15 Hz and 20 Hz (three and four times the first frequency), respectively. Note that the string is not really stationary (at rest) in these stationary states: the string vibrates, but the overall vibrational pattern doesn't change or move along the string.

Schrödinger's equation predicts just such stationary states for the electron field that surrounds the H-1 nucleus. These states are "stationary" in the same sense that the waves of Figure 7.2 are stationary: the matter wave does not move from place to place, although it does vibrate at each point. So the stationary states of the hydrogen atom represent situations in which the electron (which, remember, is a field spread around the nucleus) doesn't change its location. These states have well-verified experimental implications and represent a triumph for Schrödinger's equation and for the entire quantum field approach to electrons and other quanta. How could an electron not be a wave in a field when it has stationary states that are directly analogous to the stationary states of all other wave systems, such as the standing waves on violin strings?

The standing matter waves of an electron are more complex than standing waves on a string, however, because they happen in three dimensions. Figure 7.3 shows two-dimensional drawings of the overall shapes of seven of these stationary states. Each view is a different possible quantum state of the electron in a hydrogen atom. I've labeled the seven states (a), (b), (c), (d), (e), (f), and (g). These are portraits of the electron within hydrogen in the same sense that the five "snapshots" of Figure 6.3, or the graph at the right-hand side of that figure, are portraits of a diffracted quantum. These seven standing waves of matter are analogous to the four standing waves of the string in Figure 7.2. Each drawing shows an electron in a stationary state around a proton at the center of each diagram (but not shown). The figure shows Ψ^2 at various locations—the probability that, at any particular location, the electron will interact with whatever other fields might be present. Regions of higher probability (larger Ψ^2) are darker. The gradations between lighter and darker should vary smoothly rather than, as in this simplified diagram, abruptly. For the full three-dimensional view of each state, rotate the diagram mentally around the z-axis, shown. Thus, states (a), (b), and (e) are spherically symmetric fields, state (c) is a dumbbell, and state (d) is a doughnut.

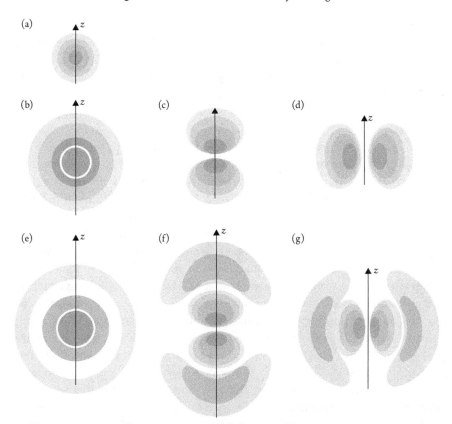

Figure 7.3 (a–g) Portraits of a quantum. Pictured are seven of the many possible stationary states of the electron in a hydrogen atom. Each diagram (a), (b), etc., pictures a possible stationary state for the electron. The figure shows Ψ^2, the probability that, at each spatial point, the electron will interact with other fields that might be present. Regions of higher probability (larger Ψ^2) are darker. For the full 3-dimensional view, mentally rotate each diagram around the z axis.

Note that the electron doesn't orbit the nucleus like a planet orbiting the sun. The stationary states don't move around. These waves just fit around the nucleus, the way the standing waves in Figure 7.2 just fit on the string. Although the probabilities Ψ^2 are unchanging, the matter field Ψ vibrates at the appropriate frequency for each state, just as the string of Figure 7.2 vibrates.

Schrödinger's analysis of the hydrogen atom was published in 1926. It proved more accurate than Niels Bohr's earlier theory of the hydrogen atom (Chapter 4), according to which the electron orbits the proton in one of many possible circular, planetlike orbits and emits light by jumping instantaneously from a higher energy orbit to a lower energy orbit. Both Schrödinger and Bohr received Nobel Prizes for their work: Bohr in 1922 and Schrödinger in 1933.

The mathematical analysis leading to these stationary states assumes the atom is isolated from the rest of the universe. The principle of conservation of energy (Chapter 3) says such a system's energy must not change. It's not too surprising, then, to learn that each stationary state has a specific and predictable energy that doesn't change over time, but this doesn't mean the electron's position or velocity is predictable. Figure 7.3 shows, in fact, the extent of its position indeterminacy.

Don't get the misimpression that the electron is at rest in these stationary states. Although the electron's state Ψ doesn't change (except that Ψ vibrates the way that each of the four states of the string in Figure 7.2 vibrate), the electron's speed is highly random and unlikely to be close to zero. Think of the field as being in rapid motion but in such a way as to maintain an unchanging density at each point in space, like a flowing stream in which the overall shape doesn't change but the water moves downstream. In state (a), for example, the average speed of the electron is 2000 kilometers per second—about 1% of light speed. When the electron is in this state, the average distance of the electron from the proton is about 5×10^{-11} m, and the indeterminacy in this distance is also about 5×10^{-11} m. And yet, despite large indeterminacies in speed and location, the electron's energy is predictable and unchanging. This is because the electron's energy comes in two varieties: kinetic energy of the electron's motion and electrical energy caused by EM forces by the proton on the electron. These two energies balance each other in just such a way that their sum—the total energy—remains predictable and unchanging.

We can learn something about these energies by examining the diagrams. A gravitational analogy helps. When you lift a rock from the floor to a table, you increase the rock's ability to do work (it's energy) because it could now fall from the table and, for example, compress a large sponge resting on the floor, doing work during the compression process. So a raised rock has more energy than a rock on the floor, and a rock has more gravitational energy when it's farther from Earth. In the same way, our electron has more EM energy when it's farther from the proton, because the electron is attracted toward the proton in the manner that the rock is attracted toward Earth. So states lying closer to the nucleus should have *lower* energy. Mathematical analysis of Schrödinger's equation bears out this classical analogy: The smallest state, state (a), has the lowest energy. It is also the simplest state: a simple sphere having low density (low probability of interacting) toward the outside and high density in the center, nearest to the proton. The other spherical states, (b) and (e), are larger and hence higher in energy. State (b) has an interesting small gap at a certain radial distance from the proton; the electron has no probability of interacting at this distance, although it can interact with high probability inside or outside this distance. A gap like this is a wave interference effect, similar to the dark lines in Figures 2.3 and 6.2.

It happens that states (b), (c), and (d) all have the same energy. The difference between them arises from different angular (i.e., orbital) motions. A spherically symmetric state such as state (b) corresponds classically to an electron that doesn't go around the nucleus at all, but instead moves back and forth in a line straight through the nucleus (electrons are leptons that don't experience the strong nuclear force). Despite the straight-line nature of this motion, the quantum state is spherically symmetric because the orientation of these possible straight lines is entirely random, and therefore the probability distribution is spherically symmetric.

State (d) represents the opposite extreme—namely, an electron with lots of angular motion and moving in a nearly circular orbit. This doughnut-shape state is similar to the orbits of the electron in Bohr's planetary atom. State (c), then, is intermediate between state (b) and state (d), and is analogous to a highly elliptical planetary orbit. States (e), (f), and (g) represent another set of three states, all of which have the same energy but different angular motions. State (g) is a double doughnut (for the *single* electron!), whereas state (f) is a double dumbbell.

State (a) is dubbed the *ground state*, in the sense that it is the stationary state having the lowest energy, like a rock that is on Earth's surface and, accordingly, has low gravitational energy (because it's on the ground—get it?). It's analogous to the upper drawing in Figure 7.2, which has the lowest frequency and thus the lowest vibrational energy. If our isolated atom happens to be in its ground state, it will surely remain there because there are no lower energy states for it to fall into, and there is no outside energy source to excite it to a higher energy state. It's the existence of such a lowest energy state, you'll recall, that guarantees the stable existence of atoms by preventing the electron from falling into the nucleus.

All other states have higher energy and are dubbed *excited states*. If our isolated atom happens to be in an excited state, it must, theoretically, remain there forever because Schrödinger's equation predicts the stationary states never change. However, *in the real universe, no system is truly isolated*. This is because the vacuum state of every quantum field extends throughout the universe. So our atom can never really be isolated, but instead must be in constant contact with all the universe's vacuum fields! The vacuum EM field, in particular, affects atoms because atoms contain electrically charged quanta.

When we include possible interactions between an excited hydrogen atom and the vacuum EM field, quantum theory predicts (and experiment confirms, or I wouldn't bother telling you this tale) that excited hydrogen atoms eventually transition into lower energy states. Furthermore, because energy must be conserved, every transition from a higher to a lower energy state is accompanied by an energy release in the form of an EM wave—a photon—that carries away the energy lost by the atom during its transition.

In summary, interaction with the vacuum causes an excited atom to transition to a lower energy state while emitting a photon. All atoms create photons by this process of transitioning—quantum jumping—between the stationary states of atoms. Let's look at these quantum jumps more closely.

Quantum Jumps

The tale just told, of the origin of light, has enormous implications for understanding atoms, molecules, and chemistry, and thus for engineering, communication, medical, and other applications. Recall (Chapter 5) that a photon's energy, measured in joules, is always Planck's constant multiplied by the photon's frequency. To conserve energy, transitions from one atomic state to a lower energy state must create photons whose energies equal the difference between the energies of the two states. Using Schrödinger's equation, physicists can calculate the energies of the stationary states of the hydrogen atom. By simply subtracting the energy of, for example, state (b) in Figure 7.3 from the higher energy of state (f), one finds the energy of the photon emitted in the transition from the state (f) to the state (b). One can then find the frequency of that photon. In other words, from the energies of these states of hydrogen, we can predict the frequencies present in the hydrogen atom's EM spectrum.

This predicted spectrum should match the spectrum emitted by a dilute gas of hydrogen atoms. In a dilute gas, each atom is relatively far from its neighbors. If, on the other hand, the atoms are too close together, each atom is distorted by its neighbors and will emit a spectrum that is altered by this distortion.

The dilute gas experiment is a favorite of student physics labs. The standard procedure uses a "discharge tube" (a thin tube about half a meter long) filled with hydrogen gas and sealed shut, with short metallic "electrodes" sticking out of each end. One electrode is attached, using a wire, to a battery's positive pole, and the other to the negative pole; chemical forces from the battery then push electrons from the negative wire onto the negative electrode and pull electrons off the positive electrode and onto the positive wire. These charged electrodes establish an electric field (but not yet any electric current) throughout the gas. At sufficiently high battery voltages, this electric field becomes strong enough to pull electrons off the atoms of the gas, turning the atoms into ions (charged atoms) and establishing an electric current (a flow of electrons).[5]

Electrons then travel through the gas, onto the positive electrode, through the battery (pushed by the battery's chemical forces) and onto the negative electrode, back into the gas, and onward around the circuit. The flowing electrons collide continually with H atoms, exciting them into their various higher energy states. During every second, many of these excited H atoms transition spontaneously into lower energy states, creating photons with all the frequencies of the many possible electron transitions in the H atom. So the tube emits light of many colors. Neon signs follow the same principles, using neon or other gases instead of hydrogen. When a prism separates the different frequencies of this radiation into differently directed beams of light, we can observe the entire spectrum of the atoms.

When physicists perform these kinds of experiments, the frequencies of the observed spectra match the theory's predictions, showing that Schrödinger's equation describes something real in nature. This match between theoretical predictions and a wide range of experiments is a substantial triumph for quantum physics. As noted in Chapter 1, the accuracies, to many figures, of these matches is often astonishing.

Because Schrödinger's equation was derived under certain simplifying assumptions, its accuracy is even better when these assumptions are removed. One assumption is that the electron in hydrogen is nonrelativistic—meaning, it moves so much slower than light that we can neglect the effects of Einstein's theory. In reality, even in the ground state, the electron moves at about 1% of light speed, implying a small but detectable relativistic correction. Another assumption is that the atom experiences no magnetic fields. In reality, every atom contains magnetic fields created by the nucleus and by the motions of electrons. Another assumption is that the numerical values of the atom's possible energies are not influenced directly by the ever-present vacuum EM field. Paul Dirac and others worked out a quantum theory that incorporates Einstein's special theory of relativity and corrects all these inaccuracies. This theory's predictions are among the most accurate in all of science. Much of quantum physics' high reputation stems from the accuracy with which it predicts atomic and other spectra.

Like the imaginary quantized child's swing of Chapter 2, an H atom's energy is restricted to a set of allowed values. It's a sterling example of quantum physics' digital nature. Figure 7.3 pictures the quantum states corresponding to several of these allowed values. Figure 7.4 graphs (on some quantitative energy scale measured, perhaps, in joules) the lowest five energy levels. The energy gap between the lowest two energies is 1.6×10^{-18} joules, and the other gaps are smaller still.[6] The energies of the seven states shown in Figure 7.3 are indicated in the figure. Every energy gap (e.g., from E_5 to E_3) represents the energy carried away by the photon emitted by a hydrogen atom transitioning between these levels. The hydrogen atom has a plethora of other, higher, allowed energy values corresponding to other stationary states.

Let's look more closely. Suppose an H atom transitions from state (c) with energy E_2, to the ground state (a) with energy E_1. Because of quantum indeterminacy, we expect this process to be random. Assuming the atom starts from state (c), quantum physics (the Schrödinger equation) predicts that the probability of the atom being detected in state (c) declines gradually from 100% at the initial time to lower values at later times, whereas the probability of the atom being detected in the ground state increases from zero. The theory further predicts

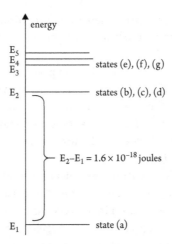

Figure 7.4 The lowest five energy levels for the electron in a hydrogen atom. The states (a) through (g) referred to in the diagram are shown in Figure 7.3. Actual quantities of energy, in joules, are not given because only relative values are important for our discussion. A hydrogen atom emits light by transitioning between these energies.

that a transition to the ground state is accompanied by a photon whose energy equals $E_2 - E_1$. Although the time of transition is not predictable, such statistics as the "average lifetime" are predictable. For example, the transition from either state (c) or (d) into the ground state has an average lifetime of 1.6 nanoseconds (1.6×10^{-9} seconds). This is the time, on average, during which the atom remains in the excited state. Such billionths-of-a-second ("nanosecond") lifetimes are typical within atoms. On the other hand, state (b) has an average lifetime of about one-eighth of a second before transitioning into the ground state—a hundred million times longer than the (c)-to-(a) or the (d)-to-(a) transition. A relatively long-lived state such as (b) is said to be *metastable*.

Don't get the misimpression that the transition from, say, state (c) occurs *during* a time span of 1.6×10^{-9} seconds. Transitioning is similar to radioactive decay; the transition itself is instantaneous (i.e., it occurs in zero time), but the time at which that transition occurs is indeterminate and is 1.6 nanoseconds long on the average. The atom remains in state (c) for an average 1.6 nanoseconds, and the (c)-to-(a) transition occurs as an instantaneous quantum jump. For many physicists, such an instantaneous jump is surprising. How do we know it's instantaneous?

Instantaneous quantum jumps within atoms were directly confirmed in 1986 by three independent groups.[7] Each group observed the light from a *single* atom as the atom transitioned into the ground state from *either* a typical excited state with a lifetime of a few nanoseconds (such as (c) to (a), as mentioned previously), *or* a nontypical metastable excited state with a lifetime of several seconds (such as (b) to (a), as mentioned earlier). They continually stimulated the atom with laser light so that as soon as the atom transitioned into the ground state, the laser boosted it back up into one of the two excited states. The atom was far more likely to transition into and out of the nanosecond state than into and out of the metastable state. Typically, the atom made zillions of transitions into and out of the nanosecond state during a period of many seconds, during which it flickered (because it was emitting photons one after the other) so rapidly that it appeared to remain lit for the entire period. Occasionally the atom would happen to be excited, instead, into the metastable state, where it remained for many seconds. During such a metastable phase, the atom no longer flickered; it remained dark. Then, it jumped randomly into the ground state and resumed transitioning into and out of the nanosecond state. So the light from the atom was on (but flickering) for a few seconds, then off for a few seconds, then back on again. The transitions between on and off were, as far as the experimenters could determine, instantaneous—showing no sign of a gradual transition from one situation to the other.

Apparently, *quantum jumps in atoms are real and instantaneous*. It must be understood, however, that the electron does not move instantaneously from one location to another location. These quantum jumps within atoms refer to transitions between two configurations of a spatially extended matter field. The matter field switches suddenly from one of the stationary states of Figure 7.3 to another. As mentioned before (Chapters 2, 5, and 6), the instantaneous, or discontinuous, or digital, nature of quantum jumps arises from the unity of the spatially extended quantum. The entire extended quantum must change everywhere, all at the same instant.

Are Quanta and Their States Real?

This question embarrasses me. This is, after all, a physics book, and physics is supposed to deal with physical reality, making the above title superfluous. But quantum physics deals with subtle and (some say) paradoxical matters. Quanta are tiny, fleeting, changeable, fragile, unpredictable, odd, made of invisible fields, and nearly impossible to observe individually; it's easy to conclude we're just making them up. One cannot examine an atom as though it were a shoe. Physicists don't doubt the existence of shoes, but some are dubious about quanta too small to see and so delicate as to change their nature when examined.

It's an old discussion. At a dinner party, a doubter once asked Ernest Rutherford, who discovered the atomic nucleus in 1911, whether he believed that atomic nuclei really existed. Rutherford's reply: "Not exist—not exist! Why I can see the little beggars there in front of me as plainly as I can see that spoon!"[8] But doubters persist, then and now.

Few physicists these days doubt the real existence of atoms, electrons, photons, and other quanta, but some experts doubt the reality of our specific descriptions of quanta. They have two reasons for skepticism.

The first is the measurement issue. Some properties of quanta seem to exist only when they are observed or "measured" by those who have the technical capacity to do so. For example, an electron passes through a single narrow slit and approaches a viewing screen, spreading all over the screen just before impact, then suddenly shrinks to atomic dimensions on impact as shown in Figure 6.3. The interaction with the screen imparts a rather precise—to within atomic dimensions—position to the electron, something it didn't have before impact. Similarly, measuring a quantum's velocity confers on the quantum an approximate velocity. Are position and velocity, then, real properties of quanta or are they only figments of the measurement process?

Einstein was in the realist camp. He wanted things to have properties, regardless of whether they were observed. His colleague, Abraham Pais, reported, "We often discussed his notions on objective reality. I recall that during one walk Einstein suddenly stopped, turned to me and asked whether I really believed that the moon exists only when I look at it."[9] Does mere observation create the observed properties, and are such properties then physically real?

You have seen that measurement does, indeed, create some observed properties. On the other hand, *measurement is an objective activity having nothing necessarily to do with human observation, or laboratories.* "Measurement" has a long history in quantum physics, starting with the views of Niels Bohr and Werner Heisenberg in the mid 1920s, views that became known as the *Copenhagen interpretation* of quantum physics (Chapter 10). To be precise about such nebulous concepts as the position or velocity of an atom, Bohr and Heisenberg focused specifically on what happens when humans attempt to measure such quantities, and the entire theory then became bound up with the notion of human measurements.

But it's clear today that quantum measurement is an entirely natural process that refers to any macroscopic impression made by quantum phenomena and is not related necessarily to human observations. A single electron impacting a viewing screen creates a more precise position for the electron regardless of whether any human observes or records the event. As noted in Chapter 1, a lot of pseudoscience has been concocted from the supposed connection between quantum physics and human consciousness,[10] but there is no evidence of any such connection. The quantum universe would be unchanged even if humans didn't exist.

A second, related, reason many physicists are skeptical about the reality of our descriptions of quanta is the notion that quantum states are merely "epistemic," not "ontological"—philosophical terms for "having to do with knowledge" and "having to do with reality," respectively. This is a kind of generalization of the first argument, concerning measurement. These physicists believe quantum states are useful for making predictions, but are not real in the way that stones and spoons are real.

Any self-respecting scientific theory should be able to predict experimental results. Given a directly observable macroscopic process that is said to establish an electron in some state Ψ, we should be able to predict any macroscopic (i.e., observable) effects that occur as a consequence of this state of the electron. The epistemic view argues that such macroscopic consequences are all we can know, and that talk of "an electron" or "the quantum state Ψ" and so forth, however useful it might be in predicting observable results, exists only in theorists' minds. But the epistemic view does not dismiss quantum theory. All physicists agree that quantum states are essential for predicting macroscopically observed results of experiments. The epistemic view argues that such states should not be regarded as really existing the way stones and spoons exist.

The debate focuses on the quantum state Ψ. Does it really exist, or is it, as mathematical physicist Jeffrey Bub asserts, "simply a credence function, a bookkeeping device for keeping track of probabilities?"[11]

Despite the historical claim that physics deals with physical realities, a surprising number of quantum foundations experts incline toward the epistemic view.[12] For example, to the question "Are quantum states real," quantum expert David Mermin responds, in a 2009 article published in *Physics Today*, "Quantum states are calculational devices and not real properties of a system."[13] He argues that such uncomfortable issues as quantum nonlocality (Chapter 9) and measurement (Chapter 10) are best dealt with by acknowledging the purely epistemic nature of quantum states, which "forces one to formulate the sources of that discomfort in more nuanced, less sensational terms. Taking that view of quantum states can diminish the motivation for theoretical or experimental searches for a 'mechanism' underlying 'spooky actions at a distance' or the 'collapse of the wavefunction'—searches that make life harder than it needs to be."

The remainder of Mermin's article exemplifies, in my opinion, a danger of this view. Because quantum states are simply states of quantum fields, Mermin next argues that quantum fields are not real, but are merely "useful mathematical tools." This leads him to question the reality of Maxwell and Faraday's classical EM field, which is, after all, the same entity as the quantized EM field but simplified by ignoring quantization. He concludes that "classical electromagnetic fields are another clever calculational device." Most physicists, however, view classical fields as properties of space itself. This leads Mermin to the further conclusion that "space and time ... are not properties of the world we live in but concepts we have invented to help us organize classical events."

Space and time are not real? Apparently there's a slippery slope in any argument that quantum entities are not real. It becomes difficult to find entities that *are* real, because, after all, everything is supposed to be ultimately quantum. Thus, one soon doubts even the reality of Rutherford's spoon, and we arrive at the subjective idealism of Bishop Berkeley (Chapter 4), a philosophy that postulates there is no external reality at all and *everything*, including stones and spoons and even ice cream, exists only in human minds. Most of us would say, as

Samuel Johnson did by kicking a presumably real stone, that this is ridiculous. Most physicists accept the reality of space, time, and fields. If one accepts the reality of fields, it's hard to see how one can reject the reality of quantized fields and thus the reality of various states of a quantum.

A significant new argument emerged in 2012. Matthew Pusey and two colleagues proved that the epistemic interpretation of quantum states implies experimental predictions that contradict the predictions of standard quantum physics.[14] Because scientists agree that the quantum predictions are correct, an interpretation that is inconsistent with those predictions is of no value.

Pusey's proof is based on standard quantum principles and relies only on a primitive reality assumption that both the realist and epistemic camps should accept—namely, the assumption that every quantum system *has* a real physical state, although that state might be unknown and quite different from the state assigned by conventional quantum physics. In other words, Pusey assumes that each electron, for example, is in some sort of actual physical situation. If electrons exist at all, this should be a safe assumption. However, one quantum father—namely, Niels Bohr—seems to have disagreed. He stated famously, "There is no quantum world. There is only an abstract physical description. It is wrong to think that the task of physics is to find out how nature is. Physics concerns what we can say about nature."[15]

Bohr's view goes even beyond the nonrealist views of the epistemic camp, by claiming there are no quantum states at all—in fact, "there is no quantum world," a statement that apparently implies quanta such as electrons, photons, atoms, and molecules, along with their states, do not exist at all. If this is indeed what Bohr meant (and Bohr has always been famously difficult to decipher concerning the foundations of physics), then I think it's safe to say that most physicists today would disagree. At any rate, Pusey's proof rules out some, but not all, nonrealist interpretations.

The details of Pusey's proof are subtle and technical, but the conclusion is simple: He shows that, if two observers with different "states of knowledge" assign different quantum states to a particular quantum system in a specific experimental situation, then at least one of those two states must lead to incorrect (in the sense of disagreeing with the predictions of quantum physics) physical predictions. So every quantum system has, at all times, a *unique* quantum state that agrees with the experiments that can be performed on the system, one that cannot be different for different states of the observer's knowledge, but is different for different states of reality. The state of a quantum system is not a matter of opinion.

Chapter 5 argues that, in the double-slit experiment, for example, something certainly comes through the slits. In fact, this "something" must come through both slits even when only a single quantum is in play, because it interferes with itself at the viewing screen (e.g., by avoiding the dark lines in the emerging interference pattern). Like the stone that Samuel Johnson kicked, this is an experimental fact that can't be brushed off as merely "epistemic." Such facts imply that single quanta are spatially extended fields, and quantum states are real configurations of these fields.

My own view is that the arguments of Maxwell and Einstein, based on EM radiation and conservation of energy, for the reality of classical EM fields show fields are real. This reality is not changed merely by adding the condition that the fields are quantized.

Overshadowing all these details is the central notion that, as Carl Sagan put it, "Extraordinary claims require extraordinary evidence."[16] The claim that our most fundamental theory does not describe reality flies in the face of all previous scientific history and is extraordinary by any definition. We dare not adopt the nonrealist view without overwhelming evidence. It is one purpose of this book to demonstrate there is no evidence of this sort—no evidence that quantum physics is anything other than a consistent theory of the real, objective world.

How Quantum States Change

A quantum's state can change in two distinct ways. Both are illustrated by Figure 6.3, showing an electron passing through a single-slit experiment. Figure 6.3a–d shows a smoothly changing matter field emitted by the source, moving toward the slit, emerging from the slit, and spreading widely as it proceeds toward the viewing screen. For nonrelativistic material quanta, this smooth evolution is described by Schrödinger's equation. Then, in Figure 6.3e, something different happens. When interacting with the screen, the electron collapses instantaneously to atomic dimensions. This unpredictable change is a quantum jump.

Schrödinger's equation applies to the motion of nonrelativistic material quanta as long as there are no quantum jumps. This motion is similar to the smooth motion of ripples on a lake. If, at the beginning of an experiment, the electron's state is as shown in Figure 6.3a, then Schrödinger's equation implies the state at later times will be as shown in Figure 6.3b–d. This evolution is deterministic in the sense that the initial state (a) certainly leads to states (b), (c), and (d).[17]

The collapse, or quantum jump, shown in Figure 6.3e, where the electron interacts with the screen and creates an observable mark, is quite different from the smooth change described by Schrödinger's equation. This is a quantum measurement—a process during which a quantum causes a macroscopic change—namely, a visible flash on the viewing screen. As noted earlier, this measurement has nothing necessarily to do with humans or human observation. Such visible flashes, created when a microscopic quantum strikes the right kind of solid material, happen all over the universe all the time with no assistance from humans.

I'll try to sort out the controversial measurement process in Chapters 10 and 11. We'll find that discontinuous quantum jumps happen because of two necessary circumstances. First, a microscopic quantum must become "entangled" with some other quantum system. Schrödinger has said that quantum entanglement is "*the* characteristic trait of quantum mechanics" (the emphasis is Schrödinger's). In my opinion, the full significance of entanglement is still largely undiscovered. Entanglement entails one of the oddest and far-reaching phenomena of the quantum world: nonlocality (Chapter 9). As we see in Chapter 10, entanglement causes the microscopic quantum's state to exhibit "definite but indeterminate and reversible outcomes" (that's a mouthful that I'll sort out later), analogous to a coin that has been flipped but has not yet been looked at.

Second, the other quantum system must be sufficiently macroscopic that the second law of thermodynamics applies to processes occurring within it, which allows the entangled state to "decohere" (another mouthful, explained in Chapter 11) in an irreversible manner, so that only one of its possible definite outcomes actually occurs.

This chapter has presented experimental evidence for instantaneous quantum jumps in atoms from one allowed atomic state to another, based on detection of a photon emitted in these transitions. Quite unlike the smooth, predictable Schrödinger evolution of a single quantum such as an electron, a quantum jump is a discontinuous and indeterminate alteration of a system's quantum state that occurs when the system interacts with another system. If one of the systems is macroscopic, and if the change of this macroscopic system's quantum state results in a macroscopic impression such as an observable flash, the interaction becomes a quantum measurement. The entire measurement process, involving collapse and the random selection of a specific outcome, is a bone of contention for quantum physicists. Chapters 10 and 11 propose a solution to this quandary.

8

Superpositions and Macroscopic Quanta

Have you ever noticed two sets of ripples in still water passing through each other? If not, please let two drops of water fall, a few centimeters apart, into a bathtub of still water, and watch. The experiment demonstrates *wave superposition*: Two waves can be present in the same body of water at the same time, and can even pass right through each other, without disturbing each other. Each set of ripples represents one possible state of the water, and the experiment shows that the superposition of the two is also a possible state of the water. Using more drops you can see the superposition of any number of sets of ripples.

Nearly every type of wave can exhibit superposition.[1] Quanta, being waves in fields, are no exception. Indeed, quanta exhibit the following *superposition principle: If a quantum can be in any one of several different states, then it can be in all of them simultaneously.* For example, because the electron in a hydrogen atom can be in any one of the seven states shown in Figure 7.3, the single electron can also be in a superposition of any or all of them.

This simple wave principle leads to some surprising possibilities. Although superposition is natural for water waves and other classical waves, there is oddness in the notion that an individual quantum can exist as a superposition. For example, a single electron could exist in a state comprising a wave packet moving to the right (Figure 7.1a) superposed with a second wave packet moving to the left. This electron would move in two opposite directions at the same time! This is not odd for water waves, but it's odd for a quantum because a quantum such as an electron is an individual—and indeed highly unified—thing. The individual things around us—my coffee mug, for example—don't move in two directions simultaneously. A water wave, on the other hand, is not regarded as a "thing," but as a "shape" of the water, so we are not surprised when it moves in two directions simultaneously. But at the quantum level, mere shapes—ripples—in quantum fields become things, and this leads to surprises.

Superposition applies to every quantum. A water molecule could move in two directions simultaneously, arriving at two quite different places. This is not something you see in your everyday world. You don't see baseballs or tables in two places at the same time. Why not? I'll discuss this in this chapter and expand on it in Chapter 11.

Quantum superposition has been known since the 1920s. It follows logically from Schrödinger's equation, the structure of which is similar to other equations, such as Maxwell's equations and the equations for the movement of water and other fluids, that predict waves. In fact, in view of the similarities between quanta of matter and quanta of radiation (compare Figures 5.1 and 5.4), it would be odd indeed if Maxwell's equations for radiation permitted superposition, but Schrödinger's equation for matter did not.

The particle-oriented physics of Isaac Newton, on the other hand, does not obey a superposition principle.[2] Because superposition is a universal wave characteristic, and because it's long been known that quantum states superpose but classical Newtonian states do not, it's surprising that the founders of quantum physics puzzled over whether quanta were waves or particles. If they superpose, they are clearly waves.

Quantum Superposition: Being in Two Places at Once

Returning to the storefront window of the opening pages, let's reconsider the interferometer (Figure 8.1). Looking at it one photon at a time, a single photon strikes beam splitter 1, which is simply a repositioned version of a half-reflecting, half-transmitting window. If beam splitter 2 is absent, this photon "chooses" randomly to impact either detector 1 or detector 2, with 50–50 probabilities for each. The fact that the light is detected one photon at a time, at one or the other detector but never both, verifies that light is quantized (bundled)

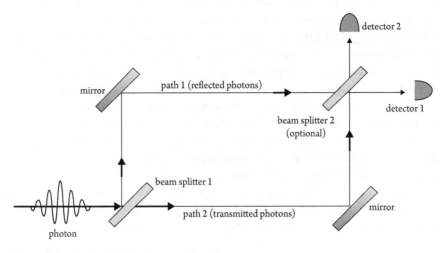

Figure 8.1 The Mach-Zehnder interferometer.

into whole photons and is not divided into parts by the beam splitter. You've seen (Chapter 6) that the statistics of this experiment demonstrate perfect randomness. This process, extended to zillions of photons, enables Alice to see her own image in the storefront window via the reflected path 1 whereas Bob inside the store also sees Alice's image via the transmitted path 2. There is no evidence here of superposition—at least not yet.

If beam splitter 2 is inserted, something quite different happens. As we saw in the opening pages of this book, the two paths are now remixed (something that doesn't happen in a storefront window where the two beams never meet) and either detector can click, regardless of which path the photon takes. The detectors no longer provide data about which path the photon takes. In this case, something unexpected happens: As we'll see below, *the experiment shows that the photon takes both paths*.

That's surprising. Here's how we know. With beam splitter 2 inserted, slightly lengthening or shortening either path alters the relative number of photons impacting each detector—something that does not happen when beam splitter 2 is absent. Let me say that again, in more detail: Suppose the experiment begins with the distances from the first to the second beam splitter along path 1 and along path 2 precisely equal. Surprisingly, the experimental result is then that *all* the photons go to detector 1 and *none* to detector 2![3] There are more surprises: If either path is lengthened slightly by a small fraction of the photon's wavelength, a fraction of the photons do go to detector 2. For example, if path 1 is lengthened slightly (without changing path 2), detector 1 clicks on most trials, and detector 2 clicks on the remaining small fraction of trials. As path 1 is further lengthened, the fraction impacting detector 1 decreases and the fraction impacting detector 2 decreases until, at the point where path 1 has been lengthened by one quarter of a wavelength (recall that photons have wavelengths), each detector clicks on 50% of the trials. Figure 8.2 shows how the fractional number of photons impacting detector 1 varies as the path-length difference between path 1 and path 2 increases, beginning from zero and ranging up to one entire wavelength (still a very short distance if it's a visible photon).

As you might expect, the experiment exhibits quantum randomness: For any particular setting of the difference between the two path lengths, quantum physics determines only the fractional number of impacts at each detector, as shown in Figure 8.2, but it does not determine the outcome for individual photons. The process is random, like a coin flip, and quantum physics determines only probabilities, as shown in the figure.

For clarity, it's worth emphasizing that, to verify the graph of Figure 8.2 experimentally, a large number (several tens, at least) of single-photon trials must be conducted at each setting of the path-length difference because the figure shows only probabilities. It's just like a coin flip; to show the odds really are 50–50, you've got to flip the coin many times.

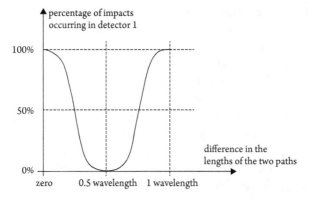

Figure 8.2 Evidence of quantum superposition in the experiment shown in Figure 8.1 with the second beam splitter inserted. The fractional number of photons impacting detector 1 for various differences in length between path 1 and path 2. This fractional number is the probability that any single photon will be detected at detector 1. As we'll see below, this graph shows that each photon travels both paths, interfering with itself at the detectors.

This is all quite odd. To see just how odd, let's compare the two cases when the difference in path lengths is zero (so that all photons impact detector 1) and when this difference is 0.5 wavelength (so all photons impact detector 2). This change from a path difference of zero to a path difference of 0.5 wavelength can be achieved *either by lengthening path 1 or by shortening path 2 by half a wavelength*. Thus, every photon responds to changes in either path 1 or in path 2. So *every photon must incorporate data about both paths*. After all, if any of the photons followed only one of the two paths, it would not change it's behavior when the *other* path was altered. The only reasonable description of this situation is "each photon travels by both paths." Each photon is in a superposition of being on both paths simultaneously.

This odd situation might be hard to believe, but it's not hard to visualize. Imagine the photon, pictured in Figure 8.1 just before it enters the interferometer, traveling simultaneously along both path 1 and path 2 within the interferometer—just like a water wave that separates into two superposed directions of motion. I will refer to these as the two *branches* of the superposition, referring to them separately as *branch 1* and *branch 2*.

To make sense of superposition, you must visualize quanta as waves in fields. The common picture of quanta as tiny things, like billiard balls only smaller, contradicts the extensive experimental evidence of superposition. If a quantum is like a billiard ball, it cannot travel two branches simultaneously. In fact, the notion that an individual quantum can be in two places simultaneously led Einstein, in 1927, to make perhaps his earliest recorded objection to quantum theory.[4]

If a superposed quantum is actually *detected* in one of its two branches, it must vanish instantaneously from the *other* branch, which could be kilometers away. Einstein argued this violates relativity's injunction against instantaneous action across a distance. However, you'll see later (Chapters 9 and 10) that quantum superposition actually does not violate the theory of relativity, because relativity forbids only the transmission of energy or information at speeds faster than light, and the simultaneous vanishing of a quantum from one branch and detection in the other branch does not involve such transmission.

When you accept the notion that a quantum is a wave, the explanation of the interferometer experiment is simple. Each quantum simply spreads out to travel both paths, just as a water wave can travel in two directions simultaneously. Based on this picture, the explanation of the experimental result of Figure 8.2 is simply that the two paths interfere when they are mixed together at beam splitter 2, just as water waves interfere when they pass through each other (Figure 2.1). When the path difference is zero, it happens that the two waves interfere constructively at detector 1 and destructively at detector 2, and when the path difference is 0.5 wavelength they interfere destructively at detector 1 and constructively at detector 2. For intermediate settings, we get partial interference. Because there is only one photon at a time in the interferometer, *each photon interferes only with itself*. If you think of a photon as a tiny particle, this is perplexing. But if you think of it as a wave, it makes sense.

With beam splitter 2 in place, this experiment is much like the double-slit experiment with no detector at the slits. In both cases, each quantum travels (more accurately, it "spreads" as a field) along both interferometer paths to the two detectors, or through both slits to a particular point on the screen. In both cases, we observe interference of the two branches of this superposition, constructive when crest meets crest or valley meets valley, and destructive when crest meets valley.

The tendency to think of quanta as tiny but ordinary objects, like a sand grain only smaller, is even stronger when the quantum is an atom or other composite material object. But atoms are waves too, and they can be superposed, as can large molecules made of many atoms, as we'll see later.

All of this provides a new perspective on what happens when beam splitter 2 is *absent*. After all, when a single quantum arrives at beam splitter 1, it's hard to see how the quantum could have any foreknowledge of whether beam splitter 2 is present or absent, because the quantum has not yet reached that location. It's reasonable to conclude that the quantum spreads along both paths regardless of the presence or absence of beam splitter 2.

In fact, quantum theory predicts that the quantized EM field (in the case of photons) or the quantized matter field (in the case of electrons, atoms, and so on) goes both ways, as a superposition, at beam splitter 1. This implies that, with beam splitter 2 absent, the quantum travels both paths, reaches both detectors, and *then* collapses when it interacts with just one or the other detector rather than both. In fact, there is direct theoretical support for the hypothesis that all quantum jumps, such as jumps between energy levels in atoms, occur only when they are detected macroscopically, and there are proposals to test this hypothesis directly and experimentally.[5]

Returning to Alice's reflection in the window, it follows that every reflected photon detected outside by Alice, and every transmitted photon detected inside by Bob, *goes both ways as a superposition* and collapses to a single location only when it reaches Alice and Bob. This provides additional food for thought the next time you pass a storefront window.

It's surprising but true that most of the big quantum issues are involved in this everyday example. It all goes to show that quantum physics, including most of its wierdness, is all around you.

Macroscopic Quantum Phenomena

During the 1920s, Niels Bohr, Werner Heisenberg, Max Born, and others established the first widely accepted view of the strange new theory of the quantum,[6] a view later known as the *Copenhagen interpretation* in honor of Bohr's institute in that city. Bohr and his colleagues, however, did not consider it an "interpretation" but simply quantum physics.[7] Until the 1950s, when apparent paradoxes caused other views to crop up (Chapter 10), Copenhagen was assumed to be the only acceptable quantum interpretation. A key feature was Bohr's insistence that a classical apparatus is always required for observations of quanta, because humans must be able to observe the outcome directly in a macroscopic device. Such an interpretation inappropriately (in the view of many physicists, including me) interjected humans into the foundations of the theory. Bohr perceived the quantum realm to be microscopic and thus accessible to humans only indirectly; this quantum realm was, he thought, distinct from the directly accessible macroscopic realm, where classical physics applies. Bohr proclaimed, in effect, that there was a quantum-to-classical boundary, and that the quantum rules did not apply in the classical realm.[8]

But because big things are made of small things, we might expect quantum physics alone to be able to explain the ordinary world we see around us (Chapter 11). It's a common but not universal opinion among physicists, and one that I share, that the physical universe is fundamentally quantum. If so, then everything you see around you, from dust grains to stars, is a macroscopic quantum phenomenon. Certainly many familiar chemical, biological, and nuclear phenomena have a quantum origin. Interactions between light and matter; the differences between solids, liquids, and gases; and the stable structure of solid materials all come down to quantum physics. This means we might expect to find macroscopic phenomena that demonstrate obvious quantum features. In the next paragraph, and in the following two sections, I'll tell the tales of three such phenomena.

If you followed my suggestion in Chapter 4 and performed a single-slit experiment using your thumb and forefinger as a narrow slit, a well-lit surface as the light source, and your retina as the screen, you have already observed directly a macroscopic phenomenon arising from a fundamental quantum principle: the wave nature of each photon. As you know, the narrow interference lines running parallel to your fingers are made by zillions of photons, each one of them acting independently of the others and interfering only with itself. The set of interference lines mirrors the quantum state of each individual photon; it is a picture of the quantum state of each photon just before it interacted with your retina. The slit, as well as it's image on your retina, are a millimeter or so wide, and definitely macroscopic. The experiment demonstrates millimeter-wide photons.

Lasers

"Light Amplification by the Stimulated Emission of Radiation" is our second example of macroscopic quantum phenomena. Lasers (you can tell it's a really important acronym because the word is now written in the lowercase) are made of a "lasing material" with atoms that have a long-lived, or metastable, excited state. An external energy source excites a large fraction of such atoms into this state. Einstein, in 1917, predicted that any atom in an excited energy state could be stimulated to quickly emit its energy by bathing the atom in radiation with a wavelength that matched the atom's emitted radiation. Lasers, invented in 1960, put this principle to work. When the external energy source builds the lasing material's population of metastable excited atoms up to the point that more atoms are in their excited state than are in their unexcited state (this is called a *population inversion*), the collection of metastable atoms begins to radiate more photons than it absorbs. A kind of chain reaction then occurs rapidly. The radiated photons stimulate other metastable atoms to radiate, their photons stimulate still more atoms, and the entire laser is soon filled with photons, all of them arising from identical atomic transitions. Typically, the laser is a tube filled with the lasing material, with a fully reflecting mirror at one end and a partially transmitting mirror at the other end. Photons that happen to be moving parallel to the tube's axis then reflect many times back and forth, stimulating other atoms to radiate, before escaping through the partially transmitting end. Quantum physics predicts *stimulated emission creates photons identical to the stimulating photon in every way, having identical wavelengths, directions of motion, and phases*. The term *identical phases* means the photons' crests and valleys are in step with each other, so they always reinforce each other constructively rather than interfering destructively. Think of zillions of identical photon wave packets (Figure 7.1) all moving in the same direction, all with their crests and valleys perfectly aligned.

Thus the zillion photons emitted by the lasing medium move in unison to form a large wave—a big quantum state that's like a single giant photon—with macroscopic effects. Like a well-disciplined army, the photons move in step, creating effects that are similar to single-photon effects, but are much more energetic and, hence, observable macroscopically. A normal flashlight creates a beam of photons with different quantum states that quickly spread apart, and the beam becomes less intense with distance. A laser's photons share a single quantum state and thus remains narrow over great distances, one of its most useful properties.[9]

Superconductivity

Our third tale of the macroscopic quantum is superconductivity. You learned in Chapter 5 that superconductivity occurs when a low-temperature metal transitions into a new kind of quantum field in which electrons form into pairs that can move freely through the metal without experiencing EM forces. The pairs experience no electrical resistance, allowing electric currents to persist for years without batteries—a powerful property for applications such as superconducting electromagnets. Although I suppose nothing really lasts forever, theoretical estimates for the lifetime of a persistent current can exceed the 14-billion-year age of the universe.[10] How does this highly quantum phenomenon work?

Ordinary electrical currents arise in metals because one or two electrons per atom are free to drift within the body of any metal, so that small EM fields (created externally by batteries) force these "conduction electrons" to flow easily through the surrounding "lattice" (or regular array) of atoms. As they flow, the negatively charged electrons interact electrically with the lattice, transferring energy away from the electrons and into the lattice's natural thermal vibrations. So the current runs down, unless there is a battery to maintain it, while the metal warms up—a phenomenon called *electrical resistance*.

The atomic lattice plays a crucial role in superconductivity. Interatomic forces, acting like tiny springs strung between the atoms, hold the lattice together, enabling it to support mechanical vibrational waves that carry thermal energy and sound through the metal. Because the atoms forming the lattice must obey the quantum rules, these lattice vibrations are quantized. This quantization results in quantized bundles of vibrational energy, called *phonons*, that move through the lattice and interact with electrons. So electrons in a metal interact not only with the photons of the quantized EM field vibrations, but *also with the phonons of the quantized lattice vibrations*.

At sufficiently low temperatures, electron–phonon interactions take an unexpected turn. At these temperatures, *pairs* of electrons turn out to have lower total energy than *individual* electrons. Because nature always prefers to "fall" into the lowest available energy state (think of rocks sliding downhill), the electrons then bind together in pairs. These pairs are unified quanta, composites of two electrons in the same way that an atom is a composite of smaller quanta. Just as an atom has very different properties from its constituent quanta, *these composite pairs have fundamentally different properties from unpaired electrons.* The paired electrons are a surprisingly large 100 nanometers apart; this is 500 times larger than the spacing between the atoms. Thus, the quantum states of the pairs overlap enormously and form a single, fluidlike macroscopic state.[11]

The new properties are best described in terms of so-called *bosons* and *fermions*.[12] All quanta can be grouped into these two broad and radically different categories. The primary difference between them is that bosons prefer to gather together in a single quantum state whereas fermions prefer isolated solitude. Photons, for example, are bosons, which is why lasers are possible. But *individual* electrons are fermions, so they can't cooperate in the way that a laser's photons cooperate. This situation changes radically when the temperature is sufficiently low for electrons to form into pairs, because *each electron pair is a boson.* So the physics of a metal changes radically below the superconducting transition temperature.

An unpaired conduction electron in a metal has a continuous range of energies available to it, enabling the electron to exchange energy easily with the lattice, creating electrical resistance and warming the lattice as described earlier. But when two electrons bind, the pair has a quite different energy structure. The pair falls into a quantum ground state of lowest energy, much as the electron in a hydrogen atom falls into its ground state. Analogous to state (a) in Figure 7.4, this state is separated from higher energy states by an energy gap. But unlike Figure 7.4, this gap is quite small. Furthermore, the energy states above this gap are not paired states, but instead are normal conducting electron states. So this lowest energy state is delicate. If you excite the pair with more than a certain minimum energy, the electron pair "rises" into a normal conducting state, where the electrons lose energy quickly to the lattice. The paired state, on the other hand, is extremely stable against losing energy to the surrounding lattice, because (like an atom in its ground state) it cannot drop to a lower energy.

When the temperature of the metal has dropped below the transition temperature, the paired electrons' newly acquired boson status allows them to fall into the same quantum state; *all the conducting electrons form pairs and fall into the same ground state.* These pairs cannot lose energy to the lattice because they are already in their ground state, and they cannot gain energy from the lattice as long as the temperature remains sufficiently low that random thermal fluctuations cannot overcome the small energy gap separating them from the normal conducting state. So the paired electrons cannot interact with the lattice at all, and they move through the metal with zero resistance.

Summary: Just as the electron matter field is a fermion field whose quanta are individual electrons, superconducting electrons form a boson field whose quanta are electron pairs. At low temperatures, these pairs are isolated energetically from the lattice and experience zero resistance. That's how superconductivity works.

Superconductivity is pretty special. Like photons in a laser beam, every pair is locked into precisely the same quantum state, with precisely the same energy and motion. The matter field of each pair spreads broadly across thousands of atoms so that the total field for all the pairs becomes a single macroscopic fluid that fills the entire superconductor with a single macroscopic quantum state. It's like laser light; small ripples (individual quanta) come together to form a large wave. Supercurrents are macroscopic yet highly quantum, often comprising zillions of pairs moving as one through many kilometers of wire.

Strikingly quantum objects are not limited to microscopic sizes.

Macroscopic Superpositions

Although it's not hard today to find macroscopic quantum phenomena, it's difficult to demonstrate macroscopic quantum *superposition*. Lasers and supercurrents are macroscopic, but they are not superpositions. Neither the laser's individual photons nor the superconductor's individual electron pairs are in two macroscopically distinguishable quantum states at the same time. The macroscopic nature of lasers and supercurrents stems simply from having a large number of quanta in absolutely identical states.

The problem of creating large-scale superpositions is important, for the practical reason that future quantum computers would necessarily involve complex superpositions extending over distances far larger than an atom and involving computational elements called *qubits* that would themselves be in superposition states. Contemporary computers use "classical bits" such as small electrical currents that have two states (perhaps "off" and "on") dubbed 0 and 1. Each bit can be in *either* one *or* the other state. Quantum computers will use qubits that can be in quantum superpositions of *both* 0 and 1, giving them phenomenal computing power for certain applications.

Macroscopic superposition is also important theoretically because it's related to the famous quantum measurement problem, according to which a quantum measurement could supposedly put a macroscopic object such as Schrödinger's cat into a superposition of paradoxically different states such as alive and dead. Chapter 10 describes this conundrum and suggests a solution.

A third reason for demonstrating macroscopic superpositions is that their utterly nonclassical implications would counter Bohr's dictum of an important distinction between the microscopic quantum world and the macroscopic classical world. There are abundant signs today that Bohr was wrong. Definitively resolving this would help clarify the quantum foundations.

The remainder of this chapter tells four remarkable tales of nearly macroscopic superpositions that have been achieved experimentally.

Swinging In Two Directions at Once

The 2012 Nobel Prize in Physics was awarded for a pair of experiments involving quantum superposition, one led by David Wineland and the other by Serge Haroche. Both experiments demonstrate intricate manipulation of individual quanta. Wineland's experiment, described here, traps electrically charged atoms—*ions*—observing and controlling them with photons. Haroche's experiment, described in Chapter 11, takes the opposite approach, observing and controlling trapped photons by sending atoms through the trap. Both experiments involve fundamental interactions between light and matter, and represent steps toward building a superfast computer based on quantum principles. Their work has also led to a 100-fold improvement in the accuracy of the measurement of time.[13]

Wineland's experiment, which he calls "a 'Schrödinger cat' superposition of an atom,"[14] manipulates a single beryllium ion, made of four protons and five neutrons in the nucleus with three electrons orbiting the nucleus; these electrons are in states similar to the states of the electron in the hydrogen atom discussed in Chapter 7. This ion's net electric charge makes the ion easily manipulated by EM fields. Using lasers to manipulate it and magnets to contain it, the ion can be trapped within a small region. The trap is designed to cause the ion to vibrate (or oscillate) back and forth. I'll describe this state in the following paragraph and then explain how Wineland coaxes the ion into a fascinating superposition of two states.

Because the ion is a composite object, the ion's quantum state has two distinct aspects. First, *the ion's center* oscillates back and forth within the trap, in a path 80 nanometers long. The quantum state of the ion's center is a wave packet similar to that shown in Figure 7.1 and is about 7 nanometers long. So the ion's center forms a 7-nanometer-long matter field swinging back and forth along an 80-nanometer path. Second, *the three electrons within the ion each have their own "electronic state,"* with each one analogous to a hydrogen atom's single-electron states (Figure 7.3). The ion's electronic state is much smaller than its overall wave packet, being only about 0.1 nanometer across (a typical size for an atom). To put this another way, the indeterminacy of the electrons relative to the ion's center is 0.1 nanometer whereas the indeterminacy of the center itself is 7 nanometers, and the whole thing oscillates to and fro along an 80-nanometer path.

Just as an electron in a hydrogen atom has a series of quantized energy levels, both the oscillatory state and the electronic state of the ion have a series of quantized energy levels. Using lasers, Wineland slowed the ion all the way down to its lowest-energy oscillations and its electronic ground state.

Wineland then executed an intricate series of four steps, each involving a specific short pulse from a laser that illuminates the ion briefly and is timed and tuned precisely to its specific task. In step 1, the ion's electronic state is split into a superposition of two branches, one branch having a higher electronic energy than the other. Step 2 uses a laser pulse with a frequency that affects only the higher energy branch. This pulse provides oscillatory energy *to that branch only*, setting it vibrating to and fro along the full 80-nanometer path. In step 3, a single laser pulse exchanges the two superposed branches, putting the energetically oscillating branch into the lower energy electronic state and the minimally oscillating branch into the higher energy electronic state. In step 4, a pulse similar to the step 2 pulse excites the minimally oscillating state into energetic oscillations that are "out of phase" with the other energetically oscillating state. The lower-energy branch then swings "to and fro" while the higher-energy branch swings "fro and to" along the 80-nanometer path! The two superposed branches pass right through each other (like water ripples, or ghosts) as they swing back through the center. It's as though my Dad was pushing a low-energy Hobby (my grade-school nickname) to and fro on my backyard swing, and also pushing a higher-energy Hobby fro and to, simultaneously!

To demonstrate that the experiment actually did achieve a superposition of a single object executing two opposite motions simultaneously, Wineland next brought the superposition's two branches together and showed they then interfere in just the expected manner; this is analogous to a single photon or electron in the double-slit experiment interfering with itself.

At 80 nanometers, this superposed motion is far from macroscopic, but it's a thousand times larger than an atom. Such distances are often described as *mesoscopic*.

A Superposed SQUID

We saw earlier that superconductivity can definitely be a macroscopic quantum phenomenon, but I'll let you decide whether the superconducting superpositions described in our next tale are properly described as macroscopic.

A superconducting quantum interference device (SQUID) is made from a tiny ring (0.14 millimeter across in one experiment—the width of a human hair) made of superconducting metal. Superconducting currents comprising billions of electron pairs can circulate within such a ring. All these pairs are in a common quantum state with a wavelength that "fits" exactly around the entire ring to form a standing wave, implying there are either one or two or three, and so on, complete wavelengths around the ring's circumference. Each possibility represents a distinct quantum state of the supercurrent—a nice example of quantization. When a supercurrent is established in such a ring, it's extremely difficult to alter its quantum state because one cannot just increase or decrease the wavelength by a small amount and still have a wave that fits.

But this superconducting ring is not yet a SQUID. To make a SQUID, one replaces a small segment of the ring with an insulating (nonconducting) segment

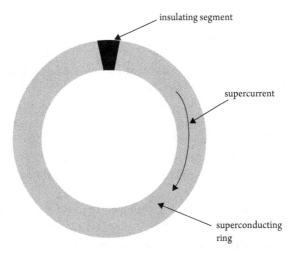

Figure 8.3 Schematic diagram of a superconducting quantum interference device.

(Figure 8.3). A process similar to alpha tunneling (Chapter 6) allows electron pairs to tunnel through such a barrier, so the supercurrent can continue despite the nonconducting barrier. But now, with the barrier in place, and also assisted by an externally created magnetic field, it's possible to alter the state from, say, five waves encircling the ring to two waves. One can then put the ring into a superposition of two or more such states, leading to the possibility of interference between different states. Now we have a SQUID with interference effects that can be put to such uses as magnetic measurements for magnetic resonance imaging and, rather unexpectedly, oil prospecting.

In fact, it's possible to create such a superposition in which the two superposed components circulate in different directions around the ring, one clockwise and one counterclockwise. Two independent research groups achieved this in 2000.[15] This represents considerable finesse: a single quantum state in which a supercurrent comprising a billion electron pairs circulates simultaneously in two opposite directions around a ring. The current and the submillimeter size of these SQUIDs are tiny by macroscopic standards, but enormous by atomic standards.

Large Superposed Molecules

Next, here's a tale of superposed mesoscopic molecules. A University of Vienna group led by Anton Zeilinger managed, in 1999, to perform the double-slit interference experiment with C_{60} molecules and, in 2002, with C_{70} molecules.[16] Every double-slit experiment demonstrates superposition, because to create an interference pattern, each quantum must pass through both slits and interfere with itself, so the quantum must be in a superposition of coming through both slits.

C_{60} and C_{70} molecules are often dubbed *buckyballs* because they are near-spherical shells, similar to soccer balls, made of carbon-atom structures reminiscent of Buckminster Fuller's architecture. In the 2002 experiment, each C_{70} molecule is rather massive, containing more than a thousand protons, neutrons, and electrons. Recall that such massive objects have shorter wavelengths than less massive objects. The C_{70} molecule's wavelength is only 1 picometer (one trillionth of a meter)[17]—100 times smaller than the size of a typical atom.

Such a tiny wavelength presents a problem, because the standard double-slit experiment requires the width of the slits, and the spacing between them, to be not much larger than the wavelengths under investigation. It would be nearly impossible to construct the tiny slits necessary to demonstrate interference for C_{70} molecules. So Zeilinger's group used an indirect method based on a subtle interference phenomenon called the *Talbot–Lau effect*[18] to perform the experiment using slits about 1 micrometer (one millionth of a meter) wide—a more practical slit size. It would take us too far afield for me to describe this method here, but Zeilinger convincingly demonstrates C_{70} molecules interfering with themselves, and shows the molecular wavelength agrees with theoretical expectations. The ingenious Talbot–Lau interferometer represents an enormous extension of our ability to measure interference at the tiny wavelengths that characterize objects having mesoscopic mass.[19]

Bigger things (literally) are possible. In 2011, a University of Vienna team—headed this time by Markus Arndt and including scientists from Germany, the United States, and Switzerland—reported a new leap in quantum interference experiments. Using the Talbot–Lau effect, the team demonstrated interference of several different large organic molecules. The largest of these had the chemical formula $C_{168}H_{94}F_{152}O_8N_4S_4$, contained 430 atoms, and was some 8 times more massive and 6 times larger than the C_{70} molecule.[20] In the future, such experiments might be performed using protein molecules, or viruses.

Will we eventually get to cats? Probably not, but who knows?

A Superposed Diving Board

Can we watch, with the naked eye, a superposition moving in two ways simultaneously? Our final tale of the superposed quantum concerns a tiny but macroscopic mechanical device, made of trillions of atoms, that can be put into specific quantized states of motion and into a superposition of such states. A team at the University of California in Santa Barbara, led by Aaron O'Connell and Andrew Cleland, put it together. It's the first visible human-made device with mechanical motion that exhibits superposition.[21] It behaves in many respects like a single atom.

Imagine a tiny diving board made of a flat, thin metal strip 40 micrometers (0.04 millimeter) long.[22] That's as long as a fine human hair is wide, and my naked eye definitely cannot see it, but maybe yours could. The Santa Barbara team caused this strip to vibrate, but not in the bending-up-and-down way that diving boards vibrate. It vibrates by expanding and contracting, like your upper body when you breathe.

The strip is made of a "piezoelectric material." This means that, if you send an alternating electric current (one that is oscillating rapidly between forward and backward) through it, the strip vibrates ("breathes") at the same rate as the current oscillates. Furthermore, if the strip is attached to an electrical circuit and if the material is then caused to mechanically vibrate, an alternating electric current flows through the circuit. So the strip transforms alternating currents into mechanical vibrations and vice versa.

This strip is fastened down at its base (like a diving board) and is attached electrically to a "qubit"—a device that can be in either of two quantum states that computer engineers dub 0 and 1. The team's qubit is a SQUID for which the "0" and "1" states are superconducting quantum states. Zero is a lower energy state and 1 is a higher energy state. The team can control and measure these SQUID states. The strategy of the experiment is then to use the SQUID to manipulate the strip into specific quantum states of motion.

So how does this doohickey work? Vibrating mechanical systems have specific quantum states, each with a specific vibrational energy. Heisenberg's principle dictates that the system cannot be in a state of rest, so it must vibrate even while in its ground state. All systems become increasingly classical as their energy increases, so if we want to demonstrate strong quantum effects, we should put them into their lowest quantum states.

A major stumbling block for previous experiments was that such low-energy quantum states require extreme cooling, because otherwise thermal vibrations would excite the system to higher energy quantum states and quantum effects would be less pronounced. Experimenters wanted a strip with a ground state that oscillated at a high frequency (and thus high energy) in order for the ground state to have more energy than typical thermal vibrations; thermal vibrations cannot then alter the state of the strip. But the common bending vibrations of a flat strip have relatively low frequencies, so thermal vibrations pose a problem. Part of the Santa Barbara team's genius lay in recognizing that breathing-mode vibrations in a thin metal strip occur at far higher natural frequencies, and thus higher energies. This means thermal vibrations are not such a problem, and the experiment can run at cooling levels that are reachable by standard laboratory techniques. Thus, the team could cool the strip to a temperature at which the strip had a high probability of being in its ground state of mechanical vibration.

How did the team know when the strip was in its ground state and when it wasn't? This is where the SQUID the came in. The SQUID qubit is an electric device with quantum states that are detectably different superconducting currents. Its state, dubbed 0 or 1, can be controlled and measured with a variable external magnetic field. When the SQUID is attached electrically to the strip, the team can feed quanta (bursts of electrical energy caused by a drop from the 1 state to the 0 state) from the SQUID into the strip, and can extract quanta (bursts of vibrational energy caused by the *strip's* drop from its excited state to its ground state) from the strip and feed it into the SQUID. As you'll see, this permits the team to verify those occasions when the strip is in its ground state.

That's how it works. Finally, the results: After cooling the strip, the SQUID was adjusted to extract one quantum of energy from the strip. The team then found that, 93% of the time, it was impossible for the SQUID to get any quanta out of the strip. This verified that the cooling did put the strip into its quantum ground state 93% of the time—quite an accomplishment in itself.

The team then put the SQUID into its higher energy state and swapped (using microwave energy bursts) the SQUID's quantum of energy into the strip as a single phonon of vibrational energy. By manipulating the SQUID and the strip, it was then possible to confirm that the team actually achieved a one-quantum excitation of a macroscopic object—another enormous accomplishment.

The team put icing on this cake by creating a *superposition* of the strip's ground and excited states, and verified it by observing the superposition's gradual "decay" into the ground state. These measurements are a fine demonstration of the quantum decay process, be it the decay of a radioactive nucleus or of a mesoscopic metal strip.

Here's how this works: The measurement process begins by using the SQUID to put the strip into its excited state just as described earlier. The strip in its excited state is like an atom in an excited state; the state is unstable and, eventually, spontaneously decays (transitions) to the ground state, emitting a phonon of vibrational energy. Quantum theory predicts that if the excited strip is left alone, it will decay gradually as a superposition that involves more and more of the ground state. Such a superposition is similar to a photon in a 50%-50% superposition of passing through each of the two paths of the interferometer in Figure 1.1, but the probabilities are now not necessarily 50% and 50%. Beginning from being in the excited state with 100% probability, the strip will evolve into being in the excited state with 99% probability (and in the ground state with 1% probability), then into being in the excited state with 98% probability, and so on (you get the idea). Theoretically, the decay follows a so-called "exponential decay curve," implying that the probability of being in the excited state first drops rapidly, then more and more slowly, and drops eventually to zero (Figure 8.4).

To demonstrate that their experiment matched the theory, the team showed the strip to be in the superposition indicated by Figure 8.4 at a variety of specific times. To verify the predicted superposition at a decay time of, for example,

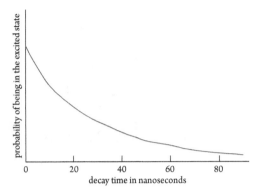

Figure 8.4 The strip's predicted exponential decay curve showing the probability, at various times, that the strip will be detected in its excited state. In the experiment, the measured probability of detecting the strip in its excited or ground state matched this curve for a variety of decay times. This agreement confirmed the quantum superposition of the strip.

10 nanoseconds, the team made a "measurement" of its superposed quantum state at 10 nanoseconds after preparing the strip in its excited state. This means they found the probability of the strip being in the excited state by performing several hundred "trials" of the following experiment: Prepare the strip in its excited state, wait 10 nanoseconds, then determine whether the strip is still in its excited state or has decayed to its ground state. The probability of being in the excited state is then the ratio of the number of trials in which it was found to be in its excited state divided by the total number of trials. These measured probabilities matched Figure 8.4 at different decay times, demonstrating that the decay followed the quantum predictions.

This discussion illustrates a second way of verifying that a quantum system is in a superposition state. The first way is by observing interference between the branches of the superposition, as in the double-slit and Mach-Zehnder interferometer experiments. The second way is by making "quantum state measurements," as described here, that show the system to occupy the branches of the superposition with the predicted probabilities for the possible measurement outcomes.

9

An Entangled, Nonlocal Universe

Quantization leads to a phenomenon so odd it caused Einstein to reject quantum physics "because it cannot be reconciled with the idea that physics should represent a reality in time and space, free from spooky actions at a distance."[1] By "spooky," Einstein meant "instantaneous across a distance."

Contrary to Einstein, just such instantaneous actions at a distance are not only predicted by quantum physics, they are now experimentally verified features of the natural world, predicted by quantum physics but known to exist even if quantum physics were someday found to be incorrect. As you know (Chapter 6), such actions are called *nonlocal*.

Quantum nonlocality is a direct consequence of the unity and spatial extension of the quantum. If a quantum changes in any way, the change must occur simultaneously across the entire quantum, for a quantum is a single unified object, without parts. And yet, each quantum occupies a spatially extended region (more precisely, a range in position and velocity), as prescribed by Heisenberg's principle, entailing that certain events occurring in different locations are correlated instantaneously.

Locality—the idea that objects are directly influenced only by their immediate surroundings—is quite a natural notion. Objects around us are influenced directly and immediately only by their immediate surroundings, and not by remotely located objects. Even a radio signal or light wave sent to us from a distant antenna or star is local in the sense that the source sends out a wave that travels to our location, and that wave then interacts locally with a radio receiver or our eye. This notion is codified in the locality principle (Chapter 6), according to which *physical processes occurring at one location should have no instantaneous effect on the real physical situation at another location.*

This chapter tells of experiments that violate this commonsense notion, how this can be, what it means, and the relation of nonlocality to quantum jumps and other phenomena.

Nonlocality and Quantum State Collapse

Quantization immediately raises questions about the locality principle. Consider a quantum passing through a single slit, diffracting, and impacting a screen (Figure 6.3). As you know, the intensity graph shown at the right-hand side of the figure represents the probability that a single electron will interact with the screen at any particular point x. The electron spreads over the entire pattern just before impact, with different probabilities of interaction at different points. Interaction with the screen then causes the electron to collapse to atomic dimensions, because the screen is made of atoms and the electron is a single quantum that cannot be detected at two places. It must collapse by interacting with a single, randomly chosen screen atom, not two or more. So the collapse happens simultaneously all over the screen and the electron collapses into one small, randomly chosen region. Einstein, in 1927, was unhappy about this. He argued that collapse violates the theory of relativity because it happens instantaneously across a distance, which seems to violate relativity's prohibition on superluminal travel.[2]

Heisenberg, in 1930, also recognized the questions about nonlocality raised by quantum state collapse.[3] He imagined one photon in a 50–50 superposition of reflecting and transmitting at a beam splitter. He noted that, although the photon's reflected and transmitted branches could be widely separated, if an experiment detects the photon in the reflected branch, the transmitted branch immediately vanishes. The reflected branch thus exerts a kind of action on the distant transmitted branch. If quantum states are real, this is a real physical change in the photon, happening instantaneously across an arbitrarily large distance, so nonlocality is built into the collapse process.

A nonrealist would argue that the disappearance of the transmitted branch represents merely an updating of knowledge and is not physical. But neither Einstein nor Heisenberg believed any such easy explanation; their intuition was that something really happens at points from which the photon vanishes.

Heisenberg noted however that "it is also obvious that this kind of action can never be utilized for the transmission of signals so that it is not in conflict with the postulates of the theory of relativity."[4] This is because the outcome is indeterminate at both locations. For example, imagine a communication system that uses a photon source and beam splitter at a central location in space to send a single superposed photon every second, with one branch going to Alice's detector on Mars and one branch going to Bob's detector on Earth. It would take some 15 minutes to communicate by radio. Could Alice manipulate her own detector in some manner to send an instant message to Bob? The answer is no because the probability of Bob detecting a photon remains unchanged no matter what Alice does; he will detect each photon with 50% probability, regardless of what Alice does. If Alice could cause the photon to either appear or not appear at her location rather than at Bob's location, then she could send Bob an instant message such as "the Yankees won the ball game" if Bob detects a photon and "the Yankees didn't win the ball game" if Bob doesn't detect a photon (Alice and Bob would have to agree on this particular "code" in advance). But she can't do such things because she can't control quantum indeterminacy. There is no conflict between the instantaneous collapse of a single quantum and relativity's ban on superluminal communication. Relativity and quantum physics delicately balance each other in this regard, without contradiction.

Entanglement

Are there possibilities for nonlocal communication when *two* quanta, instead of only one, are involved? We will find that, when two or more quanta interact, they can mix together or "entangle," and remain entangled even after the interaction is completed. Quantum physics predicts and experiments verify that, even after the quanta have separated widely, entangled quanta can influence each other instantaneously. This is surprising and has been the source of much controversy. We will find, however, that this instantaneous influence across a distance doesn't conflict with special relativity because it cannot be used to transmit information or energy. Again, there is a delicate balance between relativity and quantum physics.

Figure 9.1 illustrates schematically how two wave packets can become entangled. The gray quantum and the black quantum, initially moving rightward and upward, respectively, are initially independent of each other in the sense that you could alter one without affecting the other. Their paths then cross, causing them to interact temporarily (exert forces on each other), and then separate. But quantum theory predicts they don't separate entirely. Two quanta do indeed emerge from the interaction, but each retains some portion of the other, as indicated in the figure. Because the gray quantum is a single unified object, as is the black quantum, the *two* objects after interaction form a *single* unified whole. *The two quanta now share a unified state and are said to be "entangled."*[5]

This final situation is an entangled superposition in which both the black and gray quanta are moving in two different directions, upward and rightward. The superposition can be described as follows:

> "Black quantum moving rightward and gray quantum moving upward" superposed with "black quantum moving upward and gray quantum moving rightward."

This two-quantum superposition turns out to be remarkable. A system of two quanta with the first moving rightward and the other moving upward (the initial

An Entangled, Nonlocal Universe

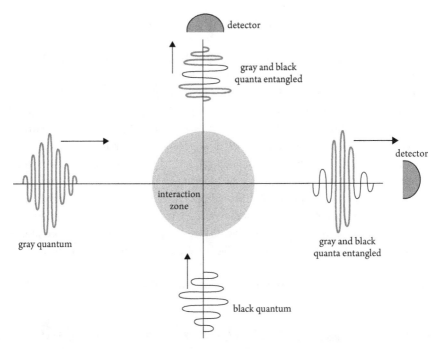

Figure 9.1 When two quanta interact and then separate, they typically remain entangled. (Thanks to Nick Herbert's *Quantum Reality* [New York: Anchor Press, 1985] for suggesting this kind of diagram.)

situation in Figure 9.1) is unremarkable, but we'll see the final entangled state is unlike anything in our macroscopic world. Entanglement is a natural consequence of the wavelike, superposable nature of quanta. This subtle state of affairs has perplexed physicists for decades and is at the root of the measurement, or Schrödinger's cat, problem (Chapter 10).

Remembering that quanta are ripples in a field, the entanglement shown in Figure 9.1 isn't surprising. Two ripples have come together and then separate in such a way that part of each incoming ripple now appears in both outgoing ripples. In view of the unified nature of each initial quantum, one now expects to find nonlocal effects. We might guess that *any alteration of one of the final quanta must entail instant changes in the other, regardless of how far apart they might be.* This would be a good guess.

This is the sort of thing experiments can check. For a few examples, entanglement and associated nonlocal effects have been demonstrated between two photons separated by 144 kilometers,[6] in a variety of systems of atoms and photons,[7] between two atoms at meter separations,[8] between two tiny diamonds separated by 15 centimeters each containing 10^{16} atoms,[9] and between two gas clouds—each made of a trillion cesium ions—at millimeter separations.[10]

An Experiment with Two Quanta

One key experiment was performed in 1990 by John Rarity and Paul Tapster in the United Kingdom, and independently by Zhe-Yu Ou, Xingquan Zou, Lei Wang, and Leonard Mandel in the United States.[11] I'll call this experiment the *RTO entanglement experiment* (for Rarity, Tapster, and Ou). I'm going to dwell on this tale for a bit, and shall return to it in Chapter 10. It illuminates entanglement much as Thomas Young's 1801 double-slit experiment (Chapter 2), to which the RTO experiment is closely related, illuminated the wave nature of light. Figure 9.2 shows one way of viewing this beautiful experiment. It begins with a pair of oppositely directed photons, A and B, that have *already* been entangled by means of a process called *parametric down-conversion*, the details of which aren't important here. What's important here is the behavior of quanta after they are entangled. Figure 9.2 shows photon A moving leftward through a double-slit arrangement, and photon B moving rightward through a second double-slit arrangement. The entangled state is described as follows:

> "Photon A passes through slit 1A and photon B passes through slit 1B" superposed with "photon A passes through slit 2A and photon B passes through slit 2B."

Note that in each branch of the superposition the two photons exit the source in opposite directions, one branch passing through slits 1A and 1B (dashed line in

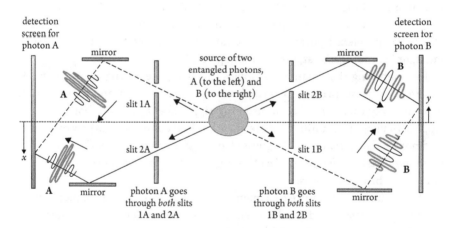

Figure 9.2 The RTO photon entanglement experiment viewed as a "double double-slit experiment." Two entangled photons, A and B, are created in the source. They travel in opposite directions, each photon passing through a double-slit apparatus, and then re-combine at a detection screen. The dashed and solid lines indicate the entanglement: the dashed 2-photon state is superposed with the solid 2-photon state.

Figure 9.2) and the other branch passing through slit 2A and 2B (solid line). The photons are moving directly away from each other along *two* superposed paths, dashed and solid. The entangled *pair* is doing two things at the same time—an "extremely quantum" situation that perfectly demonstrates entanglement.

The experiment could be described as a "double double-slit experiment." Photon A moves leftward in a superposition of passing through a pair of slits and photon B moves rightward through a second pair of slits. Mirrors bring each photon back together to interfere with itself on its own detection screen.

The actual experiment used Mach-Zehnder interferometers instead of double slits, but Figure 9.2 is logically equivalent to the RTO experiment and easier to grasp. RTO studied the patterns formed on the screens by thousands of entangled pairs. If the pairs were not entangled, each detection screen would show the familiar double-slit interference pattern (Figure 2.3). But surprisingly, neither screen shows the least trace of interference. After thousands of pairs have struck their respective screens, the pointlike impacts are found to be scattered quite randomly over each screen with no bright and dark interference lines! What's going on?

It turns out there is an interference pattern, but it's buried more deeply in the data. *The pattern lies in the correlations between the impact points of each photon pair.* Here's what this means. Recall (Chapter 6) that two random outcomes are "correlated" if knowledge of one helps to predict the outcome of the other. Assuming that the source-to-screen distance is the same for both photons, the two photons of each pair should impact nearly simultaneously.[12] So, by timing the impacts carefully (there were a few hundred impacts per second), the experimental teams could pair up each impact point *x* of photon A on the left-hand screen with the corresponding impact point *y* of A's entangled photon B on the right-hand screen. The data then consisted of a sequence of paired impact points *x* and *y*.

To understand the physics, it's helpful to visualize widely separated screens and an observer at each screen, as is indeed the case in some entanglement experiments. Some new words are useful: The unpaired data for photon A alone or photon B alone are called *local data* because they can be gathered immediately on impact by a *local observer* who observes just one screen, whereas the paired data are called *global data* because they can be gathered on impact only by *two* local observers who must then compare notes (perhaps by communicating via a light beam if they are very far apart).

As just stated, the local data show, surprisingly, that photon A impacts randomly all over its screen with no sign of interference, and the same goes for photon B. The global data are also surprising. If one studies, say, 1000 global pairs, all with the *same* impact point *y* on the right-hand screen but randomly *varying* values of *x* on the left-hand screen, one finds the 1000 impact points on the left-hand screen *form an interference pattern* and the center of this pattern is at a point *X* that is precisely equal to the position *y* at which the *other* photon impacts the right-hand screen! That is, $X = y$ (note that the x-axis points downward whereas the y-axis points upward).

For example, given that photon B impacts its screen at, say, $y = 3$ millimeters on the *right-hand* screen, the quantum state of the paired photon A forms an interference pattern centered on the point $x = 3$ millimeters on the *left-hand* screen; A's impact point must be consistent with this pattern.

That's odd. Photon B impacts at some entirely random point that turns out to be, say, $y = 3$ millimeters, and somehow the *arbitrarily distant* photon A instantly "knows" that it "should" impact at a point x that fits an interference pattern centered at $x = 3$ millimeters. Because B's impact point is indeterminate right up until the impact occurs, how can photon A "know" instantaneously where photon B impacted? And of course this works in reverse. Given that photon A impacted at, say, $x = 5.3$ millimeters, photon B "knows" instantaneously that it must impact at a point that fits an interference pattern centered at $y = 5.3$ millimeters. *There is an interference pattern, but it shows up only in correlations contained in the global data.* It doesn't show up in the local data. This will turn out to be the key point in the resolution of the measurement problem (Chapter 10).

In the ordinary double-slit experiment with single photons, each photon's state is a superposition of coming through both slits; the photon carries data about the path difference between these two branches and interferes appropriately with itself on impact at the screen. In the RTO experiment, each photon carries data about the location of the *other* photon and adjusts its impact point accordingly.[13] Each photon interferes with the *other* photon instantaneously and across an arbitrary distance! In principle, the two photons could impact screens in different galaxies, and these results would be unchanged.

This certainly seems nonlocal. As you'll soon see, it does indeed violate the locality principle that objects should be directly influenced only by their immediate surroundings.

The RTO experiment is actually done with Mach-Zehnder interferometers, one for each photon (Figure 9.3; compare with Figure 1.1 as explained in the caption), but the logic of the experiment is the same as for the double slits of Figure 9.2. Both interferometers have beam splitters to mix the paths at the detectors. Each interferometer has a "phase shifter," labeled x and y in the figure, that alters a path length; x sets the difference between the two paths for photon A, and y does the same for photon B. In terms of the equivalent experiment in Figure 9.2, these phase shifters set the observation points x and y on the two screens. The beauty of interferometers is that experimenters can study the relationship between impacts at specifically chosen values of x and y, *and these values can be changed quickly*. To understand the nonlocal aspects of this experiment, imagine that "Alice's" and "Bob's" phase shifters, x and y, respectively, are widely separated.

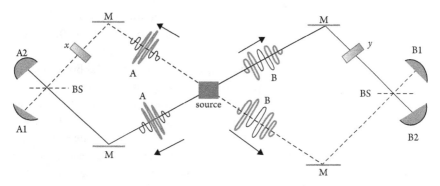

Figure 9.3 The RTO experiment using Mach-Zehnder interferometers instead of double slits as in the preceding figure. Each of two entangled photons A and B exits the source along two paths (dashed and solid lines) like the two paths of the single-photon interferometer in Figure 1.1. "M" means mirror, "BS" means beam splitter, A1 and A2 are photon detectors, as are B1 and B2. The boxes labeled x and y are "phase shifters" that enable local observers Alice and Bob to alter one of the path lengths—equivalent to altering the observed points x and y in Figure 9.2. As in Figure 9.2, the dashed line indicates one branch of the entangled superposition, and the solid line indicates the other branch.

The results: Alice's local data at detectors A1 and A2, and Bob's local data at B1 and B2, show no sign of interference or entanglement or superposition. Detectors A1 and A2 each click a random 50% of the time, and the same for detectors B1 and B2, *regardless* of the phase settings x and y. This is quite unlike single-photon superposition (Figure 8.1), where changes in the path-length difference produce the interference pattern of Figure 8.2.

The global data reveal the entanglement. As x or y is varied, the *correlations between* the outcomes at A and B vary all the way from "perfect correlation" at some settings to "no correlation" at other settings to "perfect anticorrelation" at still other settings. Here, *perfect correlation* means the two outcomes agree; either both are "1" (dashed paths, detectors A1 and B1) or both are "2" (solid paths, detectors A2 and B2). *Perfect anticorrelation* means the two outcomes disagree (either A1 and B2, or A2 and B1). *Zero correlation* means the two outcomes agree on half of the trials, and disagree on half of the trials, so that one outcome cannot be predicted from the other.

What's surprising here is that each photon's "choice" of detector 1 or detector 2 is entirely random, yet each photon always correlates it's choice appropriately with the *other* photon. How does one photon know what the other photon's choice was, when that choice is random and is not made until both impacts actually occur? And how does, say, photon A know the setting on photon B's phase shifter in order to know the appropriate degree of correlation? The future of each photon is entirely unpredictable, yet each photon "knows" what the *other* photon is doing. Once again, this certainly seems nonlocal, in just the same way the double-slit version of this experiment (Figure 9.2) seems nonlocal.

Figure 9.4 shows the degree of correlation between Alice's and Bob's photons for a fixed value, y, of Bob's phase shifter and a range of values of Alice's phase shifter. The solid part of the graph is identical to the single-photon interference graph (Figure 8.2), except that it begins at the point $x = y$ instead of at $x = 0$, and the graph shows *correlations* between Alice's and Bob's photons rather than *intensities* of a single photon's state. We might say the graph represents an *interference of correlations between two photons* rather than, as in Figures 8.1 and 8.2, an *interference of states of a single photon* (more about this in Chapter 10).

One of the beauties of the interferometer setup is that Alice can change x and Bob can change y while the two photons are in flight. Although this is not actually done in the RTO experiment, it has been done in other similar experiments (see Aspect's experiment, described later). It's strange but true that the outcomes—the correlations—then shift instantaneously to agree with the new settings of the phase shifters. Thus, if there is some subtle form of cooperation between the two photons, it can be altered locally and instantaneously by either Alice or Bob even if they reside in different galaxies.

Analysis of the RTO experiment shows that the experimental results agree entirely with the predictions of quantum physics.[14] Thus, any nonlocality uncovered in this experiment is part and parcel of quantum physics.

Next, we shall see this experiment is just as nonlocal as it sounds.

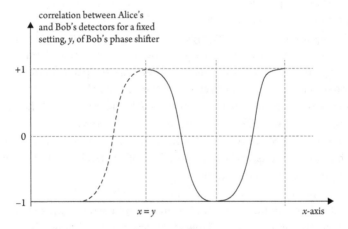

Figure 9.4 The interference of correlations. Correlations of +1 and −1 represent perfect correlation and perfect anti-correlation. The graph shows the correlation between Alice's and Bob's outcomes as Alice varies her phase shifter x for a fixed setting y of Bob's phase shifter. How does Alice's photon "know" Bob's setting y? Alice's photon adjusts instantaneously to the value of y.

John Bell's Test for Locality

It's often suggested that this instant cooperation between two distant quanta only seems nonlocal, but is actually just a result of correlations arising from purely local processes in the past, similar to the gold and silver coins sent to Alice and Bob in Amsterdam and Beijing, respectively (Chapter 6). Is RTO's apparent cooperation between distant quanta a result of undiscovered nonspooky processes?

To settle such questions, John Bell, in 1964, used standard probability methods to derive a mathematical result applicable to entanglement experiments of the RTO type. Bell defined *locality* to mean the probabilities associated with, say, Alice's outcome A1 or A2 depend only on Alice's experimental equipment settings, on Alice's specific outcome, and on past "common-cause" factors that influenced both Alice's and Bob's outcomes. If, after taking all such local and common-cause factors into account, it's still necessary to take Bob's distant outcome B1 or B2 into account to explain the degree of correlation, then the observations violate the locality principle, according to which physical processes at one location should have no instantaneous effect on the physical situation at another location.

Bell showed that locality implies a particular numerical limitation on the degree of correlation between Alice's and Bob's outcomes. Often called *Bell's theorem* or *Bell's inequality*, the precise form of this limitation is not important for this book's purposes.[15] I'll call it *Bell's locality condition*. Bell's result has nothing necessarily to do with quantum physics; it is derived entirely from normal probabilistic considerations. One reason it's so important is that *the quantum predictions for experiments of the RTO type violate Bell's locality condition.*[16] The problem is, for most values of the phase settings x and y, the quantum predictions are correlated too strongly to be a consequence of only local causes; the simultaneous outcome at the other observer's distant station must play a crucial additional role. For example, at certain phase settings x and y, quantum physics predicts, and the experiment shows, a correlation of +71% whereas Bell's locality condition implies correlations lying between 0% and +50%. So, for these phase settings, if Bob changes his phase setting y, we *must* (to explain the experimental results) assume Alice's photon readjusts it's physical state. Furthermore, because Bob could change y quickly while the photons are in flight, this readjustment of Alice's photon must occur instantaneously on alteration of Bob's phase shifter—or at least sooner than the time it would take for a light beam to reach Alice's photon from Bob's phase shifter.

Even in experiments conducted over large distances such as 144 kilometers, such nonlocal action occurs. Bell's locality condition shows quantitatively that experiments such as RTO's really do demonstrate nonlocality, and that quantum physics really does predict nature to be nonlocal.

It's important to note that the reasoning behind Bell's locality condition is based entirely on standard probabilities and does not assume any of the principles of quantum physics. This means that experiments such as RTO show directly that *the universe is nonlocal, regardless of whether quantum physics is correct*. Experiments such as RTO's rule out all local theories, and imply that *any* correct theory must incorporate nonlocality. To its credit, standard quantum physics predicts nonlocality, in agreement with experiment.[17]

This is all a stunning victory for the theory of the quantum.

Aspect's Experiment: Evidence for Superluminal Action at a Distance

RTO's entanglement experiment agrees with the quantum predictions for distant correlations, and Bell's analysis shows these correlations cannot be the result of any common-cause arrangement such as the gold and silver coin example. But are such correlations really established and altered *instantaneously* across a distance? Perhaps an EM signal or other normal local form of communication carries information to Alice's photon about Bob's phase shifter setting. An EM signal is a *local* mechanism because it travels as a wave from one spatial point to the next, with each small region of space influencing only the immediately adjoining regions, rather than acting directly and instantaneously across a distance. This experimental loophole—the possibility that only normal local cause-and-effect is occurring rather than an instantaneous nonlocal cause—is known as the *locality loophole*. The way to rule this loophole out is to make the distance between Alice's and Bob's stations large enough, and the phase shifter alteration fast enough, that *Bob can switch his phase shifter while the two entangled photons are in flight*. If we then still get nonlocal correlations (violating Bell's locality condition), they must have formed while the photons were in flight and faster than any EM signal or other local mechanism could connect the two sites.

John Clauser, working with Stuart Freedman at the University of California at Berkeley in 1972, first detected the nonlocal effects of entanglement experimentally. The team accomplished this by observing violations of Bell's locality condition in the correlations of entangled photon pairs, much like the later RTO experiment.[18] It was a remarkable breakthrough. It was also a big surprise to Clauser, who was motivated to do the experiment by his own conviction that the world was ultimately local, implying that Bell's locality condition would be *obeyed* and the nonlocal quantum predictions would be *falsified*. To his own and many other physicists' surprise, the data violated Bell's conditions and agreed with quantum physics.[19] It was a great victory for quantum physics and a fine example of science's basic value: Conclusions must be formed on the basis of evidence and reason rather than preconceived beliefs. Contrary to Clauser's original belief, the experimental facts that he helped discover indicated the universe is nonlocal.

However, Clauser's experiment suffered a possible, even if implausible, loophole, namely the locality loophole mentioned above: perhaps the apparent nonlocality is caused merely by EM waves or some other local mechanism traveling between the two photons. Clauser's phase shifters were switched only at 100-second intervals, allowing plenty of time for the observed correlations to form during photon preparation, and plenty of time for information about Alice's phase setting to somehow leak across to Bob's detectors (which were only a few meters away) in a normal local fashion.

Alain Aspect and colleagues at the University of Paris closed the locality loophole in 1982.[20] Both Aspect and Clauser performed their experiments using entangled photon "polarization states"—a quantum property into which we needn't delve—rather than entangled wave packets as in the RTO experiment. The nonlocal implications of Aspect's experiment are the same as those of the RTO experiment only with the locality loophole now closed. It was quite an accomplishment. The distance between the two sets of detectors was 12 m, for which the light travel time was only 40 nanoseconds—0.04 millionth of a second. To demonstrate the correlations were established superluminally, preparation of each photon pair and alteration of the phase settings had to occur in less time than this. By achieving a preparation time of 5 nanoseconds and by switching phase settings within 10 nanoseconds, the experiment was able to demonstrate the superluminal establishment of nonlocal correlations.

But physicists remained skeptical because of the oddness and importance (if true) of the experiment's nonlocal implications. This built-in conservatism regarding revolutionary new conclusions is crucial to science's remarkable success in extending our range of reliable knowledge. As the great planetary astronomer and science popularizer Carl Sagan phrased it, "Extraordinary claims require extraordinary evidence."[21]

Perhaps there were other loopholes that could explain the strange results. Perhaps, for example, there is a deterministic pattern in Alice's and Bob's choice of phase settings, enabling the two phase shifters to coordinate their behavior in a normal, local manner. In 1998, Anton Zeilinger and others closed this loophole in an experiment in which the two sites were separated by 400 m across the campus of Innsbruck University in Austria, allowing a light travel time of 1.3 microseconds—1300 nanoseconds. Photon pairs were sent through optical fibers to each station. The duration of each trial was far less than 1.3 microseconds, with the phase shifters reset for each pair while the photons were in flight. Phase shifter settings were selected randomly by a device based on the quantum randomness of photon beam splitters, rather than by a predetermined pattern, ruling out any undesired cooperation between the phase shifters.[22]

Yet another possible but implausible loophole is known as the *detection loophole*. This arises because, typically, only a small fraction of the photon pairs created in this experiment are actually detected. Perhaps the pairs that are detected are especially prone to violate Bell's locality conditions, leading researchers to believe that the experiment violates locality when a complete tally of all the created pairs would confirm locality. Two experiments in 2013 closed the detection loophole by using new types of detectors with significantly higher "detection efficiencies," enabling the detection of a higher fraction of pairs.[23]

In yet another experimental *tour de force*, Zeilinger's group established in 2007 a new record for long-distance entanglement. As usual, the entanglement was verified by violation of Bell's locality condition. The two stations were 144 kilometers apart. Each photon pair was created on one of the Canary Islands off the west coast of Africa, with one photon going to "Alice" on the same island and the other sent through space to another of the Canary islands where that particular photon was received by a 1-m-diameter telescope that then focused the photon on "Bob's" photon detector.[24]

Finally, at the end of 2015, three new experimental results verified the violation of Bell's locality conditions with no significant loopholes of any kind. By January 2015, the sole significant loophole was implausible indeed. Perhaps, it was argued, nature conspires to violate *both* the locality loophole *and* the detection loophole, in such a way that either violation occurs only when the *other* violation occurs. In this case, previous experiments, which tested only one or the other of these two loopholes, could fool us into thinking nature is nonlocal when it isn't. Three experiments, conducted independently, tested this compound-loopholes hypothesis by arranging to randomly switch the phase shifters while the photons were in flight (to test the nonlocality loophole) *and* detect the photons with high efficiency (to test the detection loophole).

The three experiments were published in October and December 2015.[25] According to Alain Aspect, "by closing the two main loopholes at the same time, three teams have independently confirmed that we must definitely renounce local realism [the notion that objects are only directly influenced by their immediate surroundings]. Although their findings are, in some sense, no surprise, they crown decades of experimental effort."[26]

Although experiments continue testing possible loopholes, and although some physicists still disagree, the general consensus is that *nature is nonlocal*: Objects can be directly influenced by distant events.

Is Nonlocality Ubiquitous?

Entanglement is one of quantum physics' most profound features. According to Schrödinger, "I would not call [entanglement] *one* but rather *the* characteristic trait of quantum mechanics, the one that enforces its entire departure from classical lines of thought" (emphasis in the original).[27] It would be surprising, then, if only pairs of quanta could entangle. Can three or more quanta be entangled? Can a single quantum somehow be entangled and, if so, with what? As we will see, the answer to the first question is "yes," and the answer to the second question is "yes, with the quantum vacuum." In fact, a plethora of such entanglement phenomena have been conceived theoretically and demonstrated experimentally.

It's been known since 1992 that, according to quantum physics, all entangled states—all collections of quanta whose present or past interaction causes them to share each others' quantum states, as in Figure 9.1—exhibit nonlocal behavior in which different quanta influence each other instantaneously across a distance.[28] This includes not only pairs of quanta, but also systems having three or more quanta and, surprisingly, single quanta.

It's nearly certain that entanglement and nonlocality are widespread throughout the universe. As one common example, the process of spontaneous (unassisted) emission of a photon by an atom creates an atom–photon entanglement. If the atom is among the zillions of hydrogen atoms that are isolated in intergalactic space, the atom and photon could remain isolated, without interacting with other objects, for millions or billions of years. Quantum physics predicts their entangled nonlocal relationship persists, although they might (when finally "detected" or "measured" via interaction of the photon with a detector or with another quantum) be separated across much of the observable universe.[29] Widespread, indeed.

Physicists have entangled pairs of macroscopic objects. In 2011, Ka Chung Lee and colleagues entangled the vibrational states of two diamonds, each a thin, flat square measuring 3 millimeters on a side, each containing 10^{16} carbon atoms, separated by 15 centimeters.[30]

The tale of how this was done is fascinating and instructive. Diamond is made of carbon atoms located at regularly spaced lattice points. At the beginning of the experiment, the two small diamonds were in states of no organized vibrations. Although the room-temperature atoms were all vibrating individually in a disorganized fashion, there were no phonons (quantized bundles of organized vibrational energy [Chapter 8]) present. The team then sent a single *photon* through a beam splitter, putting the photon into a superposition of two spatially separated paths or "beams," and sent one beam through one diamond and the other beam through the other diamond. When the two beams emerged from the other side of both diamonds, the emerging photon was still in a superposition state of emerging simultaneously from both diamonds, but at a lower energy (lower frequency). Thus, the single photon was in a superposition of having deposited some of its energy in both diamonds. This implied that quantized *phonon* states of each of the two diamonds had been entangled with each other! This entangled state could be described as follows:

"Diamond A has one phonon and diamond B has zero phonons" superposed with "diamond A has 0 phonons and diamond B has 1 phonon."

This is precisely the kind of entangled quantum state studied by John Bell, although Lee's team didn't attempt to verify nonlocality directly by showing experimentally that Bell's locality condition is violated.

This is related to *single* quantum entanglement. Can a lone quantum be entangled even when it is not entangled with any other quantum? You've seen that Einstein and Heisenberg were uncomfortable with the notion that the state of a single quantum, such as a photon, can collapse. Both felt that something actually happens at the many locations from which the photon *vanishes*, but they were uncomfortable with this because it seems incompatible with relativity. As we'll see, they were right on the first count; something does happen at the points from which the photon vanishes. But as we know, such vanishing doesn't violate relativity because messages cannot be sent using this effect.

Single-photon entanglement was first described in 1991 by Sze Tan and colleagues.[31] They suggested sending a single photon through a beam splitter, with the reflected and transmitted output beams going, respectively, to "Alice" and "Bob," who could be any distance apart and were equipped with their own beam splitters and photon detectors. That is, one photon passes through a beam splitter, as in Figure 1.1, but instead of going to a second beam splitter or to individual detectors, beams 1 and 2 go, respectively, to Alice and Bob, both of whom have their own beam splitters equipped with pairs of photon detectors. So paths 1 and 2 never mix, but instead go to Alice and Bob, who could be on different planets. We know, however, that only one of these observers can actually catch the single photon.

We'll see that this photon is entangled. But what's to entangle with if there is only one photon? Answer: the quantum vacuum. The photon, which moves along either path 1 or path 2, is in the following entangled superposition:

"Path 1 contains the photon and path 2 contains the vacuum" superposed with "path 1 contains the vacuum and path 2 contains the photon."

This is exactly like the entangled state of the two diamonds described earlier, but with photons instead of phonons! Both Alice and Bob actually receive something real at their beam splitters and detector pairs. One receives a photon (an excited state of the EM field) and the other receives the vacuum (an unexcited state of the EM field). This sounds like hocus pocus because it's common to think of "vacuum"—an absence of all quanta—as "nothing," so receiving it would be hocus pocus. But the quantum vacuum is quite real (Chapter 5), and not "nothing." Tan's team predicted the correlations between Alice's and Bob's observations would violate Bell's conditions for a purely local explanation and would thus demonstrate the nonlocality of a single quantum. This speaks directly to the concerns of Einstein and Heisenberg about instantaneous quantum state collapse. Tan went beyond Einstein and Heisenberg's perception that state collapse is oddly nonlocal, and predicted that this nonlocality could be detected and verified using Bell's locality condition.

There was considerable debate, following Tan's work, about whether single-photon nonlocality was physically real, and there were further proposals to test this notion experimentally. There was an indirect test in 2002[32] and a direct test in 2004.[33] The results demonstrated a nonlocal relationship between Alice's and Bob's data that clearly violated Bell's locality condition, demonstrating that entanglement between a photon and the vacuum has real nonlocal effects.

So when a quantum collapses, something happens not only at the location to which it collapses, but also at those locations from which it vanishes. At the latter locations, the possibility of a field excitation is replaced by the quantum vacuum. Tan's analysis and experiments show this to be a real nonlocal transaction between the two locations.[34]

We've seen that not only can two quanta be entangled with each other, but a single quantum can be entangled with the quantum vacuum. Larger systems, called *many-body systems*, containing three, four, or more quanta, can be entangled as well. A three-body system can contain a two-body entanglement in three distinct ways (entanglement of quanta 1 and 2, or 2 and 3, or 3 and 1), and it can also contain a three-body entanglement in which each quantum participates in the state of the other two—for a total of four distinctly different entanglements in all.

If you ponder the case of four bodies (drawing a diagram helps), you'll see it has 11 distinctly different entanglements (6 two-body, 4 three-body, and 1 four-body), and that the number of possible distinct entanglements must rise rapidly as the number of quanta increases. So things get complicated.

Every entangled state of every system of three or more quanta is predicted to have some degree of nonlocality. Bell–type locality conditions have been worked out for the various entanglements that can arise in these systems, but it's difficult to probe these systems experimentally the way that two-body systems have been probed by RTO, Aspect, and others, so it's difficult to verify their nonlocality.[35]

Most kinds of atoms are many-body systems with three or more quanta, and the standard quantum theory of atoms implies that all of them are highly entangled internally.[36] The neutral carbon-12 atom, for example, contains 6 entangled electrons, not to mention 12 entangled protons and neutrons. So entanglement and nonlocality are ubiquitous within atoms, which is why an atom behaves in so many ways like a single unified quantum; it can, for example, interfere with itself. Every highly entangled system is similar to individual atoms in the sense that the entangled system behaves in the unified, "all or nothing" manner of single quanta. The entangled photon pair of the RTO experiment (Figures 9.2 and 9.3) is, in this sense, an "atom of light."

In what sense might the universe be an entangled system? Are the cosmic background photons that come to us from the early universe (a mere 400,000 years after the Big Bang, when photons were first freed to travel through nearly empty space) entangled with material from the earlier universe? Do entangled systems really retain their entangled status forever, and regardless of distance? What are the cosmological (large-scale universe) implications of entanglement? I'm not aware of much scientific discussion about such questions.

However, a recent general result shines some light. According to mathematician Stanislaw Szarek, "There have been indications that large subgroups within [large] quantum systems are entangled. Our contribution is to find out exactly when entanglement becomes ubiquitous." [37] Szarek and colleagues found, theoretically, that for interacting systems with at least hundreds of quanta, internal entanglements are ubiquitous in the sense that any two large subsystems within them are almost certainly entangled with each other.[38] For example, in a system of 1000 quanta, any two distinct groups of 200 or more quanta are highly likely to be entangled with each other. This result indicates that the macroscopic world is highly entangled.

So it appears that ordinary objects, such as this book, contain multiple internal quantum entanglements among their parts, and that different objects are likely to be entangled with each other. Although the detailed study of entanglement in many-body systems has only gotten underway during the past few years, the world appears to be highly entangled. Entanglement-related technologies such as quantum computers, and the research prospects of uncovering significant new features of the quantum world, point toward continued explosive growth in entanglement studies. It's my guess that many tales of the entangled quantum, some of them highly significant for our understanding of the universe, remain to be discovered.

Nonlocality and Reality

Despite all the theoretical and experimental work verifying the nonlocality of entangled systems, there has always been controversy regarding whether nonlocality is physically real. Although the tide of expert opinion has recently swung in favor of reality, this cannot yet be called a consensus.[39]

Many who doubt that nonlocality is real argue that Bell's locality condition actually assumes not only locality, but also a second condition that skeptics call "realism"; they argue that the violation of Bell's locality condition points to a violation of realism rather than a violation of locality. More generally, skeptics argue that quantum physics is not about the real world but only about our knowledge of the real world. John Bell himself disagreed with this criticism, and "became terribly upset at suggestions that the proof of [Bell's locality condition] required any assumptions other than locality."[40] Furthermore, study of

Bell's proof reveals not a word concerning realism, even though Bell's paper is a response to the so-called Einstein–Podolsky–Rosen paradox which is all about the "elements of physical reality."[41]

This argument about realism and nonlocality is related to the view that quantum states don't represent reality, but are instead only bookkeeping devices to represent our own knowledge (Chapter 7). This nonrealistic view makes it easy, perhaps too easy, to explain why quantum states collapse instantaneously when a quantum impacts a viewing screen (it's because an observer's knowledge suddenly changes) and why correlations between a widely separated entangled pair of quanta can change instantaneously (same reason).

As I argued in Chapter 7, this nonrealist proposal is beyond the pale of what has, in the past, been considered scientific. Furthermore, there's no theoretical or experimental reason to adopt the nonrealist position because, strange as it may seem, nonlocality and the related notion of quantum state collapse seem to be in accord with experiment and with such general principles as special relativity's injunction against superluminal communication. The problem appears to be simply that nonlocality is odd. But there have been many odd theories that eventually received scientific consensus. What needs adjusting here is our own preconceptions.

A realistic interpretation of quantum physics entails that spatially extended quanta really can alter their entire configuration instantly. When the extended quantum field of an electron interacts with a viewing screen, the field (i.e., the electron) collapses instantaneously from "being all over the screen" to "being within an atom-size region." And as the two entangled photons of Figure 9.2 move toward their screens, they have no individual positions relative to the screen, but only relative to each other. When either member of the pair strikes its viewing screen, the other distant quantum retains its data about the first quantum's impact point and adjusts its own impact point accordingly and instantly. The reason for these instantaneous nonlocal changes is the unity of the quantum: You cannot change only part of a quantum. It's always all or nothing. The two photons of the RTO experiment are entangled in a single two-body quantum state whose extent is represented by the dashed line and solid line of Figures 9.2 and 9.3. This two-quantum system acts like a single unified quantum. It could stretch across our galaxy, yet it's a single object that can reconfigure itself, i.e. "collapse," instantaneously. It might seem spooky, but this is the physically realistic interpretation of the experiments.

Does Nonlocality Conflict with Special Relativity?

One reason for scientists' discomfort with nonlocality is that it seems to violate special relativity by providing a channel for superluminal information transfer. To see if this is so, consider the RTO experiment: both Alice and Bob see only random impacts, conveying no information, on their own detectors. As we've seen, the interference pattern lies only in the *correlations between* the two quanta, and these correlations can only be "observed" later, by gathering (at ordinary, not superluminal, speeds) data from both Alice and Bob and comparing it. Thus, *instantaneous changes of correlations can't be used for communication*, and the instantaneous nature of the change can only be discovered in retrospect.

To make this argument more concrete, suppose that in the set-up of Figure 9.3, Alice and Bob both receive the following string of 10 random 1's and 2's: 1, 2, 2, 2, 1, 1, 2, 1, 2, 2. Because they both receive the same string, the two sets of messages are apparently correlated positively. Suppose, instead, Alice changed her phase shifter to cause Bob to receive messages that were anticorrelated with hers. In this case, Bob would have received 2, 1, 1, 1, 2, 2, 1, 2, 1, 1. Locally, Bob can't tell the difference; in both cases, he simply receives a random string of messages. Bob can detect changes in Alice's phase shifter only by gathering *Alice's* data and comparing it with his. *Local observers cannot detect changes in the nonlocal correlations.* So Alice can't use such correlations to send information to Bob.

Quantum physics provides nonlocal correlations that can be altered instantaneously across any distance, while carefully protecting special relativity's injunction against superluminal information transfer. Einstein's special relativity "survives by the skin of its teeth," as Brian Greene puts it in *The Fabric of the Cosmos*.[42] But perhaps this is not so surprising. Even in low-energy situations, relativity is part and parcel of quantum physics.[43] If so, then quantum physics cannot contradict special relativity.

Quantum Jumps and the Detector Effect

Chapter 7 discussed the two kinds of quantum dynamics: smooth changes described by Schrödinger's equation and instantaneous quantum jumps such as those that occur when the electron of Figure 6.3 impacts its viewing screen. An electron's impact on a viewing screen is one example of a "measurement," defined in Chapter 2 as any macroscopic event (the flash on the screen) caused by a quantum process (the electron's interaction with the screen's atoms). This section explores the relation between measurement, entanglement, and quantum jumps.

Let's return to the double-slit experiment,[44] and compare results for two previously presented setups, which I'll call (a) and (b), and a third new setup, which I'll call (c). In (a), only one slit is open. As you know, impacts on the screen then occur randomly, with no interference, across a broad region centered behind the slit, perhaps with interference fringes at the edges. Let's assume the slit is so narrow the central region spreads over the entire screen. In (b), two slits are open. As you know, impacts are then distributed in an interference pattern across the screen. In (c), two slits are again open, but at the slits a "which-path detector" is present. I'll define this as a gadget that can determine through which slit each quantum (e.g. an electron) passes while not significantly disturbing the quantum so that the quantum proceeds, undisturbed, toward the screen. Such an "ideal" (or "nondisturbing") detector is feasible, at least to any desired approximation.[45]

Even describing this setup seems to present a puzzle because you know that, with no such detector, each quantum comes through *both* slits simultaneously (Chapter 5). Quanta are subtle. Will the operation of this detector affect the experimental results and, if so, how?

Figure 9.5 pictures the three setups and results. You already know the results of setup (a), shown in the figure as (a1), which assumes slit 1 only is open; and also as (a2), which assumes slit 2 only is open. You also know the interference pattern that results from setup (b). However, the result for setup (c) might come as a surprise. Although both slits are open, *the mere operation of a detector at the slits greatly changes the outcome on the screen, destroying the interference pattern observed in the case of setup (b).* The detector enormously changes the detected

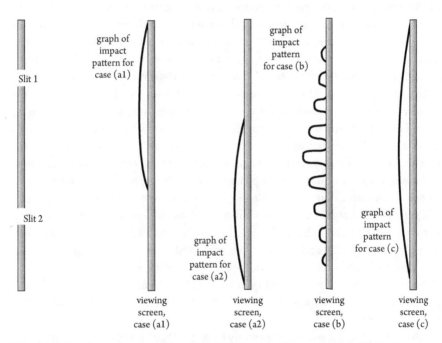

Figure 9.5 The detector effect. Case (a1) Distribution of impacts of individual quanta passing through a double-slit setup with only slit 1 open. Case (a2) The same, but with only slit 2 open. Case (b) Interference pattern of impacts with both slits open. Each quantum comes through both slits. Case (c) The detector effect: With both slits open and a "which-path detector" (not shown) present, the interference pattern jumps to a non-interfering distribution. The detector causes each quantum to come through either slit 1 or slit 2, not both.

quantum, despite our every effort not to "disturb" the quantum. I'll call this shift from pattern (b) to pattern (c) the *detector effect*. In pattern (c), the impacts are spread randomly all over the screen and are simply the sum of the two single-slit distributions shown as (a1) and (a2). *This is evidence that each quantum, rather than passing as a superposition through both slits, now passes randomly through one or the other slit.* Apparently, switching on the which-path detector causes the quantum to pass through only one slit rather than both.

The detector effect occurs quickly and apparently instantaneously upon switching the slit detector from off to on. What causes this radical change? Is it a result of interatomic forces exerted by the detector on the quantum? There are problems with such an explanation. Experimenters can reduce any such forces—for example, by moving the detector far from the slits—but the huge shift in the impact pattern doesn't change in the least. More telling, a single detector focused on only *one* of the two slits is sufficient to cause the shift, *even though quanta coming through the nondetected slit don't experience forces from the detector!* These nondetected quanta alter their impact points when the detector switches on—for example, they no longer avoid the destructive interference regions (dark lines) of the interference pattern. Furthermore, when the two slits are separated more and more widely, the effect on the undetected quanta doesn't diminish. There's some kind of long-distance effect going on here; every quantum "knows" the detector is operating, even those quanta that would seem to be unaffected by it.

Let's look closely at the detection process. The detector must respond to individual quanta passing through the slits by registering "slit 1" or "slit 2" macroscopically. So it must operate on the quantum level and must incorporate an amplifying mechanism to bring the detection up to the macroscopic level. This precisely fulfills our definition of a "quantum measurement" as a quantum process that causes a macroscopic change. The core of the detector is its microscopic interaction with the quantum—a process that must result, after the interaction, in one of two conditions of the detector. Let's label these conditions of the detector *state D1* and *state D2*, corresponding to detection of the quantum coming through slit 1 or slit 2. Even though the macroscopic detector is obviously *not* a single quantum, D1 and D2 must be considered to be quantum states because the detector carries out a quantum function. Before each measurement, the detector must also have a third quantum state, a "ready state," in which it is prepared to make a "which-slit" measurement. These three quantum states are the detector's core.

The detector must switch from ready to state D1 or state D2 when a quantum passes through. As noted earlier, the impacts then show no interference pattern, but instead spread randomly across the screen. With the help of modern electronic technology, one can correlate individual impacts with the reading of the detector for that particular quantum. One then finds the pattern made by quanta passing through slit 1 to be identical with case (a1) in the Figure 9.5, and the pattern made by quanta passing through slit 2 to be identical with case (a2).

This verifies that the which-path detector *causes* quanta to pass through one or the other slit, not both. The detector apparently causes each quantum to jump quickly from the double-slit interference state to one of the single-slit states. Let's call these single-slit states of the quantum *state Q1* and *state Q2*.

How should we describe the states of the composite system comprising both the detector and the quantum? With only slit 1 open, the obvious description of the composite system is "the quantum is in state Q1 and the detector is in state D1," and similarly with only slit 2 open. These states apply after the quantum has passed through one or the other slit, but before it impacts the screen.

What if both slits are open? Recall the superposition principle: If a quantum can be in either of two states, then it can be in both states at the same time. This suggests that the state of the composite system with both slits open is simply a superposition of both of the previously described single-slit states. We can describe this superposition state as follows:

"Quantum in state Q1 and detector in state D1," superposed with "quantum in state Q2 and detector in state D2."

The superposition principle tells us this is a possible state of the composite system, and the mathematics of quantum theory verify this actually is the state of the composite system with both slits open.[46]

This superposition state of the composite system, which I will call the *measurement state,* is central to quantum physics and is central to our consideration of the measurement problem in Chapters 10 and 11. It's an entangled state, like the entangled states of Figures 9.1 and 9.2. The crucial point is that *detections, or measurements, occur when a macroscopic detector entangles with a quantum.*

Although the which-slit detector is designed to not disturb the quantum no matter which slit it goes through, the detector effect shows that the detector changes the superposed quantum from a situation of coming through both slits to a situation of coming randomly through either slit 1 or slit 2. With the detector turned off, the single quantum spreads over both slits, making it impossible to associate individual impacts on the screen with a particular slit. The extended quantum is a single undivided "whole," occupying both slits. But, when the detector is on, *the experimenter can associate each impact on the screen with one or the other slit* by sending quanta through the slits one at a time and observing the detector reading that goes with each impact. *The which-path detector causes the quantum itself to change radically, to "collapse," so that instead of coming through both slits it now comes through one or the other slit.* The quantum is no longer a single undivided whole, occupying both slits.

This situation, with the quantum randomly exhibiting the properties of one or the other of two (or more) different quantum states, is called a *mixture.*[47] A mixture is subtly but crucially different from a superposition. It's in a gray zone between being in two states simultaneously (called *superposition*) and being either in one or the other of the two states. When a quantum is in a mixture, it has definite properties associated with either one or the other of two states, but those definite properties are indeterminate. In the double-slit experiment, we say the superposed quantum *coheres* as a single object coming through both slits, and that the detector *decoheres* this single object into an object that comes randomly through only one slit. A superposed quantum is unified or *coherent* in the sense that it is impossible, even in principle, to associate different spatial subregions (e.g., different slits) within the state with different experimental outcomes. The decohered quantum (with the detector switched on) is "incoherent" in the sense that it comes through only one slit (and is therefore not a single coherent object across both slits), but the choice of which slit it actually comes through is indeterminate until the detector records the "chosen" slit.

In summary, here's a synopsis of the curious effect of detectors: The undetected quantum comes through both slits simultaneously and coherently as a superposition. Entanglement with a which-path detector decoheres this state into an "incoherent" mixture in which the quantum comes randomly (with perfect quantum indeterminacy) through one or the other slit. It's a perfect example of how measurements affect superposed quanta, and a prelude to demystifying Schrödinger's cat.

PART 3

GETTING BACK TO THE NORMAL WORLD

10

Schrödinger's Cat and "Measurement"

A specter lurks at the heart of quantum physics. Known as the *quantum measurement problem*, it pertains, as you know, to vastly more than laboratory measurements.

I've put "measurement" in quotes in the chapter title because this widely used term can be misleading. The notion of a quantum measurement is much broader than the laboratory-based process the word seems to imply. *Measurement*—a macroscopic change caused by a quantum process—happens far more frequently outside than inside the lab. Measurement is everywhere; it's the link between the quantum and the macroscopic world.[1]

There is lots of confusion about measurements. There's no agreement regarding whether it even poses a real problem, and for those who find it to be a real problem, there's no agreement regarding whether it's been solved and, if so, how it's been solved. In the notes you'll find summaries of two recent polls of quantum experts.[2] They demonstrate a plethora of views, with most experts taking one of three positions: the measurement problem is real and unsolvable within conventional quantum physics, the measurement problem is real and solvable within conventional quantum physics, measurement is only a pseudo-problem because quantum physics is merely about knowledge and information rather than about reality.

There are two distinct parts to the quantum measurement riddle, often called the *problem of definite outcomes*, aka Schrödinger's cat, and the *problem of irreversibility*. This chapter investigates the celebrated mystery of Schrödinger's cat and suggests a resolution based on its nonlocal attributes. Chapter 11 examines the problem of irreversibility and suggests that a process known as *decoherence* resolves it. Some physicists opine that decoherence also resolves the problem of outcomes, but this appears to be mistaken.[3] It's important to note that there is no consensus among quantum physicists about the solution presented in this chapter, or about any other solution.

It's not easy to decipher what quantum theory says about measurement. *According to the analyses of many experts, quantum theory implies measurements do not have definite outcomes, but instead yield only a superposition of several possible outcomes*. If this were true, it would imply there is nonsense at the heart of quantum physics. We'll soon see why some experts say this, and later I'll suggest why these experts might be wrong.

We know from experiment, not to mention common sense, that quantum measurements result in definite single outcomes, such as "this particular photon detector clicked." Over a series of identical trials, quantum randomness implies that a particular detector might click on some trials and not on others, but measurements never yield indefinite superpositions of detectors that click and don't click during a single trial. The unpredictable outcome, when it occurs, is definite, not an indefinite superposition of two or more outcomes. The first section in this chapter describes this problem of definite outcomes in detail, and the second section presents Erwin Schrödinger's metaphor for this conundrum.

So there's a problem: Either (1) the principles of standard quantum theory need repair or (2) the standard theory is correct but needs a different interpretation or (3) the analysis leading from the standard theory to the faulty prediction is itself faulty. The third section in this chapter presents several of the most-discussed proposals for repairing or reinterpreting the theory. All have their supporters and detractors; none has garnered a consensus of knowledgeable scientific opinion.

In the last half of this chapter, I argue that the standard analysis is faulty (option 3), and present what I will call the *local state solution* of the measurement problem—a solution based on standard quantum theory. Its central idea is linked to entanglement and nonlocality; it was first presented by Josef Jauch in 1968 and was rediscovered frequently by others, including me.[4] Surprisingly, most previous analyses of measurement have not fully incorporated nonlocality, even though this phenomenon is written all over quantum measurements. The local state solution involves no unusual interpretations, and only a trivial adjustment in the standard theory. Although it has been proposed in the past, I present new arguments for it.

The Enigma of Measurement: Detecting the Quantum World

Because discussion of quantum measurement can become painfully abstract, I tie this chapter to specific examples. Recall the double-slit experiment without and with a which-path detector at the slits. With the detector off, each quantum spreads through both slits in a coherent superposition, but with the detector switched on an apparently instantaneous change of state of the entire extended quantum occurs. This puts the composite system (detector plus quantum) into the following entangled state:

> "Quantum comes through first slit and detector indicates first slit" superposed with "quantum comes through second slit and detector indicates second slit."

This *measurement state* is an entanglement of the quantum with the detector. We saw in Chapter 9 that the entanglement immediately destroys the interference, which is replaced by a simple sum of two single-slit patterns. After sending, say, 1000 individual quanta through the slits with the detector on, roughly 500 will be found to have gone through slit 1 and 500 will have gone through slit 2. As they approach the screen, the quanta are in a mixture (not a superposition) of coming either through slit 1 or slit 2.

A quantum in a mixture is analogous to a classical coin that has been flipped but not yet looked at, but there are crucial differences. As in a coin flip, the observer is ignorant of what the outcome will be. One big difference is that the predictive uncertainty in the coin flip's outcome can, in principle, be removed by gathering sufficient information about the coin's precise initial state, air resistance, the elasticity of the bounce, and so on, whereas quantum uncertainty is fundamental. You'll soon see another difference: A quantum mixture, when created by entanglement, retains subtle, nonlocal correlations that are a key clue in resolving the measurement enigma.

The which-slit measurement illustrates the general case: *When a detector entangles with a quantum, the quantum jumps into a mixture of its possible definite outcomes.* But the theoretical prediction—namely, the measurement state—seems absurdly different from the experimentally observed mixture. The measurement state, as noted earlier, is a *superposition*, not a *mixture*. How can this theoretical superposition jibe with the observed mixture? A superposition of two states means the simultaneous existence of both states. The theory seems to say we have here a macroscopic superposition in which the detector points to slit 1 and also to slit 2, implying the measurement has no definite outcome! But the experiment, as well as common sense, show that the detector points either to slit 1 or to slit 2, not both.

This is the crux of the measurement mystery. The measurement state suggests there is nonsense at the heart of the theory.

Other measurements run into the same dilemma. When a photon strikes a photographic screen, it leaves a single small mark, like one of the spots in Figure 5.1a. This mark is a measurement, with the screen acting as detector. Just as a which-slit detector collapses a photon from a superposition over both slits to a mixture over one or the other slit, the screen collapses the photon from its preimpact state of superposition over the entire screen to a mixture of impacting either grain 1 or grain 2 or grain 3, and so on, of the screen.

But according to the theory, the photon entangles with all the grains, resulting in a huge entangled superposition state:

"The photon impacts grain 1 and grain 1 darkens," superposed with "the photon impacts grain 2 and grain 2 darkens," superposed with "the photon impacts grain 3 and grain 3 darkens," and so on and so on.

The theory appears to show no definite outcome. The screen seems to be in a superposition of detecting an impact at every grain! You or I might have observed such a kaleidoscopic phenomenon late on some Saturday night, but sober scientific labs have never seen such a thing.

I hope I've convinced you that we do have a problem. I'll suggest a resolution later in this chapter. But first, let's look at a renowned dramatization of the problem, and at several of the better known attempts to fix it.

The Curious Tale of Schrödinger's Cat

It's physicists' favorite tale. As Schrödinger told it[5]:

> One can even set up quite ridiculous cases. A cat is penned up in a steel chamber, along with the following device (which must be secured against direct interference by the cat): In a Geiger counter there is a tiny bit of radioactive substance, *so* small, that *perhaps* in the course of the hour one of the atoms decays, but also, with equal probability, perhaps none; if it happens, the counter tube discharges and through a relay releases a hammer which shatters a small flask of hydrocyanic acid. If one has left this entire system to itself for an hour, one would say that the cat still lives *if* meanwhile no atom has decayed. The psi-function of the entire system would express this by having in it the living and dead cat (pardon the expression) mixed or smeared out in equal parts.
>
> It is typical of these cases that an indeterminacy originally restricted to the atomic domain becomes transformed into macroscopic indeterminacy, which can then be *resolved* by direct observation. That prevents us from so naively accepting as valid a "blurred model" for representing reality. In itself it would not embody anything unclear or contradictory. [Emphases in the original.]

Schrödinger's psi-function is, of course, the quantum state. The measurement situation is clear: The radioactive atoms are the quantum system, and the cat is a detector of the state of the atoms. The atoms and the cat are in the entangled measurement state:

> "The atoms are undecayed and the cat is alive" superposed with "the atoms are decayed and the cat is dead."

Schrödinger's reference to "macroscopic indeterminacy" makes it clear that he believes the theory to predict a superposed cat, both dead and alive. This apparent prediction of quantum physics, that the detector (a cat, in Schrödinger's example) is in an indefinite superposition of two macroscopically distinct states, is the measurement problem.

One expects that the atom–detector interaction would put both into a mixture in which the atoms either don't decay or do decay, and the cat either lives or dies. But Schrödinger points out that quantum theory appears to predict something quite different—namely, the entangled state "nondecayed atoms and live cat" *superposed* with "decayed atoms and dead cat." This seemingly describes a cat that is both alive and dead. This can't be right. If every quantum measurement put macroscopic objects into superpositions, we would observe such oddities all around us all the time.

Schrödinger realized something was wrong—after all, he called his tale a "quite ridiculous" case. But nearly everything we know of the quantum world comes from measurements, so if something's wrong with our analysis of measurements, it's a huge problem. It cannot just be swept under the rug—although it often is just swept under the rug. Quantum foundations expert Maximilian Schlosshauer asks whether the measurement problem might be "a dire warning that something is irrevocably rotten at the very core of quantum mechanics, something that could prompt this theoretical edifice to collapse at any moment, like a house haphazardly erected on swampy grounds."[6]

The measurement problem has long been the chief barrier to consensus on quantum fundamentals, splitting physicists into different camps. Some conclude that measurement poses an insurmountable inconsistency, resolvable only by fundamentally altering the theory; others conclude that quantum physics must be reinterpreted in some specialized and often esoteric manner. We'll look at both approaches.

Absent a consensus, most physicists sweep this very real problem under the rug by taking a shut-up-and-calculate attitude. From a pragmatic point of view, this is understandable because the theory works perfectly provided one doesn't ask too many questions about logical consistency. But the quandary obviously needs to be resolved. There is an embarrassing glut of proposals to resolve the measurement problem by reinterpreting or changing the theory. The following sections present tales of the five best-known suggestions.

Copenhagen

The *Copenhagen interpretation*[7] comprises the scientific and philosophical principles developed during quantum theory's founding during the late 1920s and espoused by Niels Bohr, Werner Heisenberg, Max Born, and others. It's the granddaddy of interpretations, amalgamating the sometimes differing views of many pioneers, most of whom maintained close contact with Bohr's Institute for Theoretical Physics in Copenhagen. Although the term is used frequently, there is no definitive statement of the Copenhagen interpretation, and different authors present different and even conflicting views of it. In fact, one careful analysis concludes that it is not a coherent interpretation at all because even it's three primary adherents didn't agree on its essentials. For example, Bohr insisted on the necessity of describing macroscopic measuring instruments in terms of purely non-quantum, classical, concepts, while Heisenberg and Born disagreed. For Bohr, only classical physics was intuitively accessible and therefor about "real" things, while Heisenberg argued that, for physicists, the quantum language was intuitively meaningful and described real things and processes.[8]

Despite such ambiguities, these views of Bohr, Heisenberg, Born, and their colleagues was the earliest general attempt to *understand* the quantum world in conceptual terms as contrasted with pure (or perhaps I should say "mere") mathematics, and is often described as the orthodox interpretation. Its general outlines form the basis of most quantum physics textbooks; it's so prevalent today that many physicists view it as the only serious way to look at quantum physics. The Copenhagen interpretation was influenced by the logical positivism of the 1920s and 1930s, a science-oriented philosophy that emphasized "instrumentalism"—the notion that scientific concepts should be formulated as specific laboratory operations. Thus, this interpretation focuses on laboratory measurements—macroscopic observations of flashes of light on a viewing screen, the position of an electronically controlled pointer, and so on.

According to Bohr's views, we should refrain from ascribing physical reality to the quantum world because we cannot observe it directly. Thus, the Copenhagen interpretation holds that quantum physics does not describe objective reality, but deals only with hints of a necessarily mysterious microscopic world.

For Bohr, quantum states did not represent objective reality, but instead represented only our own knowledge of reality—a subjective view that deviates strikingly from the preceding two centuries of objective science. Although Heisenberg inclined toward a more realistic view, Bohr considered quantum states to represent neither real fields nor real particles, but rather a record of the probabilities that a rational observer should ascribe to the outcomes of measurements based on the observer's information. In his famous lecture in Como, Italy, in 1927, Bohr emphasized that quanta are neither waves nor particles, but rather "abstractions, with their properties being definable and observable only through their interaction with other systems"—systems such as classically describable measuring instruments.[9] Bohr didn't desire to move beyond such abstractions for, as a logical positivist, he assumed the readings of these classical instruments formed the most direct and reliable description of nature. Bohr injected an enduring note of subjectivity into quantum physics, according to which quanta have no properties, or perhaps do not even exist, except when they are being measured—a nonrealistic view that persists among many physicists.

The Copenhagen interpretation interprets the square of the quantum state Ψ of, say, an electron, when evaluated at some particular point x in space (compare with Figure 7.1), as the probability that the electron "will be found at x." Because the electron is said to be found "at x," we must assume it is a tiny particle rather than a field. You've seen that this particle interpretation, if maintained throughout the double-slit experiment, for example, leads to contradictions. The Copenhagen interpretation avoids such contradictions by rejecting such questions as "What was the electron doing just before it was measured?" as meaningless. It views collapse of the quantum state, such as occurs when a quantum strikes a viewing screen, as a subjective nonphysical phenomenon representing a sudden updating of our knowledge resulting from the registration of new information—like the gold coin and silver coin example in Chapter 6—rather than a change in some real physical entity.

In a word, the Copenhagen interpretation is an effort to fix quantum physics by interpreting it nonrealistically.

Non-local Hidden Variables

Louis de Broglie presented an alternative to the Copenhagen interpretation in 1927, and David Bohm developed it into the *pilot-wave model*[10] in 1952. Chapter 6 described this model and it's application to the double-slit experiment. The pilot-wave model aims to fix quantum physics by avoiding quantum randomness and getting back to a deterministic theory similar to classical physics. More than an interpretation, it's a proposed new theory of microscopic physics because it postulates physical entities not found in standard quantum physics: Newtonian point particles moving in a non-Newtonian manner, with one such particle assigned to each of the system's electrons, protons, etc. The system's standard quantum field Ψ, which obeys the Schrödinger equation, guides or "pilots" these particles. This model resolves the so-called *wave–particle duality quandary* by declaring both waves and particles to be real and to exist simultaneously. The model is deterministic; given the initial positions and velocities of all the particles and given the initial state of the field, the state of the particles and the field are predictable at any other time. Each particle is always at one or another specific point, there are no superpositions of particles, and the field follows the smooth evolution prescribed by Schrödinger's equation.

The pilot wave Ψ acts instantaneously on all the particles across great distances, giving the model a strongly nonlocal, quantum-jumping character quite unlike classical physics. You saw in Chapter 6 that this model reproduces the statistical predictions of standard quantum physics for nonrelativistic situations (low energy, low speeds, and no creation or destruction of quanta). However, as might be expected for a theory in which causes (the pilot wave) act instantaneously across distances, it seems to conflict with special relativity and to be incapable of reproducing the experimentally demonstrated predictions of relativistic quantum field theory for higher energy situations. Standard quantum physics, on the other hand, is less flagrantly nonlocal and manages not to contradict relativity.

In a purely theoretical sense, the pilot wave model was a breakthrough because it had been thought that, even with the help of hidden variables, no purely deterministic model that reproduces the predictions of quantum physics was possible. Bohm disproved this notion by creating a deterministic hidden variables theory that does in fact reproduce the predictions of quantum physics. As we saw in Chapter 6, the hidden variables in this model are the initial positions and velocities of the presumed Newtonian particles. But the model is not widely accepted by physicists precisely because it doesn't offer new predictions beyond those of quantum physics, because its extreme nonlocality seems non-physical, and because it has not proved capable of extension to high-energy physics.

On the other hand, there is no measurement problem in the pilot-wave model. The measurable entities are simply Newtonian particles that can strike detectors, causing them to register or not, just as in classical physics. It's not surprising that John Bell, who outspokenly criticized what he considered to be quantum physics' unprofessional vagueness, especially its confusion about measurement, looked kindly on this theory.[11]

A Spontaneous Collapse Mechanism

You've seen that when a macroscopic apparatus measures the state of a superposed quantum, the theoretically predicted measurement state appears to be an unwieldy "Schrödinger's cat" superposition of the quantum-plus-apparatus. Even the macroscopic apparatus appears to become superposed. This suggests adding a new mechanism to quantum physics to force superpositions, if they ever reach macroscopic proportions, to collapse quickly and spontaneously. Schrödinger's cat would then collapse quickly into either an alive or dead state, fixing the measurement problem.

Three Italians, Giancarlo Ghirardi of the International Centre for Theoretical Physics in Trieste, Alberto Rimini of the Department of Physics at the University of Pavia, and Tullio Weber of the Department of Physics at the University of Trieste, have hypothesized just such a mechanism, called the *GRW model* after the authors' initials.[12] They postulate, as a new principle of nature, that, from time to time, every quantum makes a spontaneous jump into a collapsed state. The time and place of such a jump are random.

The GRW model has practically zero effect on single unattached quanta and doesn't change the predictions of quantum physics for systems of a few or even a few thousand quanta, but it has a big effect on macroscopic superpositions. The reason is that the authors choose their collapse mechanism wisely, as follows: When a superposed quantum makes a "GRW jump," the quantum's preceding superposition state collapses instantaneously to a state that exists (is nonzero) only in a small region around a single spatial point x.[13] This point is random, with probabilities equal to the square of the quantum field Ψ evaluated at that point, as in conventional quantum physics. So Ψ is more likely to collapse to the vicinity of points where Ψ^2 is large and less likely to collapse to the vicinity of points where Ψ^2 is small. As for the time of the collapse, the model specifies that all future times are equally probable, and that a jump of a single quantum occurs, on average, only once every 100 million years! So you'd have to be extremely patient if you wanted to see an isolated electron collapse. Even within a fairly hefty quantum such as a C_{70} molecule with more than a thousand protons, neutrons, and electrons, one of these quanta will, according to the model, decay spontaneously only every 100,000 years on average.

The beauty of the GRW model appears when we consider macroscopic superpositions, such as Schrödinger's cat, that are predicted to occur as a result of a measurement. Macroscopic objects contain zillions of quanta. A kilogram (about 2 pounds) of iron contains 10^{25} iron atoms, or 10^{27} quarks and electrons. According to the GRW model, this implies that, in a kilogram of iron, an average of one quantum collapses every trillionth of a second!

One quantum among 10^{27} still might not sound like much, but even a single collapse has huge implications for large macroscopic superpositions such as, presumably, Schrödinger's cat. Here's why: In the GRW model, large superpositions are quite fragile. If just one quantum in one branch of a macroscopic superposition (in either the live cat or the dead cat, for example) collapses, the GRW mechanism causes the entire superposition to collapse instantaneously wherever it is separated by a macroscopic distance from the atom that collapsed. So the GRW model does just what you'd want it to do; it forces macroscopic superpositions to collapse to a small portion of a single branch and has essentially no observable effect on smaller systems.

Problem fixed? Well, there are a couple of objections. First, it's been difficult to find a satisfactory relativistic version of the GRW model. Thus this model has not succesfully included high-energy phenomena, such as electron-positron pair creation and others occurring regularly at the Large Hadron Collider and in high-energy astrophysics. The efforts of several researchers to fix this problem have been only partially successful.[14]

More important, no experimental evidence for the GRW model has been observed—no collapses that could not be explained in terms of entanglement. Furthermore, we saw in Chapter 8 that recent experiments have observed near-macroscopic superpositions with no sign of the GRW mechanism kicking in to cause collapse. Before long, perhaps, experimental evidence will rule this interesting model in or out.

Human Minds Collapse the Quantum State

The pioneering mathematical analyst of quantum theory, John von Neumann, devotes much of his groundbreaking 1932 book *Mathematical Foundations of Quantum Mechanics* to quantum measurements, with "measurement" understood in the restricted, laboratory sense.[15] He recognized that quantum states change by means of the two mechanisms we explored in Chapter 7. One was the normal evolution obeying Schrödinger's equation. But something different happens if the system is measured. For von Neumann, a measurement is a "discontinuous, noncausal, and instantaneously acting" intervention.[16]

He postulates a process during measurements that amounts to the standard practice used by working physicists ever since: Replace the predicted superposition with a random choice between one of the other of its branches. For example, replace

"decayed nucleus/dead cat," superposed with "nondecayed nucleus/alive cat,"

with a random choice between either "decayed nucleus/dead cat" or "nondecayed nucleus/alive cat." This is what we observe, but there is a question about this postulate's logical consistency with other quantum principles, especially Schrödinger's equation.[17]

Writing in 1932, von Neumann wasn't fully aware of the problem of definite outcomes. Instead, he worried about the following consistency question: What if, instead of replacing the superposition with the corresponding mixture (as von Neumann had postulated), we decided instead to *measure* the superposition? For instance, in the double-slit experiment with a single superposed electron, we've seen that the result of measuring the electron with a which-slit detector is as follows:

"Electron comes through slit 1, apparatus indicates slit 1" superposed with "electron comes through slit 2, apparatus indicates slit 2."

If we now follow von Neumann's suggestion and measure *this* superposition, the result of this second measurement would be the following:

"Electron comes through slit 1, first apparatus indicates slit 1, and second apparatus indicates that the *first* apparatus indicates slit 1" superposed with "electron comes through slit 2, first apparatus indicates slit 2, and the second apparatus indicates that the first apparatus indicates slit 2."

Do we get the same experimental predictions if we replace this second outrageous prediction with its corresponding mixture as we would get by replacing the first outrageous prediction with *its* corresponding mixture?

von Neumann's analysis shows that the theory is consistent in this sense. That is, one can locate the "von Neumann cut" that replaces the superposition at either the measurement of the quantum or at the measurement of the first detector, without altering the predictions.

This makes for some nice mathematics and it leads to the notion of a "von Neumann chain" of measurements. Here is an example: A sensor detects an electron coming through one or the other slit, an electron amplifier boosts the sensor's signal so that it creates a small electric current, this current moves a macroscopic pointer to indicate either slit 1 or slit 2, an image of the pointer's position falls on the retina of a scientist's eye, and a signal travels from the scientist's eye to the scientist's consciousness telling her the pointer points either to slit 1 or slit 2. It makes no difference where, along this chain of causal effects, we make the arbitrary decision to replace the outrageous superposition with the collapsed state. Hence, we can choose to make this replacement within the conscious mind of the scientist who makes the measurement.[18]

Thus was born the *consciousness interpretation* of quantum physics. The entire matter is a little obscure in von Neumann's analysis, but he seems to suggest that the outrageous superposition can be replaced *legitimately* with the collapsed state at the level of the observer's subjective consciousness. His stance appears to be that something unknown happens within the human consciousness that uniquely collapses quantum states. Once *this* collapse occurs, other collapses occur all the way back down the von Neumann chain of events to collapse the original superposition.

Regardless of how von Neumann actually viewed this analysis, there's no doubt that Nobel Prize–winning quantum physicist Eugene Wigner argued during the early 1960s that consciousness plays a fundamental role in the quantum measurement process. His essay "Remarks on the Mind–Body Question"[19] relies strongly on von Neumann's work. Wigner argues that a conscious observer of the outcome of a quantum measurement must not herself be in the outrageous superposition state, but rather in only one of these states, so that any conscious observer must collapse the von Neumann chain of superpositions. However, Wigner abandoned this view in 1970.

Despite Wigner's change of opinion in 1970, the consciousness hypothesis continues to be taken seriously.[20] As critics often point out, it's vulnerable to the following kinds of objections: Which systems have sufficient consciousness to collapse a quantum state? Only research physicists? Children? Other animals? Is a laboratory required? Were quantum states not collapsing during 4.5 billion years of prehuman history? Do quantum states not collapse elsewhere in the universe? The objections are best summed up by the comment of Wojciech Zurek, a leader of the decoherence approach to measurement (Chapter 11), that "the boundary between the physical universe and consciousness ... is a very uncomfortable place to do physics."[21]

The hypothesis that consciousness causes quanta to collapse can be checked experimentally. Because there are people, some of them scientists, who believe this notion, such an experiment is worth doing. In 2006, Roger Carpenter and Andrew Anderson at the University of Cambridge's Department of Physiology, Development, and Neuroscience performed an experiment that captures the essence of Schrödinger's cat but, mercifully, without sacrificing a cat. The idea is to split information about the experimental result between two observers in such a way that neither one can know the outcome. The observers learn the outcome at a later time by sharing the information. The question is then: Did the collapse occur at the time of the experiment or when the observers became conscious of the outcome?

Carpenter and Anderson arrange a closed box equipped with a small sample of radioactive radium, two raised hammers, a Geiger radiation detector, an electronic timing device, and shielding material that allows the detector to be exposed to the radium for only 0.19 second—an interval that guarantees there will be a 50–50 probability of detecting an alpha quantum. An alpha decay during the 0.19-second "window" releases hammer 1, the fall of which allows a red ball to be released and roll into an output box within the larger box. If no alpha decay occurs during the time window, hammer 2 releases electronically, allowing a black ball to roll into the output box. Hammers 1 and 2 represent the dead and alive cat, respectively.

However, observer A has previously set up the apparatus so that the ball provides either true or false information and has obtained an electronically supplied "truth card" affirming "true" if the balls are loaded to represent the outcome correctly (red ball = decay, black ball = no decay), and "false" if the balls are loaded to represent the outcome falsely. Observer A uses a random number generator to determine randomly whether the setting will be true or false. Observer B then runs the experiment and removes from the output box a ball the color of which purports to represent the outcome, but doesn't know whether the provided information is true or false. Thus, following each run of the experiment, observer A has a card reading *true* or *false*, and observer B has a red or black ball, but neither knows the outcome until the card and ball are compared.

Observer A then returns to the lab and, while observing his truth card, asks observer B to report which ball was received. At this point, the observers become conscious of the experimental outcome. The observers then open the box and observe the experimental outcome directly. The two experimenters performed 18 repetitions of this experiment, with each experimenter performing the duties of observer A on half of the trials and the duties of observer B on the other half. On every trial, they found that, on looking inside the box, the hammer that fell agreed with the conclusion they drew from the truth card-and-ball combination—a combination that was determined entirely by the earlier experiment. So the collapse (the fall of the hammer) must have occurred when the experiment occurred, not later when the observers became conscious of the outcome.

It's worthwhile to do this experiment, but in my humble opinion one would have to have rocks in one's head not to have predicted this outcome. My hat is off to Carpenter and Anderson, who reported this, with a straight face, in a peer-reviewed journal.[22]

Every Measurement Creates Other Universes

The *many-worlds interpretation*[23] dispenses with the question of quantum state collapse by assuming it doesn't occur, despite considerable evidence (quantum jumps in atoms, double-slit experiments using which-path detectors) that it does occur. To understand many worlds by way of a simple example, consider an electron that impacts a viewing screen. As we've seen, the standard theory predicts the electron entangles with the zillions of atom-size regions of the screen and collapses into one of these regions. In many worlds, the electron doesn't collapse; instead, *reality splits up into zillions of separate realities, in each of which the electron impacts a different atom-size region.*

That's breathtaking. Just as the electron impacts the screen, its quantum state is a huge electron-plus-screen superposition with zillions of branches. In the first branch the electron ends up in the first atom, in the second branch it ends up in the second atom, in the third branch it ends up in the third atom, and so on. Up to this point, many worlds agrees with the standard theory. The standard approach assumes that detection by the screen causes this entangled state to collapse into just one of its branches, such as the one in which the electron ends up in atom number 5,139,428,371 and not in any of the other atoms. This interpretation, however, asserts that all these zillions of branches actually exist and persist forever.

But in this case, why do we not see zillions of impacts? This interpretation's answer: The universe itself splits into zillions of separate universes, in each of which only one of the branches occurs! In the universe that we happen to inhabit, the electron ends up in atom number 5,139,428,371, but you and everything else are somehow simultaneously in those other universes too, and in those universes the electron ends up in the other regions. So the many-worlds scheme requires as many universes as there are atoms in the screen—just for this single experiment and just for this single electron. Send a second electron through this experiment and you've created another big batch of universes. This interpretation might be economical in assumptions (by eliminating the collapse), but it's extravagant in universes.

One reason some people like many worlds is that it does away with quantum randomness. No single outcome is chosen in preference to the others. They all occur, but in different universes. Many worlds says it's a mistake for you to think

"But I really exist in only one of these worlds," because you really exist in all of them, although you happen to be experiencing only one of them.

This idea might seem more plausible today because of the "cosmic inflation hypothesis" developed during the 1980s by Alan Guth and Andrei Linde.[24] Inflation posits that, beginning a trillionth of a trillionth of a trillionth of a second after the initiation of the Big Bang, a certain "inflationary quantum field" pushed the universe into a short burst of exponentially accelerating expansion that proceeded at speeds much faster than light for a very brief time. This hypothesis has considerable experimental support in observations of the cosmic background radiation emitted by the Big Bang.[25] The inflationary hypothesis suggests that inflation continues forever in at least some regions of the universe, producing an infinite number of other universes—a possibility that many astrophysicists take quite seriously. However, it's hard to see how the inflationary multiuniverse has anything to do with the many-worlds interpretation, an interpretation that appears to have little in common with the Big Bang.

Yes, some physicists take this seriously.[26] This enormous splitting of the universe is supposed to happen every time a photon, electron, or other quantum becomes entangled in a measurementlike superposition. Such entanglements occur constantly, all over the universe, not to mention all over the other universes presumed by many worlds, for all past time. So there are quite a few universes by now, all going about their separate and distinct quantum evolutions and all constantly splitting. As many-worlds theorist Bryce DeWitt puts it, "every quantum transition taking place on every star, in every galaxy, in every remote corner of the universe is splitting our local world on Earth into myriads of copies of itself."[27]

For some physicists, including me, this is a bit much. As you might guess, many worlds has its critics. Occam's razor councils us to choose the simplest resolution of the measurement problem, and the many-worlds proposal is not what most people would call simple. Furthermore, it's hard to see how many worlds could be either verified or disproved experimentally, although theorist David Deutsch has suggested some idealized thought experiments.[28]

The Local State Solution of the Problem of Definite Outcomes

Perhaps the measurement problem is solvable within normal quantum physics, interpreted in a conventional realistic fashion wherein quanta and their states are objectively real features of the natural world, without alterations or special interpretations such as those put forth in the preceding five sections. The remainder of this chapter is an adventure into just such a suggested resolution of Schrödinger's cat. Regardless of whatever physicists' ultimate consensus about this suggestion might be, this discussion should deepen your understanding of quantum physics. The suggested resolution neither alters nor reinterprets the standard quantum principles. It instead uses those principles to show that the entangled measurement state (the Schrödinger's cat state) actually predicts outcomes that are random but nevertheless definite, just as we expect. This agrees with experiment and, as we will see, is entirely non-paradoxical. [29]

This section presents an argument first given (as far as I can determine) in 1968 by Josef Jauch in his comprehensive monograph on quantum foundations[30] and proposed independently several times since then.[31] Unfortunately, this *local state solution,* as I shall call it, has been generally criticized or ignored, although it is compelling, straightforward, and follows logically from standard quantum physics with no specialized interpretation. The subsequent section verifies this solution by means of a second independent argument showing that the local state solution is a necessary consequence of special relativity's ban on superluminal communication. The argument is based on the nonlocal nature of the measurement state. It demonstrates that standard quantum physics and special relativity together *require* the local state solution. Because some experts have found fault with Jauch's solution, I've placed an optional appendix at the end of this chapter that answers the primary objections.

As a typical measurement example, let's return to a single electron passing through a double-slit experiment with a which-path detector at the slits. As you know, the detector and electron entangle to form the measurement state:

> "Electron comes through slit 1 and detector indicates slit 1" superposed with "electron comes through slit 2 and detector indicates slit 2."

How can we make sense of this state? Experimentally, the electron is in a *mixture* of coming through *either* slit 1 *or* slit 2 (Chapter 9), whereas the theoretical measurement state just mentioned is a *superposition* that seems to describe an electron coming through both slits and a detector indicating both slits, with no definite outcome.

There's much more to this measurement state than meets the eye. For starters, this state is a superposition—a state of affairs that is about as nonclassical and counterintuitive as you can get. But it's not a simple superposition of two states of a single system, like the photon that follows both paths in a Mach-Zehnder interferometer; it's a superposition of two situations involving *two* systems: electron and detector. These two systems are entangled, just as the two quanta in Figure 9.1 are entangled. Furthermore, like all entangled states, the measurement state establishes a nonlocal relationship between the two systems. No matter how large the distance between them, the electron and the detector are connected instantaneously to each other in a way that violates the locality principle, the idea that objects are directly influenced only by their immediate surroundings. Given all of this, it's not surprising that the measurement state is hard to sort out and that it's true nature is not what one might have expected at first glance. It perplexed even Schrödinger—which isn't surprising, because he died before physicists learned anything about nonlocality beyond its bare existence, which was discovered in the 1935 EPR paper discussed in Chapter 6.[32]

Here's one straightforward feature of the measurement state that is known to some physicists but often ignored in discussions of quantum measurement: Standard quantum physics implies that, when a quantum and a detector are entangled in the measurement state, *neither the quantum nor the detector, individually, are in a superposition*. This is proved easily by assuming the composite system is described by the measurement state with either the quantum or the detector in a superposition; it's not difficult to derive a contradiction.[33] Thus, according to standard quantum theory, *the measurement state is actually inconsistent with either the quantum alone, or the detector alone, being in a superposition*. This conclusion is exactly what we expect and is not at all paradoxical!

Many physicists are aware that the detector (a cat in Schrödinger's example) is not in a simple superposition of indicating both outcomes, and the quantum (a radioactive nucleus in Schrödinger's example) is not in a simple superposition of both quantum states. Instead, it's the *composite* system—cat plus nucleus in Schrödinger's example—that is in a superposition of some sort. But exactly what does this mean, especially if the two "subsystems" are entangled and thus in a nonlocal relationship in which the subsystems could in principle be arbitrarily far apart without in any way altering their role in the measurement process? It's an odd situation: The detector is not individually superposed, and the quantum system is not individually superposed, and the detector could be arbitrarily far from the quantum without altering the relationship, yet somehow the detector-plus-quantum seem to be superposed. If the detector and the quantum were on two different galaxies, then what could it possibly mean to say that the detector-plus-quantum are superposed? We'll soon find out what this means, by learning precisely what is superposed in the measurement state superposition. The answer turns out to be remarkably simpler than we might have supposed.

When discussing entanglement (Chapter 9), I distinguished between immediately recorded *local data* concerning only one or the other of two entangled systems and *global data* that includes correlations between both. The important physical distinction is that global data must be obtained *after* the immediate recording of data, by gathering and comparing data, at light speed or less, from *both* local systems. In the case of an electron in a double-slit experiment that incorporates a which-slit detector, we must make the same distinction between the *local state of the electron*, the *local state of the detector*, and the *global state of the composite system.* The local state of the electron describes only the electron—what an observer of the electron observes, ignoring correlations with the detector. Similarly, the local state of the detector describes only the detector, ignoring the electron.

What does standard quantum theory tell us about these two local states? Once again, the answer is straightforward and calculated easily from standard quantum physics.[34] This calculation shows that the local states are *mixtures*—not superpositions—of their possible outcomes! But quantum physicists know quite well that the local state[35] of a system is precisely what an observer of that system observes. In other words, standard quantum physics predicts that an observer of the electron in the double-slit experiment with a detector at the slits will find the electron to be in a mixture of *either* coming through slit 1 *or* coming through slit 2. Furthermore, the which-path detector will be found to be in a mixture of either indicating slit 1 or indicating slit 2. *Although the composite system is in a superposition of some sort, an observer of the electron alone or of the detector alone observes a mixture.* Once again, this is not paradoxical; it's just what we expect.

As you know, experiments confirm this prediction (see Figure 9.5).

This prediction tells us something new, because it tells us what observers actually observe. An observer of the detector observes a detector indicating either "slit 1" or "slit 2," not both; an observer of the electron observes it to come through either slit 1 or slit 2, not both. In the same manner, an observer of Schrödinger's cat observes a cat that is either alive or dead, not both; the same observer observes a nucleus that is either decayed or undecayed, not both. There is no ambiguity about this theoretical prediction: What a local observer observes must be what really happens. If you observe a cat that is really alive, there is no way the cat can in any sense be in some situation of being both dead and alive. This prediction has been known since at least 1968. Unaccountably, it has not been taken sufficiently seriously.

So quantum theory is in complete agreement with experiment! When we make the important physical distinction between the local states and the global state, the famous problem of definite outcomes does not actually arise! Jauch noted all of this in 1968, and others, including me, have noted it since—to little avail.[36]

Yet, the belief persists that the measurement state implies a superposed cat. It's as though you are staring at a live cat and a quantum foundations expert says to you, "You're mistaken. That's not a live cat. That cat is in a superposed state of being both dead and alive." It reminds me of Monty Python's parrot routine. John Cleese purchases a parrot, but it turns out to be dead and nailed to its perch. When Cleese takes the dead parrot back to the pet shop to complain, the shop owner dogmatically insists that the parrot is not really dead. A hilarious argument ensues.[37]

This resolves the problem of definite outcomes! Josef Jauch seems to have understood all this in 1968:

> We see that both states [the state of the quantum, and of the detector] have become mixtures.... There is no question of any superposition here.... Moreover, we have a measurement since the events in [the detector] and in [the quantum] are correlated.... Thus the paradox of Schrödinger's cat can be resolved when it is reformulated entirely in physical terms.[38]

A host of experiments supports this solution, and the theory predicts it correctly.[39] What more could one ask? Yet quantum foundations specialists have long claimed this conclusion is questionable. This chapter's optional appendix discusses the arguments of these critics, and the answers to those arguments.

Jauch's conclusion is clear, and correct, but it raises an obvious question that has not, to my knowledge, been answered: What, then, *is* superposed in the measurement state? A complete resolution of Schrödinger's cat must answer this question. The next section sorts this out.

Why Quantum Physics Requires the Local State Solution

Because experts dispute or ignore Jauch's five-decades-old resolution of the measurement problem, this section presents a second argument for the local state solution—one that is independent of Jauch's argument.[40] This argument, based still on standard quantum theory, concludes that the measurement state *must* exhibit itself as mixtures with definite, observed properties because, otherwise, this state could be used for instantaneous signaling, violating special relativity. This argument also provides a surprisingly simple, even obvious, answer to the question of what is superposed in the measurement state. Like the resolution of a murder mystery in which the killer was "hiding in plain view" all along, we'll discover that the resolution is all too obvious.

Nonlocality is written all over the measurement process in general and the double-slit experiment in particular. First, with both slits open and no which-path detector, an interference pattern forms on a downstream viewing screen, showing that each quantum comes through both slits—a nonlocal situation because of the arbitrarily wide separation of the slits. Second, with both slits open and a which-path detector present at only one slit, experiments show that a noninterfering mixture forms. Every quantum conforms to this pattern, including those coming through the *undetected* slit—a nonlocal effect. Third, when each quantum impacts the viewing screen, it collapses instantaneously and nonlocally to an object of atomic dimensions. Behind all these nonlocal effects lies the unity of the quantum.

It's been known for decades that entangled states, such as the measurement state, are highly nonlocal.[41] Many analyses of quantum measurement pay scant attention to this nonlocality because, in an experiment such as the double-slit experiment with a which-slit detector, the detector is normally adjacent to the slits and the nonlocality isn't obvious. Nevertheless, detector and quantum are entangled in the nonlocal measurement state, and fundamental physics must reflect this although it may have no practical effect on the experiment at hand.

There is, in fact, at least one double-slit experiment that highlights strikingly the detector's nonlocal relationship to the detected quantum. In 1991, X. Y. Zou, Lei Wang, and Leonard Mandel at the University of Rochester, New York, performed a double-slit experiment that used an entangled pair of photons.[42] One photon went through a double-slit setup while *the other photon acted as a distant detector for the first photon.* The first photon passed through the two slits and impacted a viewing screen while the second photon traveled away from the slits along two superposed paths to a separate detector that could be arbitrarily distant from the first photon. The ingenious geometry of these paths ensured that by either (1) inserting or (2) not inserting a barrier along one of the second photon's two paths, this second photon could either (1) detect or (2) not detect the slit through which the first photon came. So the second photon acted as a distant and optional which-path detector for the first photon.

As expected, during those trials in which the barrier was inserted in one path of the second photon, the distant first photon impacted the screen in a mixed-state pattern of coming through one or the other slit; and during those trials in which the barrier was not inserted, the distant first photon impacted the screen in a double-slit interference pattern—a striking demonstration of the nonlocal, "wholistic" nature of measurement. When I first read the paper describing this remarkable experiment, I practically fell out of my chair. Nonlocality is shocking. If the first photon were sent into space, and the decision to insert or not insert the barrier were made on Earth when the first photon was halfway to the next star, the distant photon would presumably jump instantaneously between the mixture and the superposition depending on whether the second photon performed or did not perform a which-path detection.

The experiment of Zou and colleagues provides strong evidence that measurements have everything to do with the nonlocal effects of entanglement.

The resolution of the measurement problem lies in sorting out the entangled measurement state. The RTO experiment (Chapter 9) affords the perfect tool for this. Each of RTO's two entangled photons is a microscopic which-path detector for the other photon, much as in the experiment of Zou and colleagues. This is because RTO's two photons are entangled in the same manner as a quantum and its detector are entangled in the measurement state. Because physicists can manipulate photons more agilely than macroscopic detectors, we can learn much more about measurements from the RTO experiment than from the double-slit experiment or Schrödinger's cat. Furthermore, as I've mentioned before, the measurement problem really has two quite independent components. The first is Schrödinger's cat. The second, which I'm saving for Chapter 11, is the "irreversibility problem" associated with the macroscopic mark, such as the audible "click" of a detector, that any true measurement must make. In the RTO experiment, entanglement occurs at the outset of the experiment whereas the macroscopic mark doesn't come until the end, making it possible to study the effect of entanglement alone without the added confusion of the irreversibility problem.

The RTO experiment certainly demonstrates the measurement state's nonlocal character. If we think of one of the entangled photons as a which-path detector for the other photon, the experiment shows that a quantum and its detector can be separated widely and yet incorporate information instantly about their mutual correlation. Measurements act instantaneously across a distance.

Here is the crux of the measurement problem: Given that the measurement state is a superposition, precisely what is superposed? Is Schrödinger's cat superposed? The decaying nucleus? We've seen that the answers to these two questions are no. Are states of the combined cat-plus-nucleus superposed? As noted earlier, this sounds plausible. Nevertheless, it cannot be correct. Treating the composite cat-plus-nucleus as a single system would ignore a crucial aspect of the physics—namely, the entanglement between the *two* systems—an entanglement that is the whole point of the measurement process.[43] Measurement entails correlations between the measured system and the measuring apparatus. If we could treat the composite system as though it were a noncomposite system, it wouldn't be a measurement at all.

As you know, the RTO experiment demonstrates that, when two systems are entangled, quick changes in the phase of either system alter the correlations between them *instantly*. This, in turn, implies *these correlations must be unobservable—invisible—at the location of either system* because, as discussed in Chapter 9, if they were observable, then alterations in phase could be used to send instant messages from one local observer to the other local observer, which is in violation of special relativity. So the pattern observed on either screen *must* be a mixture.

This is the deep reason why the double-slit interference pattern must collapse to a mixture as soon as a which-path detector is turned on (i.e., as soon as the entanglement is established). In the entangled state, neither photon nor detector can show the least sign of interference, because such a sign could be used, in an appropriate entanglement experiment, to send instant messages. *The collapse from superposition to mixture follows logically from the nonlocality inherent in the measurement state plus special relativity's ban on superluminal signaling.* If a superposed cat showed up, it could be used for instant signaling.

Summarizing: I've demonstrated in this and the preceding section, using two independent arguments, that the entangled measurement state exhibits itself as mixtures of both subsystems.

But the entangled measurement state of the combined system is a superposition. What *is* superposed, if not cats and not nuclei and not cats-plus-nuclei? We have already seen the answer, in connection with the RTO experiment: Figure 9.4 shows the evidence that, when two systems are entangled as in the measurement state, *correlations* are superposed. In fact, the measurement state superposition can be perfectly pictured by imagining, in Figure 9.3, system A to be an apparatus such as an electron detector or Schrödinger's cat, and system B to be a quantum such as an electron passing through a pair of double-slits or a radioactive nucleus. In the measurement setup, both phase shifters x and y are fixed at zero so that the detector is exactly "in phase" with the quantum. That is, the detector states $A1$ and $A2$ are perfectly correlated with the quantum states $B1$ and $B2$. The RTO experiment surpasses the measurement state situation by allowing the phases x and y to be altered; when this is done, the experimentors find the data shown in Figure 9.4, showing inference of the two branches $A1$-$B1$ and $A2$-$B2$ (the dashed line and solid line, respectively, of Figure 9.3), demonstrating that the correlations between $A1$ and $B1$ on the one hand, and $A2$ and $B2$ on the other hand, are superposed.[44]

So the measurement state is not paradoxical or even difficult. We just need to read it correctly. The RTO experiment teaches us that the state of an electron and its which-path detector in a double-slit experiment should be read as follows:

> "The electron comes through slit 1" is correlated perfectly (with 100% probability) with "the detector indicates slit 1," and "the electron comes through slit 2" is correlated perfectly with "the detector indicates slit 2."

The simple word "and" following the comma indicates the superposition. Referring to Figure 9.3, both the dashed line pairing A1 and B1, and the solid line pairing A2 and B2, are true simultaneously. A1 is correlated with B1 *and* A2 is correlated with B2. The superposed entities are simply *correlations*, implying a non-paradoxical measurement state; both correlations are true.

To ward off misconceptions, I must add that this "local state" analysis is not offered as yet another interpretation of quantum physics. Rather, I'm arguing that the local state solution is an implication of standard quantum physics.

A Brief History of the Measurement Problem

I've spent far more words on the measurement problem than it deserves, because it's solved trivially by looking, as Jauch did, at the local states. Measurement should not have been, and should not remain, a problem. It's instructive to look at how it got this way.

The problem of definite outcomes, first recognized by Schrödinger in 1935, appears to have been first solved by Jauch in 1968, but his solution wasn't recognized by the physics community. Jauch's solution has still not been accepted, although it is frequently rediscovered. This unfortunate history has gone hand in hand with the sluggish evolution of our understanding of entanglement. Although Schrödinger intuited correctly in 1935 that entanglement is *the* (his emphasis) characteristic quantum trait, entanglement's physical consequences were little understood at that time. He believed quantum physics predicted his famous cat to be in an indefinite superposition—an understandable mistake at that early time—thus initiating the problem of outcomes. In 1935, Einstein, Podolsky, and Rosen (Chapter 6) pointed out the counterintuitive nonlocal predictions that arise for correlated pairs of quanta, but didn't pursue further the unusual properties of such systems. During the 1930s, Einstein and Schrödinger were dissatisfied with entanglement because it seemed to violate relativity's speed limit on the transmission of information—an important misconception that was not corrected until 1987.[45] None of this stimulated much published work in the 1920s and '30s, during which much of standard quantum physics was established.

World War II brought most work on quantum foundations to a halt, and after the war the job-preserving admonition was to "shut up and calculate" rather than philosophize about quantum foundations. It wasn't until 1964, when John Bell published his test for locality and showed that quantum physics violates that test, that a new vista was opened on entanglement and nonlocality.[46] One might have expected Bell's work to create a scientific sensation, but instead it was published in an obscure journal and largely ignored even by those who did happen to learn about it. Such a paper should have led quickly to experiments regarding whether nature actually violates Bell's test, but the experiment was performed only in 1972 and was little noticed for years after that. Alain Aspect did notice, and in 1982 carried out a more definitive experiment demonstrating the superluminal change of correlations that violated Bell's locality condition.[47] We are still debating its significance.[48]

In 1968, Jauch solved the problem of definite outcomes by the obvious strategy of asking what a local observer observes. His answer was straightforward: The detector is observed to be in a mixture, and so is the quantum. Jauch thought this local state solution (as I have called it) resolved the Schrödinger's cat problem, and he concluded his discussion by stating, "Thus the paradox of Schrödinger's cat can be resolved when it is reformulated entirely in physical terms."[49] His book does not mention entanglement or nonlocality.

Jauch did not follow up on his 1968 insight. In 1973, he published an inconclusive and brief essay that seems to backtrack on his 1968 claim of resolving the problem of outcomes.[50] Without mentioning his 1968 analysis, the 1973 essay states, "The measuring process exhibits features which are apparently inconsistent with the Schrödinger type evolutions." He regards this as evidence that quantum physics itself is contradicted by experiment. Jauch ends by referring vaguely to the Copenhagen interpretation as a possible solution. The article shows that Jauch did not understand the nonlocal implications of entanglement. In 1973, this was not surprising. As we have seen, the entangled and nonlocal nature of the measurement state is crucial to understanding the problem. Entanglement's creation of local states with definite properties and global states with superposed correlations causes the instantaneous quantum jump, from a superposition of states to a mixture of local properties, that we observe. Unfortunately, Jauch died in 1974.

To grasp the measurement state, it's necessary to understand its nonlocal correlations and their interference, but there was no consensus about nonlocality until at least 1982 (Aspect's experiment). By that time, a plethora of quantum reinterpretations and revisions had already appeared, precisely in order to get around the measurement problem. Many quantum foundations experts became invested intellectually in one or the other proposed resolution and assumed the standard tools of quantum physics could not solve the measurement problem in a logically consistent manner. Jauch's straightforward proposal was largely ignored and forgotten. Since that time, the focus has been on reinterpretations and revisions rather than on solving the problem within standard quantum physics.[51] Anyone proposing to solve the problem within standard quantum physics must deal with decades of encrusted misconceptions.

Optional Appendix: Rebutting the Objections

Many experts opine that the measurement problem represents a fundamental inconsistency that cannot be fixed without changing or reinterpreting the quantum principles, and that the local state solution in particular does not resolve the problem. I'll present the three most significant of these arguments and rebut them.

First and foremost, detractors cite the problem of definite outcomes: The state we actually observe is either a nondecayed atom and live cat or a decayed atom and dead cat whereas the theoretical measurement state represents (according to the detractors) both situations simultaneously.

The preceding discussion refutes this argument. The theory predicts not only the global state but also two local states. The local predictions are what an observer at the cat or at the nucleus actually witness. These definite local outcomes are either an alive or dead cat and either a nondecayed or decayed atom. These predicted mixtures of observable properties are exactly what we see. Furthermore, direct calculation shows that when the cat and the nucleus are entangled in the measurement state, neither the cat nor the nucleus is in a superposition state. The superposition is present only in the global state (i.e., in the correlations between the definite local states). These correlations are not observed directly by local observers and are evident only after data are collected from both local sites and compared.

Furthermore, the global state's correlations can, in principle, be altered by either local observer, instantly altering the local state of the other, possibly distant, local observer. Thus, the global correlations must be shielded from direct local observation lest we transgress on special relativity's injunction against instant signaling. This implies the nucleus and cat *must* be described by mixtures, not superpositions.

Conclusion: *The measurement state is a superposition only of correlations between the nucleus and the cat, not a superposition of the nucleus or of the cat.* The nondecayed nucleus is correlated perfectly with an alive cat *and* the decayed nucleus is correlated perfectly with a dead cat. Thus the outcomes are definite and the measurement state is not at all paradoxical.

A second objection is that the theoretically predicted and experimentally observed mixture is "improper" and thus unacceptable.[52] A "proper" mixture, as the term is used in quantum physics, is one that arises from simple human ignorance of the actual quantum state and is not a quantum notion. The same notion arises in classical thermodynamics, where the precise classical state of a complex system is represented by probabilities simply because humans cannot feasibly observe or calculate the precise positions and velocities of zillions of molecules. In fact, we apply ignorance-based probabilities to a classical situation whenever we describe a flipped coin, before it's looked at, as having a 50–50 chance of being heads or tails. In quantum physics, a proper mixture would be a situation in which a system is in a specific quantum state, but we don't know which state.

An "improper" mixture occurs for purely quantum reasons. It arises from entangled quanta whenever we observe just one of the quanta, much as you might observe just one die when throwing a pair of dice. With fair dice, observation of only one die will show the probability of each outcome to be one sixth. This result—one sixth for each outcome—is the "reduced probability distribution" for the single die arising from the probability distribution for two dice.[53] In the same spirit, an observer of just one of a pair of entangled quanta observes a reduced probability distribution for the one quantum. Both quantum theory and experiment show that this reduced probability distribution turns out to be a mixture of single-quantum properties, not a superposition.

Far from being really improper, so-called *improper mixtures* that arise when one views only one member of an entangled pair are, in fact, at the heart of quantum randomness! When a detector measures a quantum, the two entangle in the measurement state; we then find the measured quantum to be in a mixture of definite outcomes with prescribed probabilities. This is how nature creates quantum randomness. It is predicted correctly by quantum theory's so-called "improper" mixtures.

Figure 9.5 furnishes an instructive example. When a superposed quantum emerges from a pair of double slits, and a which-path detector observes the quantum, the detector entangles with and thus decoheres (removes the interference pattern from) the quantum. The entanglement causes the quantum to jump from the superposition evidenced by Figure 9.5b into the mixture evidenced by Figure 9.5c—a mixture of being either on path 1 or on path 2. This is a textbook example of how quantum randomness arises: Without measurement, the quantum comes through both slits; with measurement, the quantum comes randomly through one or the other slit. The randomness arises from the entanglement caused by measurement.

Schrödinger's cat is another example. The entanglement puts the nucleus into a mixture of being nondecayed or decayed, and the cat into a mixture of being alive or dead—further examples of quantum randomness.

The distinction between proper and improper mixtures is useful and appropriate, but the choice of words is not. Better terms would be *mixture* and *local mixture*. A quantum is in a "mixture" of quantum states if it is really in one or the other state but we don't know which. A quantum is in a "local mixture" of properties if a local observer of the quantum observes definite but indeterminate properties whereas a "global observer" (one who tallies the outcomes of both local observers) observes that the quantum participates in a global state that correlates the quantum's properties with some other quantum's properties. There is nothing improper about local mixtures; they express the uncertainty inherent in quantum physics.

The third objection is a mathematical problem called *basis ambiguity*.[54] The objection is that the local state solution is ambiguous mathematically in certain special cases. In the case of Schrödinger's cat, this special case occurs at only one instant during the atom's radioactive decay. That instant is the atom's half-life, the time at which the atom has precisely a 50–50 chance of decaying.[55] Suppose this half-life is 30 minutes. If we ask about the health of the cat at times shorter than 30 minutes, there's no problem about basis ambiguity; if we ask about the cat at times longer than 30 minutes, there's no problem about basis ambiguity. But if you want to know what happens at precisely 30 minutes, there's a problem with the theory; something special, called *basis ambiguity*, happens.[56]

The endnotes explain the problem and its technical resolution.[57] The idea of the resolution is that the detector must be designed to be correlated, *in an unambiguous way*, with the state of the quantum. The equipment must be set up to guarantee that a nondecayed atom is correlated with an alive cat, and a decayed atom is correlated with a dead cat. This is all that matters, and it is neither physically nor mathematically ambiguous. Basis ambiguity is irrelevant to the physics discussion.

The physical irrelevance of basis ambiguity is further confirmed by the fact that it arises for Schrödinger's cat only precisely at the half-life. After all, there's nothing distinctive physically about this particular instant in the life of an atom. A radioactive atom's half-life is simply a useful statistic, like the median age of US citizens (36.9 years). Nothing special happens when a person reaches this age, and nothing special happens to a nucleus when it reaches its half-life, regardless of whether or not the nucleus has become entangled with Schrödinger's cat.

11

The Environment as Monitor

> The moving finger writes; and, having writ,
> Moves on: nor all thy Piety nor Wit
> Shall lure it back to cancel half a Line,
> Nor all thy Tears wash out a Word of it.
> —"The Rubaiyat" of Omar Khayyam[1]

I hope you are by now convinced that quantum physics describes the microscopic world entirely consistently and with unparalleled experimental accuracy. But still, there's one *slight* problem: Quantum physics seems to fail utterly in describing the world around us!

We never see tables or teapots, not to mention ice cream, as wavy, space-filling fields that are possibly here and possibly there and possibly both here and there. Tables and teapots don't quantum jump, nor does ice cream. Classical physics may be inaccurate at the microscopic level, but it does explain how such ordinary objects move in response to forces. If quantum physics describes the microscopic world correctly, and if the macroscopic world is made of microscopic objects, then the quantum principles should lead ultimately to tables and teapots. This chapter explains, at least in part, how.

Quantum physics began with Max Planck's hypothesis that eventually led to the quantum. The crucial quantum principle is the universe is made of these highly unified, extended bundles of field energy. To the extent that any given phenomenon depends on the spatially extended field nature of quanta, the phenomenon should be considered as "quantum." To the extent that quanta can be represented by pointlike objects, the phenomenon can usually be considered "classical." But this classical–quantum boundary is a useful metaphor, not a law of nature. Most physicists, me included, regard the world as fully quantum.

The clearest expression of the extended field nature of quanta is the principle of superposition. Because electrons, photons, atoms, and so on are simply disturbances in fields, these objects superpose just as disturbances in the surface of a tub of water superpose. Several states can all be present at the same time.

This chapter describes how the process called *decoherence* converts these wavy superpositions into particlelike mixtures.

The quantum measurement problem must be part and parcel of any discussion of how quantum physics explains our normal world. After all, quantum measurements—quantum phenomena that cause macroscopic changes—are the bridge from the micro- to the macroworld. Schrödinger's cat is a dramatic example. The microscopic decay of a radioactive atom triggers a device that can kill a macroscopic cat. Schrödinger launched the measurement problem when he noted that the quantum rules seem to imply something we never see: a macroscopic superposition—namely, a cat that is both alive and dead. Chapter 10 suggested a resolution of this problem of definite outcomes.

But there's more than that to the measurement problem. Macroscopic processes are irreversible. The moving finger moves on. "Things run down." The second law of thermodynamics demands it. So quantum measurements must be irreversible, even though there appears to be no trace of irreversibility in the microscopic world. As we'll see, decoherence solves this mystery.

The Problem of Irreversibility

Chapter 10 showed how entanglement between a quantum and its measuring device converts superpositions into mixtures with definite properties, resolving the Schrödinger's cat problem. But entanglement cannot be the complete solution to the measurement problem because it occurs entirely within the evolution prescribed by Schrödinger's equation, and this evolution turns out to be "reversible" because it can be run backward—at least in principle; equivalently, it proceeds at an unchanging entropy level.[2]

Chapter 3 described reversible and irreversible processes in terms of entropy and the second law of thermodynamics. We discovered that most real macroscopic physical processes are irreversible. For example, in a box partitioned into a compartment filled with hot gas and a compartment filled with cold gas separated by a wall that allows energy flow, energy is overwhelmingly likely to flow from hot to cold rather than the reverse. This one-way, or irreversible, heating causes the hot gas to cool and the cold gas to warm until the two sides reach a common temperature, at which time the box is in thermal equilibrium. During the process of coming to equilibrium, the microscopic disorganization or "entropy" of the gas increases.

As you know, this example demonstrates the second law of thermodynamics: The total entropy of all participants in any macroscopic physical process is overwhelmingly likely to increase; decreases in total entropy seldom occur, and can be neither sustained nor controlled. This implies the one-way nature of macroscopic physical processes. For all practical purposes, the entropy of the universe increases, and cannot decrease, during macroscopic processes.

Schrödinger's equation, the fundamental description of quantum motion, can be shown to be entirely reversible.[3] If it really is a quantum world, and if all systems therefore proceed according to Schrödinger's equation (or its relativistic analogs), how can the world's entropy increase? The apparent contradiction is most evident in quantum measurements. After all, by definition, every measurement involves a macroscopic change of some sort, and such changes must obey the second law, so entropy increases. Furthermore, we can observe this entropy increase as a permanent (i.e., irreversible) mark made by the measurement. But when a quantum is measured, the quantum and its detector obey Schrödinger's equation as they entangle to form the measurement state, so it seems that entropy doesn't increase. How can we resolve this conundrum?

A variation on an experiment first performed in 1922 by German physicists Otto Stern and Walther Gerlach demonstrates the connection between quantum measurements and the second law.[4] I'll focus only on the features of this experiment relevant to the irreversibility of measurements. Here's the experiment.

A horizontal stream of silver atoms passes between a pair of magnets (Figure 11.1). The entering atoms, at the left, have all been prepared previously in identical states that I'll call the "zero state." There's no need for us to concern ourselves with the precise nature of this initial state.[5] The magnets are shaped to create a so-called "inhomogeneous magnetic field" in the space between the magnets, as shown in Figure 11.1. This field separates the stream in such a way that if a detection screen is placed "downstream" from the magnets as shown, atoms

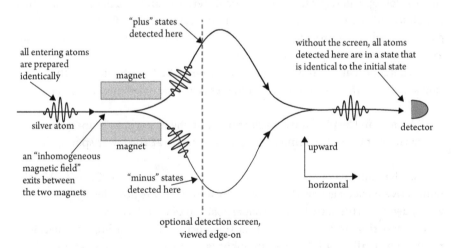

Figure 11.1 A variation on the Stern–Gerlach experiment. Identically-prepared superposed atoms pass through an "inhomogeneous magnetic field." With a detection screen positioned as shown, a random 50% of the atoms strike the screen at the upper impact point, and 50% at the lower point. But when the screen is removed and additional magnets (not shown) are inserted to bend the two streams back together, a detector finds all the atoms to be in the state in which they were initially prepared (at the left).

make visible impacts on the screen at two different spots, one above and the other below the original direction. Individual atoms impact randomly at one or the other spot with a 50% probability—a striking example of quantum indeterminacy because the atoms were all prepared identically. Examination of the two beams shows that atoms striking the upper spot are no longer in the zero state but are in a different state called the "plus state" whereas the atoms striking the lower spot are in a new state called the "minus state." So the magnet-plus-screen combination acts as a detector to determine which atoms are in the plus state and which are in the minus state. This is entirely analogous to a double-slit experiment with a which-slit detector putting quanta into the "slit 1 state" or the "slit 2 state" before the quanta strike a viewing screen.

There's more. If one removes the detection screen and instead installs some appropriately chosen magnets (not shown in Figure 11.1) at certain points along both paths, one can bend each stream back onto its original horizontal path, as shown. On studying the atoms in the converged stream, a perhaps surprising result emerges; none of these atoms are in the plus state and none are in the minus state. Instead, every one is in the zero state from which it started!

So this experiment, without the detection screen, is reversible. But with the screen, the experiment is obviously irreversible because the atoms make an irreversible mark when they strike the screen. It follows that any irreversibility and entropy increase in this experiment are *entirely* due to the impact on the screen, because we have seen that, without the screen, the experiment is reversible. Thus the macroscopic detection makes all the difference, and it is here that we must search for all the irreversible effects of measurement. Specifically, mere entanglement, such as occurs when a which-slit detector operates in the double-slit experiment, is not responsible for irreversibility.

The detection at the screen works just as described in Chapter 10. The detection screen interacts with each atom to form the entangled measurement state:

"Atom is in the plus state, flash appears at the upper spot" superposed with "atom is in the minus state, flash appears at the lower spot."

This is the controversial Schrödinger's cat state analyzed in Chapter 10 (the radioactive nucleus is now the silver atom, and the cat is now the screen), where we argued that this state is only a superposition of correlations.

This experiment demonstrates that the macroscopic recording of a measurement is no small detail. No measurement is complete until it makes a mark on the macroscopic world, and making such a mark must, because of the second law, involve an entropy increase, implying irreversibility. As John Wheeler repeatedly stressed, "No elementary quantum phenomenon is a phenomenon until it is . . . *brought to a close by an irreversible act of amplification*"[6] (my emphasis). If the measurement is recorded by the audible click of a detector, for example, the irreversible mark is a sound wave spreading out in all directions into the air around the detector. This sound wave warms the air a little and eventually disperses (vanishes, for all practical purposes) into a large volume of air. There is no way nature is going to spontaneously gather up every last bit of this wave's dispersed energy (while necessarily cooling the air), reverse the entire dispersal process, and use this energy to restore the detector and the air to their states before the click. In fact, the reversed process is prohibited by the second law because it would reduce the total entropy of the universe!

We conclude that microscopic quantum processes, including entanglement, remain reversible as long as they remain microscopic, and that the irreversibility of measurements must be rooted in the macroscopic recording process. The next section looks at a macroscopic recording process that goes on all over the universe all the time.

How Environmental Decoherence Collapses Superpositions

A pebble lies on a sunny beach, immersed in an environment that includes atmospheric molecules, photons from the sun, cosmic rays from stars, and even photons from the Big Bang. During every second, many such quanta interact with the pebble, reflecting or otherwise *scattering* off the pebble in every possible direction. The scattered photons must carry away data about the pebble's orientation, structure, and color, because otherwise the pebble could not be seen. Such natural "measurement" processes occur all the time, regardless of the presence or absence of humans to consciously observe the gathered data.

To study the quantum features of such a natural measurement, instead of a pebble let's consider an atmospheric atom that is in a highly nonclassical state of being superposed at two or more macroscopically separated locations. We know that such a superposition can be created in the laboratory, for instance by passing the atom through a double-slit apparatus, but it could also occur naturally, for example if the atom passes through an opening that is sufficiently narrow to cause the atom to diffract widely. What happens when such an atom interacts with quanta in the surrounding environment? The superposed atom interacts with, say, an environmental photon in a manner analogous to the way a superposed electron coming through double slits interacts with a which-path detector: If the interaction is significantly different at the atom's two superposed locations, a photon can "measure" the atom by entangling with it in a measurementlike state. Such an entangling interaction is just like a which-path measurement as discussed in Chapter 9! The scattered photon carries away which-path data about the atom, just as a which-path detector carries away data about an electron coming through the slits. As discussed in Chapter 10, this measurement collapses or *decoheres* (removes, via a series of small environmental interactions, the interference pattern from) the superposed atom, converting its state from a superposition of being at both locations to a local mixture of being either at the first location or at the second.

In the natural environment, a single scattered photon is not likely to entirely measure (entirely decohere) the atom, because environmental quanta are not specifically organized, the way a laboratory detector is organized, to measure superpositions. A which-path detector in a laboratory is carefully constructed to respond in detectably different ways at different locations of the measured quantum, quite unlike environmental quanta, which interact randomly in all kinds of ways. Although a good which-path detector requires only one interaction to reliably distinguish between— reliably decohere—the superposed states of the detected quantum, a large number of environmental quanta must scatter from a typical superposed atom to completely decohere it and turn it into a local mixture.

Careful analysis[7] shows that decoherence of typical naturally occurring superpositions requires many environmental interactions. So decoherence by the environment involves a series of partial measurements and partial collapses, each of them instantaneous, nonlocal, and similar to the single-step measurement that occurs at the slits in a double-slit experiment with a which-path detector.

It's through this *environmental decoherence* process that everyday objects such as pebbles and this book lose their extended quantum field nature and behave classically, with no obvious trace of superposition, interference, or nonlocality. Small objects, of atomic dimensions, are less susceptible to environmental decoherence simply because fewer environmental quanta scatter from them as a result of their smaller size, and because each scattering event can cause only a tiny amount of decoherence because the entanglement involves nearly indistinguishable locations. Thus, a superposed photon from the Big Bang might travel the universe for 13.8 billion years without decohering, while a grain of sand on Earth, should it happen to show any signs of superposition, is decohered environmentally and essentially instantly because of the myriad environmental interactions it experiences. Widely extended superpositions of small objects, such as a fine dust grain superposed in two locations separated by a millimeter, are also decohered nearly instantly because the branches of such a superposition are so distinct that a single photon reflecting from one branch but not the other can turn the entire superposition into a mixture. In a similar way, a single detector at only one of two parallel slits is sufficient to turn a quantum that's initially in a superposition of coming through both slits into a mixture that comes through either one or the other slit.

For mesoscopic objects such as dust grains, environmental decoherence turns superpositions quickly into mixtures. How quick? This has been calculated theoretically[8] and measured in a few experiments. A fine dust grain is some 10^{-5} m across. If it's in a superposition of being in two places with its two superposed branches separated by this same distance so that the two branches are right next to each other, as though the grain had come through two closely adjacent slits, it would be decohered entirely by normal air on Earth in only 10^{-31} second. Even the best laboratory vacuum (which still contains plenty of air molecules) would decohere it within 10^{-14} second—a hundredth of a trillionth of a second. If this grain were in deep outer space, cosmic background radiation from the Big Bang would decohere it in 1 second. For a smaller object, such as a large molecule with a diameter of 10^{-8} m, these decoherence times are longer (still assuming it's in a superposition with two branches that are separated by the molecule's diameter): 10^{-19} second if the molecule is in normal air, 0.01 second if it's in the best laboratory vacuum, but much longer than the age of the universe if it's in deep space.

The message is that sufficiently small superpositions can survive awhile, but meso- or macroscopic superpositions are fragile and are decohered quickly by tiny environmental interactions. This is why the "quantum world" is usually identified with the microworld. When typical quantum features become meso- or macroscopic, they generally vanish quickly.

So the quantum universe appears classical at the macroscopic level because the enveloping quantum environment "monitors" every object constantly, and macroscopic objects are especially susceptible to this decoherence process. Nature is full of which-branch detectors! Note that humans aren't required in any of this—no physicists, no laboratories. Nature has been collapsing superpositions, and quanta have been losing their wavy field nature, at least since the Big Bang.

The role of the environment as an ever-present which-path monitor that turns mesoscopic and macroscopic superpositions into mixtures was first clearly recognized by Wojciech Zurek during the early 1980s.[9] The work of Zurek, his colleagues, and others has, in large part, explained how the quantum world leads to the classical world of our experience.[10]

A wide variety of experiments have demonstrated environmental decoherence and convinced physicists that decoherence really is the mechanism that converts the quantum world into nonwavy tables and teapots.[11] One beautiful example is an experiment by the University of Vienna group under Anton Zeilinger and Marcus Arndt. You met this group in Chapter 8, where we examined their Talbot–Lau interferometer technique that demonstrated, in 2002, interference in large molecules such as C_{70} and certain biological molecules. The interference showed each of these large molecules to be in a superposition of following more than one path on its journey through the interferometer.

Using the same technique in 2004, this group was able to demonstrate convincingly the "environmental" decoherence of superpositions of C_{70} molecules.[12]

I put *environmental* in quotation marks because in this experiment the environment came from within the molecules themselves rather than from an external environment. Individual C_{70} molecules passed through the interferometer, as in the 2002 experiment. But there was a new twist; the experimenters heated the molecules just before sending them through the interferometer. They expected that, with sufficient heating, the molecules would themselves emit thermal radiation in the form of visible and infrared "thermal photons," just as an electric hot plate emits thermal radiation (you can feel its warmth at a distance, and it might glow red) when heated. The radiation comes from the random thermal motion of the many atoms and other quanta within the molecule.

According to decoherence theory, each of these radiated photons should act as a partial which-path detector, collapsing the molecule's superposition state partially by carrying a certain amount of which-path data from the molecule into its surroundings. If enough such data are transferred, this which-path measurement should decohere the molecule, causing the interference pattern to dim or vanish—the signature of a superposition evolving into a mixture.

Their results demonstrated decoherence in action. With the molecules only slightly heated, the emitted photons had low energies and thus low frequencies and long wavelengths—too long to distinguish between the possible paths, which for such a massive molecule are separated by extremely small distances (recall from Chapter 6 that a quantum's indeterminacy decreases as its mass increases). But when the molecules were heated to a few thousand kelvins, the emitted photons' wavelengths became short enough to distinguish between the possible paths, so the which-path data transmitted to the surroundings was sufficient to partially decohere the superposed molecules, causing the interference pattern to partially vanish. At sufficiently high temperatures, theory and experiment showed that a mere two or three emitted photons sufficed to decohere each molecule. There was quantitative agreement between the predictions and observations. The interference pattern began decohering at just the predicted temperature, and the degree of visibility of the remaining interference was just as predicted.

The experiment revealed the step-by-step action of decoherence as data leaked, photon by photon, into the environment. It also demonstrated the extreme sensitivity of superpositions to decoherence. Just a few high-temperature photons turn a massive superposed molecule into an incoherent mixture. Superposed quanta are delicate.[13]

Decoherence and the Measurement Problem

Decoherence provides a solution of the irreversibility problem for natural environmental measurements. Decoherence was first introduced during the early 1980s to explain how our apparently classical surroundings arise from measurementlike interactions with the environment.[14] The local state argument (Chapter 10) shows that such measurements turn superpositions into entangled nonlocal measurement states with definite outcomes, resolving the problem of definite outcomes for environmental measurements. But just as for laboratory measurements, we must ask if this also resolves the irreversibility problem associated with natural measurements.

The answer is yes. Here's why. As we've seen, when a small superposed dust grain is decohered by the surrounding environment, the object undergoes a series of small entanglement-caused partial collapses. These environmental measurements are "recorded" by the many environmental photons and air molecules that interact with the grain and then disperse widely. Data about the superposition state of the grain before it decohered are now scattered randomly far and wide. Just as one cannot unscramble an egg, one cannot reversibly gather these pieces back together and reconstruct the global state of the air plus the superposed grain. This argument is especially compelling in the case of environmental decoherence, because of the environment's enormous size. This is an obvious example of the second law in action; for all practical purposes, the process is irreversible and entropy has increased.

Because there is famous opposition to introducing for-all-practical-purposes arguments into physics,[15] it needs to be noted that the second law itself is inherently a for-all-practical-purposes principle. You've seen (Chapter 3) that a box full of gas *could* evolve spontaneously into separate regions of hot gas and cold gas simply by chance, with no external assistance. The chances of this are ridiculously small, but they are not zero, and if we consider boxes containing smaller and smaller amounts of gas, these odds increase. Such for-all-practical-purposes arguments that trace back to the second law are entirely in keeping with the principles of physics. Here's a tale about that.

Joe, deep in philosophical conversation with Schmo, points out, "Entropy never decreases."

"Never?" asks Schmo.

"No, never," responds Joe.

"What? NEVER?" shouts Schmo.

Looking up, Joe shrugs. "Well, hardly ever."

So the local state resolution plus decoherence combine to resolve the measurement problem entirely for the case of natural environmental measurements. The environment measures a superposed object (molecule, dust grain, and so on) when myriad environmental quanta convert the superposition into a measurement state that now takes the form of a nonlocal entanglement between the object and a very dispersed state of the many environmental quanta. The local state argument implies this measurement state represents a mixture of definite properties, whereas the highly dispersed nature of the environmental quanta guarantees the process is irreversible.

There has, for years, been a question about what is the exact relationship between decoherence and the solution of the measurement problem. Some accounts appear to imply that decoherence alone resolves the measurement problem, but this is not true. An especially clear proof of this appears in an article provocatively titled "Why Decoherence Has Not Solved the Measurement Problem: A Response to P. W. Anderson," by Stephen Adler.[16] Anderson, a Nobel laureate, claimed in 2001 that decoherence does solve the measurement problem.[17] Adler is convincingly correct; decoherence alone does not solve the problem of definite outcomes, so it does not by itself solve the measurement problem.[18] Maximilian Schlosshauer, a quantum foundations expert who has written widely about decoherence, also states that "decoherence cannot solve the problem of definite outcomes in quantum measurement."[19]

The connection of decoherence to the measurement problem is that it resolves the irreversibility problem in the case of natural environmental measurements, and it shows how environmental interactions transform a superposed quantum into the measurement state, but it does not solve the problem of definite outcomes associated with this measurement state. The problem of definite outcomes is, however, resolved by the local state analysis, as explained in Chapter 10.

So the local state solution combines with environmental decoherence to solve the measurement problem for the case of environmental measurements. What about the case of laboratory measurements? Here, the preceding chapter resolved the definite outcomes problem, but not the irreversibility problem. Irreversibility poses slightly different problems in the two cases, because lab measurements are sufficiently controlled that the natural environment has little effect. In fact, much of an experimentalist's efforts go precisely into ensuring that random environmental interactions have no significant effect on an experiment's outcome.

Our discussion of the Stern–Gerlach experiment (Figure 11.1) suggested that the answer lies in the irreversible nature of the macroscopic detection process. This suggestion resembles the resolution of the irreversibility problem for environmental measurements: Environmental measurements are "recorded" by innumerable environmental quanta whereas lab measurements are recorded by detection screens, electronic clicks, or, perhaps, cats.

Let's focus on a single impact made on a detection screen by one electron in a double-slit experiment—one of the small spots in Figure 5.4a, for example. How was it recorded? This mark was initiated by a single electron, but to observe it macroscopically the original impact had to be detected and then amplified sufficiently for humans to see it. This is the purpose of every laboratory detector of microscopic events, and it's why an "irreversible act of amplification" (as Wheeler puts it, earlier) is essential to quantum measurements.

In the 1989 experiment that produced Figure 5.4, each electron was emitted in a single coherent state, the same state for each electron. EM fields accelerated these electrons to high speeds, about 180,000 kilometers per second, or 60% of light speed—much faster than the electron in a normal hydrogen atom, which orbits at an average speed of "only" 2000 kilometers per second. High energies were needed to make a detectable impact. Each electron went through a pair of parallel slits, then spread out into an interference state that interacted indeterminately with the screen. The interaction entangled the electron with the screen, decohering the electron and collapsing it instantaneously from its earlier superposition state over the entire screen to a locally observed mixture. In other words, the electron collapsed randomly into one or the other of many small atom-size regions in the screen.[20]

For the fluorescent screen used in the experiment, this interaction created some 500 photons that marked and amplified the location of the impact—already an irreversible process because the photons are emitted randomly in all directions. In a process called the *photoelectric effect*, each photon then struck a metal plate and dislodged an electron from it, converting each of the 500 photons into a low-energy electron. These 500 low-energy electrons were then focused into a point image that could be displayed on a TV monitor in much the same way that old (before flat screens) TV tubes operated. Every step of this process—creation of 500 photons, conversion to photoelectrons, and amplification to create the final display—creates entropy and is irreversible thermodynamically. It's the second law of thermodynamics in action.

This illustrates the general case: Laboratory detection and amplification of a quantum event necessarily involves irreversibility and entropy production. The laboratory detector, a macroscopic device made of many microscopic quanta, plays the decohering role that the environment plays in natural measurements.

To summarize: The quantum measurement problem, in both its laboratory and environmental senses, is resolved entirely by the local state solution of the problem of definite outcomes and by the irreversibility of the decoherence process.

Universal Entanglement

Microscopic "measurements," such as those in the RTO experiment, pose an interesting and enlightening case. Recall that this experiment involves two entangled photons that "measure each other" up until one or both strikes a detector. While the photons are flying toward their detectors, no macroscopic mark has yet been made; the measurement is still reversible and not yet complete. As discussed earlier, locally observable definite, but still-indeterminate, outcomes already reside in the mixed local states of the photons, although the global state is still reversible. The outcome is still undecided. A macroscopically observable impact of either photon with a detector then decoheres the global measurement state superposition (i.e., the two correlations pictured by dashed and solid lines in Figures 9.2 and 9.3) and makes the measurement irreversible, completing the measurement. This is when nature decides the outcome—an outcome that is indeterminate and reversible right up until the first impact with a detector.

An interesting state of affairs exists before that impact. Each photon flies through the laboratory while entangled with another photon a macroscopic distance away. This situation, when a quantum travels through space while entangled with another distant quantum, is common and probably universal. Photons and other quanta from the Big Bang, or a photon emitted into space by the decay of an excited atom, or any other entangled quantum that travels through space while experiencing no decohering interactions, are all in this situation. Similarly, a photon entering your eye just now arises from some quantum interaction elsewhere and must remain entangled with other quanta until some further interaction occurs. Such quanta are in local mixtures of definite but indeterminate outcomes and, simultaneously, are in an entangled global coherent state involving two widely separated quanta. For example, seventy percent of the atoms in the universe are hydrogen atoms made during the Big Bang. Most have drifted in isolation since shortly after the Big Bang and must still retain an entanglement with other primeval material.

One upshot from all this is the suggestion of universal entanglement. Every quantum might be entangled with other quanta, and some of these entanglements might extend over macroscopic and cosmological distances. It's interesting to contemplate, then, the implication that appears to arise from nonlocality. Whenever a single quantum interacts here on Earth, some other quantum entangled with it and perhaps lying a great distance away, adjusts its quantum state. Arguably, every move you make causes an instantaneous microscopic quiver across the universe.

An Experimental *Tour de Force*

The 2012 Nobel Prize in Physics was awarded for a pair of experiments involving quantum superposition, one directed by David Wineland and the other by Serge Haroche. Wineland's experiment (Chapter 8) coaxed a single atom into moving back-and-forth in two opposite manners simultaneously. I'll round out this chapter with Haroche's experiment, performed in 1996 with colleagues at the École Normal Superieure in Paris. During the course of exploring the details of measurement-caused collapse, this extravagant experiment demonstrates most of the quantum fundamentals, providing the perfect finale for these tales.[21]

Like Wineland's experiment, Haroche coaxed a single atom into an extraordinary superposition and demonstrated the reality of the superposition by interfering the atom with itself. He then measured the superposition with a "which-branch" detector. When the detector and atom entangled, the superposition collapsed into a mixture of its two branches, just as in the two-slit experiment with a which-path detector. Haroche then examined the details of the collapse by creating a detector that can vary continuously from "on" to "off." The results demonstrate that both instantaneous state collapse in the laboratory and noninstantaneous environmental decoherence arise from entanglement and are part and parcel of the same measurementlike process.

Let's walk through the experiment. Haroche used a rubidium atom, element number 37, as the quantum to be superposed and measured. Rubidium falls in the first column, fifth row, of the periodic table of elements. This means the atom has a tightly bound and stable inner set of 36 electrons arranged into four concentric spheres, or "electron shells," around the nucleus, plus one lone electron outside these inner shells that forms a fifth incomplete electron shell (remember that each electron is a field spread all around the nucleus). As you can imagine, this outlying 37th electron is manipulated easily, by chemical reactions or lasers, into different states without disturbing the four stable inner shells. Using lasers, Haroche's group maneuvered this electron away from its normal location in the fifth shell and all the way out to the *51st* of the infinite number of theoretically possible shells. This is remarkable. It's astonishing that, with the help of precise laser tuning, the single electron could be directed so precisely into the 51st shell while avoiding, say, shells 50 and 52. Furthermore, within the 51st shell there are many different possible states or "orbits," among which Haroche chose the most circular.

The 51st shell has an enormous 125-nanometer radius—2500 times larger than an ordinary atom! When such a huge atom is exposed to EM radiation, the outer electron behaves like an extended antenna linked strongly to the EM field, which is a useful property and one reason Haroche chose this state of this atom. Any atom like this, with a single electron in a giant orbit, is called a *Rydberg atom* in honor of Swedish physicist Johannes Rydberg who studied these "big atoms."

Any atom with an electron in a high-energy excited state will eventually decay spontaneously to lower energies because of the unavoidable interaction between the atom and the vacuum EM field that fills the universe (Chapter 7). Surprisingly, Rydberg atoms, with an excited electron that is in a circular orbit, are fairly stable against this process, so the atom can endure lots of experimental manipulation without decaying—another reason for choosing this state.

Having coaxed the atom into this Rydberg state, Haroche then tickled it with a brief EM microwave "pulse." If the pulse has just the right frequency and duration, it can put the atom into a delicate superposition of having its outer electron in shell 50 and also shell 51. I'll call these two superposed Rydberg branches A+ for the higher-energy branch, shell 51, and A− for the lower-energy branch, shell 50.

Achieving this superposition state is, in itself, an astonishing accomplishment! But the experiment goes much further by measuring and manipulating this superposition.

As his detector, Haroche used a so-called *high-Q cavity* containing an EM field. The cavity is a small, evacuated region bounded by mirrors that can store photons for long enough to do the experiment. A high-Q cavity is a cavity that stores photons for a long time—0.0002 second in this case—long enough for a photon to bounce two million times back and forth between the mirrors. The cavity is an evacuated 3-centimeter-long, pillbox-shaped cylinder containing, at its two flat ends, highly polished mirrors, each 5 centimeters in diameter, facing each other. A laser provides between 0 and 10 photons to the cavity. This cavity field is "coherent" in the same way that photons in a laser beam are coherent (Chapter 8). All the photons participate collectively in a single quantum state of the field, so that the field within the cavity acts as a single quantum object. In each experimental trial, a few photons are placed inside the cavity, the photon source is turned off, and the photons then bounce freely between the mirrors. This nearly perfect photon box is the key to the experiment. Figure 11.2 shows the layout. A source O emits one rubidium atom at a time; a laser pulse at B excites the moving atom into the Rydberg state A+; a microwave pulse at R_1 transforms this state into

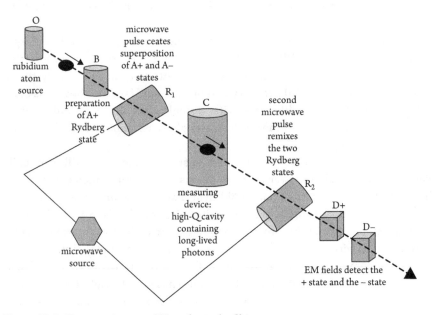

Figure 11.2 The experiment of Haroche and colleagues.

a superposition of A+ and A−; the superposed atom passes through the detecting cavity C, which measures (and thus collapses) the atom's state; in some trials a second microwave pulse at R_2 is used (for reasons described later); and detectors D+ and D− record whether the measured atom emerges in state A+ (in which case D+ records it) or state A− (in which case D− records it). Note that Haroche did not learn the outcome of each trial by observing the photons in detector C, which is the way one might expect a "detector" to be used. Instead, it was more convenient to observe the atom directly, at D+ and D−, after it was collapsed by detector C into one of the branches A+ or A−.

Experimental trials begin with C empty (no photons), so that each atom passes through C unmeasured. In this case, detectors D+ and D− demonstrate interference between the two superposed branches of the atom, just as the double-slit experiment demonstrates interference at the screen if there is no which-slit detector at the slit. This happens because the superposed rubidium atom interferes with itself, just as an electron passing through two slits (with no detector) interferes with itself. This interference verifies that each atom approaching the cavity really is in a superposition of the two Rydberg states.

Now let's introduce measuring device C into these proceedings. Just a few photons are in the cavity, forming an EM field. For the cavity to operate as a measuring device, it must transform into one field state, called $E+$, if the atom passes through the detector in state A+, and into a detectably different field state, called $E-$, if the atom passes through in state A−. Schrödinger's equation then implies that, if the *superposed* atom passes through the cavity, the atom-field interaction within the cavity puts the composite atom-plus-field system into the entangled measurement state. This is what happens in all measurement situations (Chapter 10), such as an electron passing through a double slit in the presence of a which-path detector. The result is an entangled "Schrödinger's cat" with atomic state A+ correlated with state E+ of the cavity *and* A− correlated with state E− of the cavity.

As you know, such a measurement collapses the atom to a mixture of A+ and A−. This mixture is exactly what Haroche's group found. The two detectors, D+ and D−, verified the expected results. The atom was found to be in the E+ state on 50% of the trials and in the E−state on the other 50%, with no interference effects. It's just like the double-slit experiment! With no photons in the cavity (no measurement), each rubidium atom persisted in its superposed state and demonstrated interference; with a measurement, each atom entangled with the cavity field and collapsed into one of the atom's two superposed states, becoming an incoherent 50–50 mixture showing no sign of interference.

At this point, Haroche had verified the details of measurement and collapse of the superposed quantum state in a new experimental context. Next, the experiment explored the range of phenomena between the unmeasured atom (no photons in the cavity) and the measured atom (photons in the cavity). Call it *partial measurement*.

A proper detector must distinguish clearly and unambiguously between the different states it attempts to measure. I've dubbed the detector states for Haroche's experiment $E+$ and $E-$. For a true measurement, these two field states must be distinct.[22] Haroche then explored the intermediate range between no measurement and perfect measurement by altering the separation between the two mirrored faces in the cavity in such a way as to cause states $E+$ and $E-$ to be just slightly distinct, moderately distinct, or entirely distinct. As the EM field states became more and more distinct, the interference fringes became less and less distinct, and finally vanished altogether. This highlights once again the connections between entanglement, measurement, and collapse, and supports decoherence theory by showing how partial measurement entails partial collapse.

As though all this weren't enough, Haroche added a final flourish. By means of a laser pulse at R_2, he altered the atom's superposition in such a way that detection of atomic states $A+$ or $A-$ were now correlated with two different *superposition* states of the cavity field—states that involve superpositions of $E+$ and $E-$. The point of this is that Haroche wanted to observe the gradual natural decoherence of these cavity field superpositions caused by the small imperfections that must inevitably be present in the cavity's mirrors. To do this, he sent not one but two atoms through the cavity on each trial, with the second atom following the first after a prescribed time delay. Think of them as a cat followed by a mouse. The cat puts the cavity into a superposition and the mouse then probes the state of the cavity's field to detect its expected decoherence a short time later.

By performing many trials with different time delays between the cat and the mouse, various initial superposition states of the cavity field, and various numbers of photons in the cavity, the team traced the gradual decoherence of the cavity superposition resulting from environmental imperfections. The results confirmed that the cavity field did indeed evolve into a mixture over time, demonstrating the role of entanglement and decoherence in converting the quantum world into our normal world.

The results confirmed that, as expected, decoherence is faster when the two branches of the superposition are more distinct. If you *could* finagle a cat into a superposition of dead and alive, it would decohere quickly because of the enormous difference between its branches. And decoherence is faster when the number of photons in the cavity is larger. Larger systems, if put into a superposition state, decohere more quickly. So you'll probably never encounter superposed tables, teapots, cabbages, kings, or ice cream.

You have passed through the looking glass and explored the land of the quantum. I hope the journey has been meaningful and fun.

For more than a century, much has been made of the odd and supposedly paradoxical nature of the quantum. This presumed quantum spookiness has led to an excess of attempted fixes and interpretations. Many experts have even declared the theory to be not a description of reality at all, but only a mathematical recipe that helps humans predict the results of experiments. As Niels Bohr put it, "There is no quantum world. There is only an abstract quantum description."[23] According to this hypothesis, quantum theory doesn't describe anything real at all, so there's no cause for concern about collapse of the quantum state and other odd quantum behaviors.

Such an easy resolution of the quantum quandaries amounts to giving up on science's project of understanding the realities of the natural world. It's an extraordinary claim, requiring extraordinary proof.

But there is no such proof, and there are no grounds for regarding quanta as any less real than rocks. Indeed, rocks are made of quanta. Although it has long been a conceit of humankind to imagine the universe to be centered around us, there is no reason to think that reality comprises only the kinds of things we experience in our own daily lives. The real world does not fade from existence, nor does it become incomprehensible, at distances that happen to be several powers of ten smaller than teapots. Atomic and subatomic processes are just as real as teapots and, with the help of technology, accessible to human experimentation and understanding.

From the viewpoint of the macroscopic and classical world that we are pleased to call "normal," there is certainly oddness in wave–particle duality, indeterminacy, quantum states, superposition, nonlocality, measurement, and quantum jumps. But there are no logical contradictions here, no disagreements with experiment, and nothing that should persuade us that quantum physics is about anything other than the real world. Quantum physics is either charmingly counterintuitive or maddeningly puzzling, depending on your taste, but it is entirely self-consistent and experimentally accurate. It's time to accept it with all its charms and puzzles, and stop trying to repair or reinterpret it. It's time, in other words, to relax and admit the world is not as we had thought. Nature is far more creative than we could have conceived.

Our most fundamental theory is in better shape than its detractors suppose. Quantum physics is a remarkable treasure trove of far-reaching phenomena and ideas whose surface we have probably only begun to scratch. It's time to fully embrace these ideas, incorporating them into our ways of thinking about the universe, about our planet, and about ourselves. This is a process that will engage our minds and stretch our imaginations far into the future, for quantum physics is, indeed, not what anybody could have imagined.

NOTES

Preface

1. Paul Dirac, *Quantum Mechanics*, 4th edition (Oxford: Oxford University Press, 1958), vii.
2. Freeman J. Dyson, "Innovation in Physics," *Scientific American* 199 (September 1998), 74–82.
3. Alan Mackay, *A Dictionary of Scientific Quotations* (London: IOP Publishing, 1991), 83.

Chapter 1

1. It is possible, however, to turn one photon into two photons. This difficult process, called *down-conversion*, does not really split the original photon; rather, it uses the energy of the original photon to create a new pair of photons. Furthermore, the created pair is "entangled"—meaning that it remains, in many respects, a single object, an "atom of light."
2. Surprisingly, evidence shows each photon is both transmitted and reflected at the window, as a "superposition" (Chapter 8). The superposition doesn't "collapse" to one side or the other until the photon is detected on one side or the other. Thus, either Alice or Bob, and not both, detect the photon, but the photon travels both paths and collapses only when it's detected.
3. As we see in detail in Chapter 8, this paragraph is slightly oversimplified. An individual photon striking the glass plate does not immediately "decide" whether to reflect or to transmit. Instead, *it goes both ways at the same time*. This bizarre (by our macroscopic standards) behavior is called *superposition* of the two "states" (reflection and transmission). The decision of which path the photon takes is not made until the photon is *detected*, by either Alice or Bob. The photon is said to "collapse" or to "quantum jump" on detection.
4. Although the figure doesn't show it, path 2 bends slightly downward on entering beam splitter 1. It then emerges from the far side of the beam splitter on a path that is exactly parallel to its incoming path, but is displaced slightly downward.
5. However you can't actually see a single photon; your eye must receive several closely spaced photons to sense light.
6. With beam splitter 2 in place, when either path 1 or path 2 is varied in length, an interference pattern appears at the two detectors over many trials of the experiment. This verifies that each photon interfered with itself, and for this it must have gathered information from *both* paths. So each photon goes both ways at beam splitter 1. Furthermore, this behavior cannot depend on the presence or absence of beam splitter 2, because the photon interacts with beam splitter 2 only after it has passed through beam splitter 1.
7. The only kinds of superpositions considered in this book are position-superposition and velocity-superposition. There are also other types of superpositions: superpositions of spin, of polarization, and of other quantum properties.
8. en.wikiquote.org/wiki/Werner_Heisenberg. Wikiquotes, "Werner Heisenberg," February 20, 2016.

9. More massive objects such as baseballs generally have very short wavelengths, so a superposition of significantly different positions would have to involve distances of many wavelengths. It would be hard to maintain "phase coherence" (i.e., a single, consistent quantum state) over such distances.
10. In addition to these "energy states," other states are also possible. Each of these other states is a particular "superposition" (or combination) of the energy states.
11. You've got to get the box of atoms fairly hot because hydrogen atoms join together easily to form hydrogen molecules made of two joined hydrogen atoms and have a different spectrum from atomic hydrogen. Heating prevents such bonds from forming.
12. A neon sign's glass tube doesn't feel hot because the gas inside is at such a low density that it can't transfer much thermal energy to the glass. The gas, however, has quite a high temperature.
13. G. Gerald Gabrielse, "The Standard Model's Greatest Triumph," *Physics Today* 66, December (2013), 64–65. The quoted number is actually $|g|/2 - 1$, where g represents the "g-factor."
14. https://en.wikipedia.org/wiki/Archaeoastronomy_and_Stonehenge. Wikipedia "Archaeoastronomy and Stonehenge," May 27, 2016.
15. Ian P. Bindloss, "Contributions of Physics to the Information Age," www.physics.ucla.edu/~ianb/history/, January 13, 2012.
16. Max Tegmark and John Wheeler, "100 Years of Quantum Mysteries," *Scientific American* 299, February (2001), 68–75.
17. Maximilian Schlosshauer, Johannes Kofler, and Anton Zeilinger, "A Snapshot of Foundational Attitudes Toward Quantum Mechanics," *Studies in History and Philosophy of Modern Physics* 44 (2013), 222–230.
18. Art Hobson, "There Are No Particles, There Are Only Fields," *American Journal of Physics* 81 (2013), 211–223.
19. Albert Einstein, Boris Podolsky, and Nathan Rosen, "Can Quantum-Mechanical Description of Physical Reality Be Considered Complete?" *Physical Review* 47 (1935), 777–780.
20. Alain Aspect, "Bell's Inequality Test: More Ideal Than Ever," *Nature* 398 (1999), 189–190.
21. Quoted in Michael Shermer, "Quantum Quackery," *Scientific American* 292 (2005), 34.
22. Deepok Chopra, *Quantum Healing: Exploring the Frontiers of Mind/Body Medicine* (New York: Bantam, 1989).
23. Maximilian Schlosshauer, *Decoherence and the Quantum-to-Classical Transition* (Berlin: Springer Verlag, 2007), 364–365.
24. Bruce Rosenblum and Fred Kuttner, *Quantum Enigma: Physics Encounters Consciousness* (New York: Oxford University Press, 2006).
25. Ibid., 4, 120, 179, 186.
26. Victor Stenger, "Quantum Quackery," *Skeptical Inquirer* 21.1, January–February (1997), 37–42. It's striking that this article has, by coincidence, the same title as the article listed earlier by Michael Shermer.

Chapter 2

1. Albert Einstein to Michael Besso, quoted in Max Born, *The Born–Einstein Letters 1916–1955*, trans. Irene Born (New York: Macmillan, 1971).
2. Robert L. Weber, *Pioneers of Science: Nobel Prize Winners in Physics* (Bristol: Adam Hilger, 1980), 64–66. The finest example of Einstein's critique of quantum physics is the "EPR" paper: Albert Einstein, Boris Podolsky, and Nathan Rosen, "Can Quantum-Mechanical Description of Physical Reality Be Considered Complete?" *Physical Review* 47 (1935), 777–780. This brilliant but ultimately incorrect paper challenged Heisenberg's principle. It is the first paper to demonstrate nonlocality as a prediction of quantum physics.
3. However, a quantum can (believe it or not) be in a superposition of both existing and not existing, in which case the quantum field is in a superposition of the "vacuum state" and

the "single quantum state." Note that, even in this case, the quantum is in a superposition of both "existing" and "not existing." You can't have just a fraction (such as three quarters) of a quantum.
4. Experiments have determined these sudden transitions occur across spatial regions at speeds faster than light speed and are consistent with instant transition, but it's impossible for experiments to establish that such transitions actually occur in zero time because extremely short time intervals are hard to measure.
5. Quantum physicists, especially those involved in theoretical fundamentals, often speak of *wave functions* rather than *quanta* or *quantum states*. A wave function is a mathematical description of a particular situation or "state" of a physical quantum, such as a photon or an electron. Physically, an electron *is* its wave function, and similarly for photons, atoms, and so forth. This book doesn't use the term *wave function*. See Art Hobson "There Are No Particles, There Are Only Fields," *American Journal of Physics* 81 (2013), 211–223.
6. For a good book on the hunt for and discovery of the Higgs Boson, see Sean Carroll, *The Particle at the End of the Universe* (New York: Dutton, 2012).
7. Neutrinos are created in nuclear reactors and by Earth's geological radioactivity, impacts of cosmic rays on the atmosphere, the energy processes operating at the center of every star, explosions of stars, and the Big Bang. Because they interact with matter only weakly, it takes several years before even one of the zillions of neutrinos passing through your body actually interacts with your body. They travel at nearly light speed.
8. Light moves more slowly through water or other materials, but this is because the material's atoms occasionally absorb (destroy) and then reemit (create) each photon so that a photon's average travel speed is less than light speed. Every photon, whenever it actually exists, moves at light speed.
9. More direct evidence, in the form of direct laboratory observation of quanta of dark matter, might be found at any time, perhaps by the time you read these words.
10. You'll find several good discussions of dark matter, and some discussion of dark energy, in Carroll, *The Particle at the End of the Universe*.
11. More precisely, the galactic clusters are moving away from each other. These clusters, as well as each galaxy within each cluster, are held together by gravity.
12. For the complete story by the leader of one of the two supernova observing teams, read Robert P. Kirshner's *The Extravagant Universe* (Princeton, NJ: Princeton University Press, 2002). For a brief description, see Art Hobson, "And the 2011 Nobel Prize in Physics Goes To: The Accelerating Universe," *The Physics Teacher* 12 (November 2012), 468–469.
13. Alfred Wegener's geological theory of plate tectonics is a good example. For a brief account, see "Quakes, Tectonic and Theoretical," Kenneth Chang, *New York Times*, p. WK4, January 15, 2011.
14. Here are two good books on the process of science, and the dangers of pseudoscience: Carl Sagan, *The Demon-Haunted World: Science as a Candle in the Dark* (New York: Random House, 1995); Michael Shermer, *Why People Believe Weird Things: Pseudoscience, Superstition, and Other Confusions of Our Time* (New York: W. H. Freeman, 1997).
15. This reason is called the *law of inertia*, or *Newton's first law*.
16. Dark energy might be related to the quantum vacuum field (Chapter 5). Dark energy and the quantum vacuum are discussed in Carroll, *The Particle at the End of the Universe*.
17. For a brief summary of the cosmic microwave background and cosmology's "standard model," see Adrian Cho, "Universe's High-Def Baby Picture Confirms Standard Theory," *Science* 339 (2013), 1513.
18. Darren Wong and Boo Hong Kwen, "Shedding Light on the Nature of Science through a Historical Study of Light," http://www.researchgate.net/publication/251619647, May 23, 2015.
19. Isaac Newton, *Optics, A Treatise of the Reflexions, Refractions, Inflexions and Colours of Light*, (Dover Publications, 1952 [originally published in 1704]).
20. A good textbook pursues this suggestion: Ralph Baierlein, *Newton to Einstein: The Trail of Light* (Cambridge: Cambridge University Press, 2000).

21. Robert Crease, of the State University of New York at Stony Brook philosophy department and historian of Brookhaven National Laboratory, asked physicists in 2002 to nominate the most beautiful experiment of all time. The winner was the double-slit experiment applied to the interference of single electrons (Chapter 5) rather than, as Thomas Young did in 1801, to the interference of light. Young's double-slit experiment with light was ranked number five in Crease's poll. See George Johnson, "Here They Are, Science's 10 Most Beautiful Experiments," *New York Times*, September 24, 2002.
22. Before passing through the two slits of Figure 2.2, the light must be filtered so that it is all of one color (one wavelength), and it must first come through a single narrow slit and then spread out (diffract) to pass through both slits 1 and 2. The vibrations at slit 1 are then synchronized with those at slit 2. This makes Young's experiment (Figure 2.2) the exact analog (but in three dimensions) of the water wave experiment of Figure 2.1, in which the water waves from the two openings have the same frequency and they are synchronized.
23. Earth's gravitational field also exists at other points, such as near the sun, although a rock released near the sun will fall toward the sun. Both the sun's field and Earth's field exist near the sun, but the sun's field is stronger.
24. Many leading physicists have expressed this conclusion. Here are three authoritative examples: "The basic ingredients of nature are fields; particles are derivative phenomena." By Steven Weinberg, *Facing Up: Science and Its Cultural Adversaries* (Cambridge, MA: Harvard University Press, 2001), 221. "The only way to have a consistent relativistic theory is to treat all the particles of nature as the quanta of fields, like photons.... This approach now gives a unified picture, known as *quantum field theory*, of all of nature." By Robert Mills, *Space, Time, and Quanta: An Introduction to Modern Physics* (New York: W. H. Freeman, 1994), 386. "In quantum field theory, the primary elements of reality are not individual particles, but underlying fields. Thus, e.g., all electrons are but excitations of an underlying field ... the electron field, which fills all space and time." By Frank Wilczek, "Mass without Mass I: Most of Matter," *Physics Today* 52, November (1999), 11–13. Further quotations, and a full discussion of wave-versus-particle issues in quantum physics, may be found in Art Hobson, "There Are No Particles, There Are Only Fields," *American Journal of Physics* 81, (2013), 211–223.
25. For further discussion of atoms and molecules, see Art Hobson, *Physics: Concepts and Connections*, 5th edition (San Francisco: Pearson Education, 2010), 35, 38.
26. "Science Quotes by Democritus of Abdera," Today in Science History (Todayinsci 1999–2016). http://todayinsci.com/D/Democritus/Democritus-Quotations.htm.
27. For an excellent nontechnical book on the proposed string theory unification of physics, see Brian Greene, *The Elegant Universe* (New York: W. W. Norton, 1999)
28. Bill Spargo, Recording Secretary, "Minutes of the 2107th meeting, the remarks of Robert A. Nelson," October 1999, Philosophical Society of Washington, http://www.philsoc.org/1999Fall/2107minutes.html.
29. The surprising motion of pollen particles suspended in water was discovered in 1827 by botanist Robert Brown. Much later, in 1905, Albert Einstein published a paper explaining in detail how this "Brownian motion" resulted from pollen being moved by impacts from individual water molecules. This explanation served as important confirmation that atoms and molecules actually exist.
30. Max Tegmark and John Wheeler, "100 Years of Quantum Mysteries," *Scientific American* February (2001), 68–75.
31. He also made certain simplifying assumptions, as physicists usually do when deriving general laws. The "idealized radiator" that Planck had in mind is called, perhaps surprisingly, a *blackbody* and the radiation is called *blackbody radiation*.
32. Well, there is at least one fairly obvious idea. A "countable" set of numbers, such 1, 2, 3, and so on, is much more simple mathematically than a continuum of numbers, such as "all the numbers between zero and one." A continuum includes an infinity of numbers, is not countable, and includes the "irrational" numbers that cannot be expressed as the ratio of two integers. Quantized energy levels are more simple, more "primitive," and

more "fundamental" in this sense. Maybe this has something to do with why energy is quantized.
33. This is true even of composite quanta, such as atoms (composites of quarks and electrons). Every atom's components are entangled nonlocally (Chapter 9) and thus when one component is altered, all components are altered. Entanglement causes it to act as a single object, although it is made of parts.

Chapter 3

1. There are any number of websites offering convincing visual demonstrations of Galileo's thinking. See, for example, http://demonstrations.wolfram.com/GalileosThoughtExperimentOnInertia/. Enrique Zaleny, "Galileo's Thought Experiment on Inertia," Wolfram Demonstrations Project, 2016.
2. "In the limit of long inclines," as mathematicians would say. Galileo was the first to consider such ideal, limiting situations. Herbert Butterfield, *The Origins of Modern Science* (New York: Simon & Schuster, 1965).
3. Ibid., 15.
4. John B. S. Haldane, *Possible Worlds and Other Papers* (London: Transaction Publishers, 2002; originally published in 1927 by Chatto & Windus, London).
5. Art Hobson, *Physics: Concepts & Connections*, 5th edition (San Francisco: Pearson Addison-Wesley, 2010).
6. Ari Ben-Benahem, *Historical Encyclopedia of Natural and Mathematical Sciences*, vol. 1 (Berlin: Springer, 2009).
7. Isaac Newton. Isaac Newton, *Optics, A Treatise of the Reflexions, Refractions, Inflexions and Colours of Light* (Dover Publications, 1952, originally published in 1704).
8. Isaac Newton and Gottfried Leibniz invented calculus independently in the mid 17th century.
9. Pierre-Simon Laplace, *A Philosophical Essay on Probabilities* (New York: John Wiley & Sons, 1902).
10. Brian Greene. *The Elegant Universe* (New York: W. W. Norton, 1999).
11. However, neutron stars are sometimes referred to as giant nuclei, composed mostly of neutrons. A neutron star is typically some 12 kilometers in radius, but it's held together by gravity, not by the strong nuclear force, so its classification as a "giant nucleus" is dubious. It's best classified as a star.
12. It was once thought that the isotope bismuth-209, having 83 protons and 126 neutrons, was stable; But in 2003 it was discovered that this nucleus undergoes alpha decay with a half-life of 9×10^{26} seconds—more than a billion times longer than the age of the universe. See references in Wikipedia, "Bismuth-209," December 30, 2015. http://en.wikipedia.org/wiki/Bismuth-209. Wiki also has good articles on alpha and beta decay.
13. Some physicists reject the "ability to do work" definition of energy because of difficulties in applying it to thermal energy. This is because, according to the second law of thermodynamics, only a fraction of a system's thermal energy can be converted to work by allowing the thermal energy to flow to a second system whose temperature is lower. However, that fraction approaches 100% as the second system's temperature approaches zero, and thus we can convert "all" (i.e., as close to 100% as we like) of the original system's thermal energy to work in this limiting idealized situation. Such limiting idealizations are traditional in physics, so there's no reason to allow this objection to overrule the obvious and straightforward definition of physics' most important word.
14. Do a Web search on James Prescott Joule, whose experiments were the key to understanding that thermal energy (warmth) is actually a form of energy and must be included in statements of conservation of energy. His experiments are marvelous and enlightening.
15. *Heat* is an ambivalent term. Either *thermal energy* or *internal energy* should be used instead. The only proper use of the word *heat* is as a verb related to the process of "heating," which

is the transfer of thermal energy by means of its natural flow from higher to lower temperatures. This book sticks to the term *thermal energy* and eschews *heat*.
16. This definition is good enough for this book, but it's oversimplified because it overlooks the "latent thermal energy" that resides in phase transitions. For example, when you warm a block of ice through the freezing temperature, the ice remains at 0°C while thermal energy turns the ice into water. This thermal energy is thus "latent" (i.e., not expressed as a temperature increase) in the water.
17. In fact, according to the "inflation" theory of the origin of the universe, the positive energy of the universe exactly balances its gravitational energy, which is entirely negative, so that the total energy of the universe is always zero! Search Wiki for "zero-energy universe" and note the references therein. Thus, energy did not need to be created during the Big Bang. The Big Bang simply separated the negative energy (gravitational) from the positive forms of energy. As another way to look at this, gravity (the sudden formation of gravitating systems) energized (provided positive energy to) the universe.
18. John Rigden's *Einstein 1905: The Standard of Greatness* (Cambridge, MA: Harvard University Press, 2006) tells the story of Einstein's "miraculous" year. For a wonderful nontechnical book for nonscientists and scientists about Einstein's special and general theories of relativity, see Ira Mark Egdall, *Einstein Relatively Simple: Our Universe Revealed in Everyday Language* (London: World Scientific, 2014).
19. Albert Einstein, "Ist die Tragheit eines Korpers von seinem Energieinhalt abhangig?" *Annalen der Physik* 18 (1905), 639–643. Search for a translation under the title "Does the Inertia of a Body Depend on Its Energy Content?"
20. If the pair's kinetic energy is high enough, new higher-mass quanta can be created, such as W^+–W^- pairs.
21. The "right condition" is that the photon must be near an atomic nucleus that absorbs some of the photon's momentum.
22. Because $2^{10} = 1024$.
23. If this is the second law, then where is the first law? The first law of thermodynamics is logically equivalent to conservation of energy. It says that, in any process during which a system is heated and has work done on it (with no other energy inputs or outputs), the heating done on the system plus the work done on it equals the total energy increase of the system.
24. The inventor of the inflationary theory of the Big Bang has written a nontechnical book about it: Alan Guth, *The Inflationary Universe: The Quest for a New Theory of Cosmic Origins* (Reading, MA: Addison-Wesley, 1997).
25. If the initial state, as the Big Bang began, was a single quantum state (i.e., a so-called *pure state*), then it's entropy was precisely zero.

Chapter 4

1. Albert Einstein, "Maxwell's Influence on the Development of the Conception of Physical Reality," in *James Clerk Maxwell: A Commemorative Volume 1831–1931* (New York: The Macmillan Company, 1931), 66–73.
2. Frank Wilczek, "Profound simplicity," *The New York Academy of Sciences Magazine*, October 9, 2008, http://www.nyas.org/publications/Detail.aspx?cid=1577681c-007f-4b23-bcad-7d6d32 ccca5b.
3. A. Andrew Janiak, *Newton: Philosophical Writings* (Cambridge: Cambridge University Press, 2004), 102.
4. Newton's declaration, stated in his *Principia*, reads as follows: "I have not as yet been able to discover the reason for these properties of gravity from phenomena, and I do not feign hypotheses." Frank Wilczek, *The Lightness of Being: Mass, Ether, and the Unification of Forces* (New York: Basic Books, 2008).
5. In particular, this would explain Max Born's influential description of the wave function psi (evaluated at a particular spatial point x) as representing the probability amplitude

that "a particle will be found at x." As we will see, there are no particles. The correct statement of the Born rule is that psi, evaluated at x, represents the probability amplitude that an interaction will occur at x.
6. I use the word *framework* in the sense that philosophers use *ontology* to mean "the underlying reality."
7. For a good presentation of the field concept, see Rodney A. Brooks, *Fields of Color: The Theory That Escaped Einstein*, 2nd ed. (Prescott, AZ: Allegra Print and Imaging, 2011). Brooks' book is an engaging defense of this book's notion that there are no particles; there are only fields.
8. *The Scientific Papers of James Clerk Maxwell*, vol. 1 (New York: Dover Publications, 1965), edited by W. D. Niven, 360.
9. But unlike Faraday, Maxwell believed these fields were a property not of "mere space," but rather a property of a universe-filling mechanical substance: "ether." Later, Einstein's work led physicists to drop the notion of ether and, instead, to view fields as properties of space alone, as Faraday had done.
10. By "motion of charged objects," I mean to include the spin of electrons and other charged objects. Although I don't discuss "spin" in the main text, it is a permanent quantum property of most quanta that is, in most respects, similar to the spinning motion of such macroscopic objects as a spinning top and our spinning Earth.
11. Light speed is known to better accuracy today, of course. However, the logic of our system of measurement has changed. Rather then defining a basic unit of distance, such as the meter (whose previous definition was the length of a certain number of wavelengths of the light from a certain type of atom), scientists now *define* the meter as the distance light travels in vacuum in 1/299,792,458 second. This definition fixes the speed of light in vacuum at 299,792,458 m per second, an exact number, and leaves the meter as the quantity that remains to be determined, in the future, to better accuracy. See https://en.wikipedia.org/wiki/Speed_of_light for references.
12. James Clerk Maxwell, "A Dynamical Theory of the EM Field," *Philosophical Transactions of the Royal Society of London* 15 (1865), 459–512.
13. Faraday and, later, Einstein held the view that there is no ether and that the EM field is a property of "mere space" (i.e., of space alone). Maxwell, however, continued to believe that EM waves are mechanical waves in an as-yet-unobserved material medium, the ether. The Faraday/Einstein view, that fields require no material medium, no ether, is the consensus view of physicists today.
14. In 1973, the weak force and the EM force were unified in a single "electroweak" theory. This *quantum field theory*, as it was called, established quantum fields as the basic ingredients of nature. Physicists Steven Weinberg, Abdus Salam, and Sheldon Glashow received the 1979 Nobel Prize for this work. Quantum fields are discussed much more fully in Chapter 5.
15. From Hertz's measured values of frequency and wavelength, we can figure out the speed of the wave. For example, if we have a wave with a wavelength of 2 m that has a frequency of 3 Hz (three waves emitted every second), then the wave must move a distance of 2×3 m every second, or 6 m per second. Similarly, a wave with a wavelength of 0.6 m and a frequency of 5×10^8 Hz must move at a speed of $(0.6) \times (5 \times 10^8)$ m per second, or 3×10^8 m per second.
16. Hertz used these terms in his elegant and physically insightful lecture "On the Relations between Light and Electricity," delivered to the German Association for the Advancement of Natural Science and Medicine in Heidelberg in 1889. An English translation is published in the *American Journal of Physics* 25 (1957), 335–343.
17. IR and UV are also directly (without technology such as a radio receiver) detectable by humans as radiated warmth and sunburns, respectively.
18. There are gravitational fields, dark matter fields, dark energy fields, Higgs fields, and several known types of normal matter fields, to name a few.
19. However, there are virtually no x rays or gamma rays around you because Earth's atmosphere protects us from most of this high-energy radiation that exists in outer space.

20. R. H. Stuewer, ed., *Historical and Philosophical Perspectives of Science* (New York: Gordon and Breach, 1989), 299.
21. James Boswell, *The Life of Samuel Johnson* (New York: Penguin Classics, 1986).
22. See Contemporary Physics Education Project, "Standard Model of the Fundamental Particles and Interactions" (2000), http://www.pha.jhu.edu/~dfehling/particle.gif.
23. It's called *quantum chromodynamics*.
24. See Chapters 9 and 10 of Frank Wilczek's wonderful book, *The Lightness of Being: Mass, Ether, and the Unification of Forces* (Philadelphia: Basic Books, 2008). Also see Frank Wilczek, "Mass without Mass," *Physics Today* 53, (1999), 11–13.
25. There were also other difficulties. It was expected that more intense light would cause electron emission whereas less intense light would not, because there is more energy in the more intense light. But it turned out that, for a given frequency, electrons were either emitted or not emitted regardless of intensity. Instead, the light's frequency was the deciding factor: For sufficiently high frequencies, even dim light caused electron emission. For lower frequencies, no electrons were emitted, although the light was very intense. Physicists didn't understand why frequency was the controlling factor.
26. John Rigden, *Einstein 1905: The Standard of Greatness* (Cambridge, MA: Harvard University Press, 2006).
27. Robert Weber, *Pioneers of Science: Nobel Prize Winners in Physics* (Adam Hilger, Bristol, 1988). Bohr's work is on pp. 67–69. This book provides brief authoritative accounts of the prize-winning work of all the Nobel Laureates.
28. Ibid., 91–92.
29. Ibid., 109–111.
30. Historical documentation for the work and views of Faraday and Maxwell can be found in Nancy Nersessian's beautiful book: *Faraday to Einstein: Constructing Meaning in Scientific Theories* (Boston: Martinus Nijhoff, 1984). Parts of this chapter are based on Nersessian's book.

Chapter 5

1. You can, however, cause a quantum to vanish while creating two or more other quanta, but these new quanta are not parts of the original quantum.
2. If you do an Internet search on "particles or fields," you'll find differing opinions. However, the last real champion of an all-particles view was Richard Feynman, and we'll see that even he admitted to exasperation when trying to explain the double-slit experiment in terms of particles. In recent years, nearly all quantum field experts (Steven Weinberg, Frank Wilczek, Michael Redhead, and Robert Mills, for example) have considered the universe to be made entirely of fields. For several rigorous arguments supporting this view, see my paper, "There Are No Particles, There Are Only Fields." For a qualitative overview, see Rodney Brooks, *Fields of Color* (Prescott, AZ: Rodney Brooks, 2011). The only major exceptions to the all-fields view today are those who argue that there is no microscopic physical reality at all—no particles and also no fields. The main support for such a nihilistic view is the notion that quantum physics seems to be full of fundamental contradictions. I hope to demonstrate in this book that there are no such contradictions.
3. However, Planck didn't realize at that early time the full significance of his formula, nor did he know about photons.
4. Images courtesy of Wolfgang Rueckner, Harvard University Science Center. See W. Rueckner and P. Titcomb, "A Lecture Demonstration of Single Photon Interference," *American Journal of Physics* 64 (1996), 184–188.
5. The individual points of light in the figure are small, but they are far larger than individual atoms. Each impact point represents the exposure of a single photographic "grain" of metallic silver, made of about a billion atoms.
6. David Bohm, elaborating on an idea put forward in 1927 by Louis de Broglie, suggested in 1952 an interpretation of quantum physics in which everything is made of tiny Newtonian

particles that are directed by a "pilot wave" that does come through both slits and thus is able to direct the particles to behave according to the appropriate pattern, either single or double slit. However, supporters of this theory have been unable to generalize it to include the effects of special relativity, whereas standard quantum physics was long ago extended to include special relativity. This and other interpretations are discussed in Chapter 10.
7. Paul Dirac, *Quantum Mechanics*, 4th ed. (Oxford: Oxford University Press, 1958).
8. Explanation: With a wider separation between slits, the path difference between the path from slit 1 to any fixed point on the screen and the path from slit 2 to the same fixed point increases, so the number of bright lines between the center of the screen and this fixed point must be larger, so the bright lines must be squeezed together more closely.
9. Frank Wilczek, "The Persistence of Ether," *Physics Today* 52 (1999), 11–13.
10. Richard Feynman, *The Character of Physical Law* (Cambridge, MA: MIT Press, 1965). Feynman also says, in the same lecture, "I think I can safely say that nobody understands quantum mechanics" (p. 129).
11. Daniel Clery, "The Dark Lab," *Science* 347 (2015), 1089–1093.
12. Richard Feynman, Robert Leighton, and Matthew Sands, *The Feynman Lectures on Physics*, vol. 3 (Reading, MA: Addison-Wesley, 1965), 1-1. The emphasis on "only" is Feynman's. However, physicists today might not agree this experiment contains the *only* quantum mystery.
13. The electrons must be selected to have a particular speed (a particular energy).
14. Claus Jonsson, "Elektroneninterferenzen an Mehreren Kunstlich Hergestellten Feinspalten," *Zeitschrift fur Physik* 161 (1961), 454–474. Translated version: "Electron Diffraction at Multiple Slits," *American Journal of Physics* 42 (1974), 4–11.
15. Akira Tonomura, J. Endo, T. Matsuda, T. Kawasaki, and H. Exawa, "Demonstration of Single-Electron Buildup of an Interference Pattern," *American Journal of Physics* 57 (1989), 117–120. The outcome of this experiment had been predicted decades earlier and is similar to the outcome of the Davisson and Germer experiment. Although the Tonomura experiment only confirms well-known results and predictions, it is extremely enlightening pedagogically. Strictly speaking, this is not a double-slit experiment because an "electron biprism" based on electrostatic forces, rather than two slits, is used to separate the matter field into two streams. A true double-slit experiment, using slits rather than the electron biprism and demonstrating the buildup of the interference pattern from individual electron impacts, was performed only in 2013. See Roger Bach, Damian Pope, Sy-Huang Liou, Herman Batelaan, "Controlled Double-Slit Electron Diffraction," *New Journal of Physics* 15 (March 2013), 033018.
16. The honor of the suggestion of the wave nature of electrons goes to Louis de Broglie. He imagined this idea in 1923 and elaborated on its specific quantitative consequences in papers published in 1924 and 1925. The 1927 experiments of Davisson and Germer and of George P. Thomson confirmed de Broglie's predictions. de Broglie received the 1929 Nobel Prize for this suggestion.
17. For example, there are six types of fundamental "quark" fields and six types of fundamental "lepton" fields. Protons and neutrons are made of quarks whereas electrons are one of the six types of leptons. In the low-energy (nonrelativistic) situations that we're focusing on in this book, they are described by Schrödinger's equation.
18. Robert Crease, "The Most Beautiful Experiment in Physics," *Physics World* 15 (2002), 15–17.
19. Louis de Broglie, *Recherches sur la Theorie des Quanta*, PhD diss., Paris University, 1924. The English translation is quoted in Jim Baggott, *The Quantum Story* (Oxford: Oxford University Press, 2011), 38.
20. Steven Weinberg, *Facing Up* (Cambridge, MA: Harvard University Press, 2001), 221.
21. Also see Brooks, *Fields of Color* and P. R. Wallace, *Paradox Lost* (New York: Springer-Verlag, 1996), 16–18.
22. Jean-Louis Basdevant, *Lectures on Quantum Mechanics* (New York: Springer, 2007). "It was Schrödinger who introduced the Greek letter Ψ for the wave function, which has become a tradition" (p. 34).

23. Paul Dirac, "The Quantum Theory of the Emission and Absorption of Radiation," *Proceedings of the Royal Society of London* A114 (1927), 243–265.
24. Chen-Ning Yang, "The Conceptual Origins of Maxwell's Equations and Gauge Theory," *Physics Today* 67 (2014), 45–51.
25. We won't discuss spin in this book. For a good introduction, see Leon Lederman and Christopher Hill, *Quantum Physics for Poets* (Amherst, NY: Prometheus Books, 2011).
26. The reason is that special relativity requires physics to be "symmetric under time reversal": If a certain microscopic process is possible, and if we imagine that a movie is made of this process and then run in reverse, the events viewed in this reversed movie must also be possible in our normal (not reversed) universe. Richard Feynman showed that if we imagine an electron moving backward in time, its physical effects would be precisely the same as the effects of another quantum just like the electron, only carrying a positive charge and moving *forward* in time. Special relativity then requires this "reversed" quantum actually to exist. We call it the *positron*.
27. The ALPHA Collaboration, "Confinement of Antihydrogen for 1,000 Seconds," *Nature Physics* 7 (2011), 558–564.
28. The section is based on Art Hobson, "Teaching Elementary Particle Physics: Parts 1 and 2," *The Physics Teacher* 49 (2011), 12–15, 136–138.
29. Some of the best nontechnical books explaining the standard model are, in alphabetical order by author, Baggott, *The Quantum Story* ; Sean Carroll, *The Particle at the End of the Universe* (London: Penguin Books, 2012); Leon Lederman with Dick Teresi, *The God Particle* (New York: Houghton Mifflin, 1993); Leon Lederman and Christopher Hill, *Beyond the God Particle* (Amherst, NY: Prometheus Books, 2013); Don Lincoln, *The Quantum Frontier* (Baltimore: John Hopkins University Press, 2009); Robert Oerter, *The Theory of Everything* (New York: Pi Press, 2006); Steven Weinberg, *Dreams of a Final Theory* (New York: Pantheon Books, 1992); Frank Wilczek, *The Lightness of Being* (New York: Basic Books, 2008). For a short, qualitative introduction to the standard model, see my conceptual physics textbook Art Hobson, *Physics: Concepts & Connections*, 5th ed. (San Francisco: Pearson Addison-Wesley, 2010), Chapter 17.
30. For further discussion, see Carroll, *The Particle at the End of the Universe* Also, see Wilczek, *The Lightness of Being* 95–96, and Frank Wilczek, "Mass without Mass II," *Physics Today* 53 (2000), 13–14.
31. Interestingly, the two electrons in each pair turn out to be nonlocally entangled (Chapter 9). L. G. Herrmann, F. Portier, P. Roche, A. Levy Yeyati, T. Kontos, and C. Strunk, "Carbon Nanotubes as Cooper-Pair Beam Splitters," *Physical Review Letters* 104 (2010), 026801-1–026801-4. Also see John Cartwright, "Entangled Electrons Do the Splits," Physicsworld.com, October 14, 2009, Physicsworld.com/cws/article/news/2009/oct/14/entangled-electrons-do-the-splits.
32. The electron field is a so-called *fermion field* whereas the field of paired electrons is a *boson field* that has very different properties from a fermion field. Although each fermion must occupy a different quantum state, many bosons can occupy the same quantum state, so paired electrons can "condense" into the same ground state. This condensation is responsible for superconductivity.
33. The 2013 Nobel Prize was awarded to Francois Englert and Peter Higgs for discovering the origin of the intrinsic mass of the fundamental quanta. A third person, Robert Brout, would have shared the prize had he not died in 2011; the prize is not awarded posthumously.
34. Lawrence Krauss and James Dent, "Higgs Seesaw Mechanism as a Source for Dark Energy," *Physical Review Letters* 111 (2013), 061802.
35. Brian Greene, *The Fabric of the Cosmos: Space, Time, and the Texture of Reality* (New York: Alfred A. Knopf, 2004).
36. Brian Greene, *The Elegant Universe: Superstrings, Hidden Dimensions, and the Quest for the Ultimate Theory* (New York: W. W. Norton, 1999).
37. T. S. Eliot, *Little Gidding*, 1942, http://en.wikiquote.org/wiki/Four_Quartets.

38. Gerhard Hegerfeldt, "Particle Localization and the Notion of Einstein Causality," in *Extensions of Quantum Theory 3*, ed. A. Horzela and E. Kapuscik (Montreal: Apeiron, 2001), 9–16; Gerhard Hegerfeldt, "Instantaneous Spreading and Einstein Causality in Quantum Theory," *Annals of Physics* 7 (1998), 716–725; Gerhard Hegerfeldt, "Remark on Causality and Particle Localization," *Physical Review D* 10 (1974), 3320–3321. A similar result is also proved in David Malament, "In Defense of Dogma: Why There Cannot Be a Relativistic Quantum Mechanics of Localizable Particles," in *Perspectives on Quantum Reality*, edited by R. K. Clifton (The Netherlands: Springer Netherlands, 1996), 1–10.
39. For a good book about such questions, see Greene, *The Fabric of the Universe*.
40. Astrid Lambrecht, "The Casimir Effect: A Force from Nothing," *Physics World*, September (2002), 29–32; Peter Milonni, *The Quantum Vacuum: An Introduction to Quantum Electrodynamics* (Boston: Academic Press, 1994), 54–59. The 1% accuracy was not achieved with the parallel-plate geometry, but instead was realized with two conducting cylinders placed a mere 20 billionths of a meter apart. The parallel-plate geometry has been tested, but the experiment is more difficult because it's hard to keep the plates precisely parallel, so there is less agreement between theory and experiment. The parallel-plate geometry is used here because it's easier to explain.
41. The official term for this kind of breaking down, or "analyzing into simple waves," is *Fourier analysis*.
42. William Unruh, "Notes on Black Hole Evaporation," *Physical Review D* 14 (1976), 870–892; Paul Davies, "Scalar Production in Schwarzschild and Rindler Metrics," *Journal of Physics A* 8 (1975), 609–616; John Bell and J. Leinaas, "Electrons as Accelerated Thermometers," *Nuclear Physics B* 212 (1983), 131–150.
43. According to the general theory of relativity, the effects of acceleration and gravity are identical. Thus, the gravitational field just outside a black hole should be sufficiently enormous to create a noticeable Unruh effect. Stephen Hawking, working independently of Unruh, predicted in 1974 that this effect would occur just outside the event horizon (the radius within which objects must, irrevocably, fall down into the black hole) of any black hole, and would cause the famous "Hawking radiation" that he hypothesized should come from black holes. Although Hawking radiation has not yet been observed, it's widely suspected to exist.
44. This chapter is based on Art Hobson, "There Are No Particles, There Are Only Fields," *American Journal of Physics* 81 (2013), 211–223.

Chapter 6

1. Richard Feynman, *The Character of Physical Law* (New York: Modern Library, 1965), Chapter 6.
2. Wikipedia, *Solvay Conference*, 28 October 2015. http://en.wikipedia.org/wiki/Solvay_Conference#cite_note-2.
3. S. Pironio, A. Acin, S. Massar, A. Boyer de la Giroday, D. N. Matsukevich, P. Maunz, S. Olmschenk, D. Hayes, L. Luo, T. A. Manning, and C. Monroe, "Random Numbers Certified by Bell's Theorem," *Nature* 464 (2010), 1021–1046; Cristian S. Calude, Michael J. Dinneen, Monica Dumitrescu, and Karl Svozil, "Experimental Evidence of Quantum Randomness Incomputability," *Physical Review A* 82 (2010), 022102.
4. Maxilmilian Schlosshauer, ed. *Elegance and Enigma: The Quantum Interviews* (Berlin: Springer-Verlag, 2011). See Question 5.
5. Quoted in Henry Stapp, *Mind, Matter and Quantum Mechanics*, 3rd ed. (Berlin: Springer-Verlag, 2009), 171.
6. Matthew Pusey, Jonathan Barrett, and Terry Rudolph, "On the Reality of the Quantum State," *Nature Physics* 8 (2012), 475–478.
7. The "surface" of a nucleus is that surface within which the strong nuclear binding force dominates, and outside of which the strong force falls rapidly to zero.

8. The figure and discussion follow Giancarlo Ghirardi, *Sneaking a Look at God's Cards: Unravelling the Mysteries of Quantum Mechanics* (Princeton, NJ: Princeton University Press, 1997), Section 3.7, "The Uncertainty Principle."
9. In the experiment, the pattern of electron impacts is 12 mm wide; adjacent interference fringes are 1.4 mm apart. Each electron is 12 mm wide.
10. Δx is a measure of the "spread" among a large number of values of x obtained in a long run of separate trials. It's the "standard deviation" of x—the square root of the difference between the average value of x^2 and the square of the average value of x over a long run of trials.
11. This distinction between measurement uncertainties and the minimal size of the quantum is pointed out and analyzed in Lee A. Rozema, Ardavan Darabi, Dylan H. Mahler, Alex Hayat, Yasaman Soudagar, and Aephraim M. Steinberg, "Violation of Heisenberg's Measurement–Disturbance Relationship by Weak Measurements," *Physical Review Letters* 109 (2012), 100404.
12. Atoms with very large electron states are called *Rydberg atoms*. A Rydberg electron is shaped like a thin torus (ring), reminiscent of planetary orbits. However, the electron is spread out around the entire torus.
13. As discussed in Chapter 5, the smallest distance that has any physical meaning is probably the Planck distance, which is about 10^{-35} m, in which case this would be the absolute minimum Δx.
14. Albert Einstein, Boris Podolsky, and Nathan Rosen, "Can Quantum–Mechanical Description of Physical Reality Be Considered Complete?" *Physical Review* 47 (1935), 777–780.
15. Letter from Einstein to Max Born, March 3, 1947; Max Born (ed.), *The Born–Einstein Letters: Correspondence between Albert Einstein and Max and Hedwig Born from 1916 to 1955* (New York: Walker, 1971).
16. It's not clear whether this phrase should be attributed to Richard Feynman or David Mermin. See N. David Mermin, "Could Feynman Have Said This?" *Physics Today* 57 (May 2004), 10–11.
17. Physicists will recognize the statistical aspects of this scheme as being conceptually identical with classical statistical mechanics. The difference is that, instead of using the equations of motion of classical mechanics, Bohm's theory uses the equations of motion of quantum physics.
18. Sheldon Goldstein, "Bohmian Mechanics," *Stanford Encyclopedia of Philosophy*, March 4, 2013. http://plato.stanford.edu/entries/qm-bohm/.
19. Jeffrey Bub, *Interpreting the Quantum World* (Cambridge: Cambridge University Press, 1997), 244–245.
20. Sheldon Goldstein, "Bohmian Mechanics."
21. Hrvoje Nikolic, "Would Bohr Be Born if Bohm Were Born before Born?" *American Journal of Physics* 76 (2008), 143–146.
22. This chapter is based in part on Art Hobson, "Teaching Quantum Uncertainty," *The Physics Teacher* 49 (2011), 434–437, and Art Hobson, *Physics: Concepts & Connections* (San Francisco: Pearson/Addison-Wesley, 5th ed. 2010), Chapters 12 and 13.

Chapter 7

1. I'm assuming here an ideal double-slit experiment in which the dark lines are absolutely dark.
2. This "mathematical object" is a vector in an abstract, complex, many-dimensional, vector space called *Hilbert space*.
3. Ψ can be not only positive or negative, but also "complex" (i.e., it can have an "imaginary number" component). There is nothing mysterious or imaginary about this; it's simply part of the mathematics of complex Hilbert spaces. This complex component is required to take a quantum wave's "phase" into account mathematically. If Ψ is complex, than Ψ^2

must be replaced by the positive real number $|\Psi|^2$. I must omit the complex phase from this book to communicate with nonphysicists; I instead take the wave nature of quanta into account in other, less mathematical, ways.
4. Here's why: The wave speed along the string is the same in both cases, implying the product of the wavelength and the frequency is the same (this is a standard physics result). But we can see from the diagram that the second state's wavelength is half of the first state's wavelength. So the second state's frequency must be twice the first state's frequency, or 10 Hz. Similarly, the third state has three times the frequency of the first state (which is sometimes called the *fundamental frequency*), and the fourth state has four times the frequency of the first state.
5. Before the electric field (the voltage) is switched on, the atoms will pair up into hydrogen molecules. With the field switched on, the molecules break up quickly into individual atoms.
6. We never measure the absolute numerical value of the energy of an atom or any other system. We always measure *relative* values of energy—the difference between two energy values. Thus, only energy *differences* such as $E_2 - E_1$ are physically significant.
7. James Bergquist, Randall Hulet, Wayne Itano, and David Wineland, "Observation of Quantum Jumps in a Single Atom," *Physical Review Letters* 57, (1986), 1699–1702; Warren Nagourney, Jon Sandberg, and Hans Dehmelt, "Shelved Optical Electron Amplifier: Observation of Quantum Jumps," *Physical Review Letters* 56, (1986) 2797–2799; Th. Sauter, W. Neuhauser, R. Blatt, and P. E. Toschek, "Observation of Quantum Jumps," *Physical Review Letters* 57, (1986), 1696–1698; commentary in James Gleick, "Physicists Finally Get to See Quantum Jump with Own Eyes," *New York Times*, Science section, October 21, 1986.
8. Ernest Rutherford, *The Collected Papers of Lord Rutherford of Nelson* (New York: Interscience Publishers, 1962), 305.
9. Quoted in N. David Mermin, "Is the Moon There When Nobody Looks? Reality and the Quantum Theory," *Physics Today* 38, (1985), 38–47; Mermin's reference is Abraham Pais, "Einstein and the Quantum Theory," *Reviews of Modern Physics* 51, (1979), 863–914.
10. Several scientists promoted this notion. Eugene Wigner held this view for many years and then abandoned it (Chapter 10). John Wheeler argues for a central role of the observer; see *The Ghost in the Atom*, ed. P. C. W. Davies and J. R. Brown (Cambridge: Cambridge University Press, 1986), Chapter 4, "John Wheeler," 58–69; this is a book of interviews with quantum physicists.
11. Quoted in Maximilian Schlosshauer, *Elegance and Enigma: The Quantum Interviews* (Berlin: Springer-Verlag, Berlin, 2011). Seventeen physicists and philosophers, all deeply involved in quantum foundations, respond to penetrating questions about the central issues. The quoted statement by Jeffrey Bub appears on page 94.
12. Ibid., Chapter 4, where several experts support the epistemic view. Several others, also quoted in Chapter 4, support the realist view. Also see the epistemic view of Christopher A. Fuchs and Asher Peres, "Quantum Physics Needs No 'Interpretation,'" *Physics Today* 53 (March 2000), 70–71; and responses from several readers: *Physics Today* 53, (September 2000), 11–14.
13. N. David Mermin, "What's Bad about This Habit" *Physics Today* 53 (May 2009), 8–9. This article received extensive commentary in *Physics Today* 53 (September 2009), 10–15.
14. Matthew Pusey, Jonathan Barrett, and Terry Rudolph, "On the Reality of the Quantum State," *Nature Physics* 8 (2012), 475–478.
15. Quoted in Aage Petersen, "The Philosophy of Niels Bohr," *Bulletin of the Atomic Scientists* 19 (1963), 8–14.
16. Carl Sagan, "Episode 12," *Cosmos: A Personal Voyage*, PBS, December 14, 1980.
17. There is an unspoken assumption, however, that the electron does not interact until it arrives at the viewing screen. It does not, for example, impact the partition containing the slit, but instead passes through the slit in Figure 6.3c.

Chapter 8

1. There are some kinds of waves, called *nonlinear waves*, that *do* disturb each other when they pass through each other. Quantum physics is, however, based on linear waves.
2. However, oscillating systems such as vibrating metal springs are an exception; even according to classical Newtonian physics, these systems do obey the superposition principle, i.e. they can be in two or more vibrational states simultaneously.
3. Here is the physics behind this result. As described in the text, the photon takes both paths and interferes with itself at the detectors. Along each of the two paths, 180-degree "phase changes" occur at each reflection *with the exception* of the reflection along path 1 at beam splitter 2; this particular reflection is "internal" (which produces no phase change) because, as you can see from the diagram, the reflective face of beam splitter 2 is on the *right* side of the glass plate (it's on the *left* side of the plate in beam splitter 1). If you now add up the phase changes, you'll find that each path experiences two 180-degree phase changes as they proceed to detector 1, so they arrive "in phase" and interfere constructively. On the other hand, as the paths proceed to detector 2, path 1 experiences *two* 180-degree phase changes whereas path 2 experiences only *one* 180-degree phase change, so the two arrive "out of phase" and interfere destructively.
4. Einstein made this statement at the Fifth Solvay Conference in Brussels in 1927. See Max Jammer, *The Philosophy of Quantum Mechanics* (New York: Wiley, 1974), 114–118.
5. See, for example, H. M. Wiseman and J. M. Gambetta, "Are Dynamical Quantum Jumps Detector Dependent?" *Physical Review Letters* 108 (2012), 220402.
6. As we've seen, Einstein was a prominent holdout against Bohr's views. Louis de Broglie, originator of the hypothesis that material quanta such as electrons have wave properties, along with Erwin Schrödinger, also disagreed importantly with the Copenhagen view. Nevertheless, Bohr's view of quantum physics dominated the physics community from 1927 to at least 1960.
7. Don Howard, "Who Invented the 'Copenhagen Interpretation?' A Study in Mythology," *Philosophy of Science* 71 (2004), 655–668.
8. For a nontechnical historical account of Bohr's view of quantum physics, see Henry J. Folse, *The Philosophy of Niels Bohr: The Framework of Complementarity* (Amsterdam: North-Holland Physics Publishing, 1985).
9. For a good nontechnical discussion of lasers, see Paul Hewitt's *Conceptual Physics*, 12th ed. (San Francisco: Pearson Education, 2015), Chapter 30.
10. J. C. Gallop, *SQUIDs, the Josephson Effects and Superconducting Electronics* (New York: Taylor and Francis Group, 1991), Chapter 1.
11. T. D. Clark, "Macroscopic Quantum Objects," in *Quantum Implications: Essays in Honour of David Bohm*, ed. B. J. Hiley and F. David Peat (London: Routledge & Kegan Paul, 1987), 121–150.
12. The defining difference between bosons and fermions is that bosons are quanta having integer (0, 1, 2, 3, ...) spin, whereas fermions are quanta having half-integer spin (1/2, 3/2, 5/3, ...). This definition is not important for this book, and I'm trying to avoid any discussion of "spin." See any quantum physics textbook for these concepts.
13. Eric Huss, "The Nobel Prize in Physics 2012," Press Release, October 9, 2012.
14. Christopher Monroe, D. M. Meekhof, B. E. King, and David Wineland, "A 'Schrödinger Cat' Superposition State of an Atom," *Science* 272 (1996), 1131–1136.
15. Jonathan Friedman, Vijay Patel, W. Chen, S. K. Tolpygo, and J. E. Lukens, "Quantum Superposition of Distinct Macroscopic States," *Nature* 406 (2000), 43–46; Casper van der Wal, A. C. J. ter Haar, F. K. Wilhelm, R. N. Schouten, C. J. P. M. Harmans, T. P. Orlando, Seth Lloyd, and J. E. Moolj, "Quantum Superposition of Macroscopic Persistent-Current States," *Science* 290 (2000), 773–777.
16. Markus Arndt, Olaf Nairz, Julian Vos-Andreae, Claudia Keller, Gerbrand van der Zouw, and Anton Zeilinger, "Wave–Particle Duality of C_{60} Molecules," *Nature* 401 (1999),

680–682; Bjorn Brezger, Lucia Hackernuller, Stefan Uttenthaler, Julia Petschinka, Markus Arndt, and Anton Zeilinger, "Matter-Wave Interferometer for Large Molecules," *Physical Review Letters* 88 (2002), 100404. There is a good explanation of this experiment in M. Schlosshauer, *Decoherence and the Quantum-to-Classical Transition* (Berlin: Springer-Verlag, 2007), Chapter 6.
17. This is the wavelength at the 100-m per-second speed of the molecules in the experiment. At slower speeds, where quantum effects are more pronounced, the wavelength becomes longer.
18. In 1997, John Clauser suggested using the Talbot–Low effect to observe interference. For more about this development and other interference techniques, see Markus Arndt, "De Broglie's Meter Stick: Making Measurements with Matter Waves," *Physics Today* 67 (2014), 30–36.
19. Zeilinger's experiment uses a "diffraction grating" in place of a double slit. A diffraction grating is like a double slit only with many parallel slits instead of just two, and it produces a more distinct interference pattern. In Talbot–Lau interferometry, each quantum comes only through each *pair* of adjoining slits (pers. comm., Markus Arndt, the University of Vienna team, July 3, 2014); in this respect it operates much like a double-slit experiment.
20. Stefan Gerlich, Sandra Eibenberger, Mathias Tomandi, Stefan Nimmrichter, Klaus Hornberger, Paul Fagan, Jens Tuxen, Marcel Mayor, and Markus Arndt, "Quantum Interference of Large Organic Molecules," *Nature Communications* 2.263 (2011). doi: 10.1038/ncomms1263.
21. A. D. O'Connell, M. Hofheinz, M. Ansmann, Radoslaw Bialczak, M. Lenander, Erik Lucero, M. Neeley, D. Sank, H. Wang, M. Weides, J. Wenner, John Martinis, and A. N. Cleland, "Quantum Ground State and Single-Phonon Control of a Mechanical Resonator," *Nature* 464 (2010), 697–703. *Note*: To make the presentation more understandable to nonscientist readers, I use a notation that is the reverse of the notation in the published article.
22. The diving board is a thin strip of aluminum nitride sandwiched between two thin strips of metallic aluminum connected to opposite terminals of an electrical circuit. Aluminum nitride is a strong "piezoelectric" material—i.e., it expands or contracts in response to an electrical voltage placed across it. Thus, an oscillatory voltage placed across the two aluminum strips causes the aluminum nitride strip to vibrate in the "breathing" mode, alternately expanding and contracting.

Chapter 9

1. Albert Einstein, in a letter to Max Born, dated March 3, 1947. Max Born, *The Born Einstein Letters* (London: Macmillan Press, 1971), page 158.
2. Einstein made this point in his remarks at the Fifth Solvay Physics Conference in Brussels in 1927. For a good account, see Jim Baggott, *The Quantum Story: A History in 40 Moments* (Oxford: Oxford University Press, 2011), 118–122.
3. Werner Heisenberg, *The Physical Principles of the Quantum Theory* (Chicago: University of Chicago Press, 1930), 39. [Reprinted in 1949 by Dover Publications.]
4. Ibid.
5. In more technical language, the two emerging quanta share their phases but not their amplitudes. Thus, a change in the phase of one quantum causes an instantaneous change in the other quantum's phase.
6. Anton Zeilinger, "Teleportation," *APS News*, p. 6, February 2013; R. Ursin, F. Tiefenbacher, T. Schmitt-Manderbach, H. Weier, T. Scheidl, M. Lindenthal, B. Blauensteiner, T. Jennewein, J. Perdigues, P. Trojek, B. Omer, M. Furst, M. Meyenburg, R. Rarity, Z. Sodnik, C. Barbieri, H. Weinfurter, and Anton Zeilinger, "Free Space Distribution of Entanglement and Single Photons over 144 Kilometers," *Nature Physics* 3 (2007), 481–486.

7. Jean-Michel Raimond and Michel Brune, Serge Haroche, "Manipulating Quantum Entanglement with Atoms and Photons in a Cavity," *Reviews of Modern Physics* 73 (2001), 565–582.
8. Chris Monroe, "Experiment Demonstrates Quantum Entanglement between Atoms a Meter Apart," *Physics Today* 60 (2007), 16–18.
9. Ka Chung Lee et al., "Entangling Macroscopic Diamonds at Room Temperature," *Science* 334 (2011), 1253–1256; commentary and summary, 1180, 1213–1214.
10. Brian Julsgaard, Alexander Kozhekin, and Eugene Polzik, "Experimental Long-Lived Entanglement of Two Macroscopic Objects," *Nature* 413 (2001), 400–403; Benjamin Stein, "Entanglement of Macroscopic Objects," *Physics Today* 54 (2001), 9.
11. John Rarity and Paul Tapster, "Experimental Violation of Bell's Inequality Based on Phase and Momentum," *Physical Review Letters* 64 (1990), 2495–2498; Zhe-Yu Ou, Xingquan Zou, Lei Wang, and Leonard Mandel, "Observation of Nonlocal Interference in Separated Photon Channels," *Physical Review Letters* 65 (1990), 321–324. See discussions in Art Hobson, *Physics: Concepts & Connections* (San Francisco: Pearson, 2010); Art Hobson, "Teaching Quantum Nonlocality," *The Physics Teacher* 50 (2012), 270–273.
12. There will be some lack of simultaneity because of quantum uncertainties.
13. More precisely, each photon carries information about the relation between its own path-length difference and the path-length difference of the other photon, which amounts to knowing the relation between x and y.
14. For an especially clear quantum analysis of the RTO experiment, see Michael Horne, Abner Shimony, and Anton Zeilinger, "Introduction to Two-Particle Interferometry," in *Sixty-Two Years of Uncertainty*, ed. A. I. Miller (New York: Plenum Press, 1990), 113–119.
15. For an explicit statement of Bell's locality condition and technical discussion, see Nicolas Brunner, Daniel Cavalcanti, Stefano Pironio, Valerio Scrani, and Stephanie Wehner, "Bell Nonlocality," *Reviews of Modern Physics* 86 (2014), 419–478, especially Section I. Also see Wikipedia, "Bell's Theorem" June 13, 2016. http://en.wikipedia.org/wiki/Bell's_theorem. For a readable statement and derivation of Bell's theorem see Nick Herbert, *Quantum Reality* (New York: Anchor Press, 1985), Chapter 12. John Bell discusses his work on nonlocality in *The Ghost in the Atom*, ed. P. C. W. Davies and J. R. Brown (Cambridge: Cambridge University Press, 1986), Chapter 3, "John Bell," 45–57.
16. Some physicists argue that Bell's result assumes not only the locality principle, but also "realism." However, Bell's article deriving his result makes no such additional assumption: John Bell, "On the Einstein–Podolsky–Rosen Paradox," *Physics* 1 (1964), 195–200; reprinted in John Bell, *Speakable and Unspeakable in Quantum Mechanics* (Cambridge: Cambridge University Press, 1993), 14–21.
17. For further discussion of the significance of Bell's analysis, see Nicolas Brunner, "Quantum Mechanics: Steered Towards Non-locality," *Nature Physics* 6 (2010), 842–843; Alain Aspect, "Bell's Inequality Test: More Ideal Than Ever," *Nature* 398 (1999), 189–190; Herbert, *Quantum Reality* Chapter 12.
18. Stuart Freedman and John Clauser, "Experimental Test of Local Hidden-Variable Theories," *Physical Review Letters* 28 (1972), 938–941. The difference from the RTO experiment is that Freedman and Clauser's experiment entangles and detects photon polarizations rather than photon positions and directions of motion. For this and related work, Clauser was awarded the Wolf Prize in Physics in 2010, sharing the award with Alain Aspect and Anton Zeilinger.
19. For Clauser's motivation and reaction toward his experiment, see Herbert, *Quantum Reality*, Chapter 12.
20. Alain Aspect, Jean Dalibard, and Gerard Roger, "Experimental Test of Bell's Inequality Involving Time-Varying Analyzers," *Physical Review Letters* 49 (1982), 1804–1807. Aspect discusses this work with physicist Paul Davies in Aspect and Davies, *The Ghost in the Atom*, 40–44.
21. This precise wording of this old idea was apparently first formulated by Carl Sagan (writer and host), for the Public Broadcasting Company's television series "Cosmos," Episode

12, broadcast on December 14, 1980. See Wikipedia, "Marcello Truzzi," March 21, 2016, https://en.wikipedia.org/wiki/Marcello_Truzzi#.22Extraordinary_claims.22.
22. Gregor Weihs, Thomas Jennewein, Christoph Simon, Harald Weinfurter, and Anton Zeilinger, "Violation of Bell's Inequality under Strict Einstein Locality Conditions," *Physical Review Letters* 81 (1998), 5039–5043.
23. Marissa Giustina, Alexandra Mech, Sven Ramelow, Bernhard Wittmann, Johannes Kofler, Jorn Beyer, Adriana Lita, Brice Calkins, Thomas Gerrits, Sae Woo Nam, Rupert Ursin, and Anton Zeilinger, "Bell Violation Using Entangled Photons without the Fair-Sampling Assumption," Nature 497, 227 (2013); B. G. Christensen, K. T. McCusker, J. B. Altepeter, B. Calkins, T. Gerrits, A. E. Lita, A. Miller, L. K. Shalm, Y. Zhang, S. W. Nam, N. Brunner, C. C. W. Lim, N. Gisin, and P. G. Kwiat, "Detection-Loophole-Free Test of Quantum Nonlocality, and Applications," *Physical Review Letters* 111, 130406 (2013).
24. R. Ursin, F. Tiefenbacher, T. Schmitt-Manderbach, H. Weier, T. Scheidl, M. Lindenthal, B. Blauensteiner, T. Jennewein, J. Perdigues, P. Trojek, B. Omer, M. Furst, M. Meyenburg, R. Rarity, Z. Sodnik, C. Barbieri, H. Weinfurter, and Anton Zeilinger, "Free Space Distribution of Entanglement and Single Photons over 144 Kilometers." *Nature Physics* 3 (2007), 481–486.
25. B. Hensen et al., "Loophole-free Bell Inequality Violation Using Electron Spins Separated by 1.3 Kilometres," *Nature* 526 (October 29, 2015), 682–686; Marissa Giustina et al., "Significant-Loophole-Free Test of Bell's Theorem with Entangled Photons," *Physical Review Letters* 115 (December 16, 2015), 250401-1–250401-7; Lynden K. Shalm et al. "Strong Loophole-Free Test of Local Realism," *Physical Review Letters* 115 (December 16, 2015).
26. Alain Aspect, "Viewpoint: Closing the Door on Einstein and Bohr's Quantum Debate," *APS Physics* 8 (December 16, 2015), 123.
27. Erwin Schrödinger, "Die gegenwartige Situation in der Qauntenmechanik" [The Present Situation in Quantum Mechanics], *Naturwissenschaften* 23 (1935), 807–812, 823–828, 844–849. Translated by John D. Trimmer at https://www.tuhh.de/rzt/rzt/it/QM/cat.html#note1.
28. Sandu Popescu and Daniel Rohrlich, "Generic Quantum Nonlocality," *Physics Letters A* 166 (1992), 293–297; Nicolas Gisin, "Bell's Inequality Holds for All Non-product States," *Physics Letters A* 154 (1991), 201–202. The technical condition for entanglement is that the N-body quantum state *cannot* be written as a simple product of N single-quantum states. The title of Gisin's article contains an important error: "Holds" should be replaced by "Is Violated."
29. Howard Wiseman and Jay Gambetta, "Are Dynamical Quantum Jumps Detector Dependent?" *Physical Review Letters* 108 (2012), 220402.
30. Ka Chung, M. R. Spraque, B. J. Sussman, J. Nunn, N. K. Langford, X. M. Jin, T. Champion, P. Michelberger, K. F. Reim, D. England, D. Jaksch, I. A. Walmsley, "Entangling Macroscopic Diamonds at Room Temperature." *Science* 334 (2011), 1253–1256.
31. Sze Tan, David Walls, and Matthew Collett, "Nonlocality of a Single Photon," *Physical Review Letters* 66 (1991), 252–255.
32. Egilberto Lombardi, Fabio Sciarrino, Sandu Popescu, and Francesco De Martini, "Teleportation of a Vacuum: One Photon Qubit," *Physical Review Letters* 88 (2002), 070402.
33. Bjorn Hessmo, Pavel Usachev, Hoshang Heydari, and Gunnar Bjork, "Experimental Demonstration of Single Photon Nonlocality," *Physical Review Letters* 92 (2004), 180401-1–180401-4.
34. For further details about single-quantum nonlocality, see Art Hobson, "There Are No Particles, There Are Only Fields," *American Journal of Physics* 81 (2013), 211–223.
35. Jordi Tura, A. B. Sainz, T. Vertesi, M. Lewenstein, and A. Acin, "Detecting Nonlocality in Many-Body Quantum States," *Science* 344 (2014), 1256–1258.
36. For electronic states of atoms, a primary reason for this entanglement is the requirement that the state must be antisymmetric under exchange of any two electrons. This, in turn, implies the Pauli exclusion principle that each electron within a given atom must be in a different quantum state and have a distinct set of quantum numbers. This principle explains many of the regularities seen in the periodic table of the chemical elements.

37. Guillaume Aubrun, Stanislaw Szarek, and Deping Ye, "Phase Transitions for Random States and a Semicircle Law for the Partial Transpose," *Physical Review A* 85 (2002), 030302-1 –030302-4. For a good discussion of this work, see Case Western Reserve University, "Einstein's 'Spooky Action' Common in Large Quantum Systems," *ScienceDaily*, May 28, 2013. www.sciencedaily.com/releases/2013/05/130528122433.htm. Brian Dodson, "Quantum Entanglement Isn't Only Spooky, You Can't Avoid It," June 10, 2013. http://www.gizmag.com/quantum-entanglement-ubiquitous/27836/.
38. However, this statement of "near certainty" must be qualified. The statement assumes there is not some principle of nature that selects the nonentangled states over the entangled states. The analysis simply counts the entangled and nonentangled states, and assumes the probability of each type actually occurring in nature is proportional to the numbers of states of each type.
39. M. Schlosshauer, *Elegance and Enigma: The Quantum Interviews* (Berlin: Springer-Verlag, 2011). One of the open-ended questions asked is: What do the experimentally observed violations of Bell's inequality tell us about nature? Of the 17 experts polled, seven replied it implies nature is nonlocal, five replied it only implies quantum physics is not a theory about the real world and doesn't imply nature herself is nonlocal, and five gave other replies.
40. Ibid., question 8.
41. Bell, "On the Einstein–Podolsky–Rosen Paradox." The Einstein–Podolsky–Rosen paradox, and the concept of an "element of reality," are presented in Albert Einstein, Boris Podolsky, and Nathan Rosen, "Can Quantum-Mechanical Description of Physical Reality Be Considered Complete?" *Physical Review* 47 (1935), 777.
42. Brian Greene, *The Fabric of the Cosmos* (New York: Alfred A. Knopf, 2004), this statement appears in Chapter 4, page 117.
43. As an important example, Schrödinger's equation is the nonrelativistic limit of the quantized Dirac equation of quantum field theory. For a proof, see E. G. Harris, *A Pedestrian Approach to Quantum Field Theory* (New York: Wiley-Interscience, 1972), 50–53.
44. This experiment is analyzed beautifully in Maximilian Schlosshauer, *Decoherence and the Quantum-to-Classical Transition* (Berlin: Springer-Verlag, 2007), 63–65.
45. What's needed is a device that correlates with a quantum coming through one or the other slit, without altering it. That is, if the quantum is in the slit 1 state, the quantum causes the detector to indicate "slit 1" while the quantum remains in the slit 1 state. And if the quantum is in the slit 2 state, the quantum causes the detector to indicate slit 2 while the quantum remains in the slit 2 state. Such a detector does not disturb the quantum, regardless of which slit it comes through. For further discussion, see Schlosshauer, *Decoherence and the Quantum-to-Classical Transition*.
46. We know that |Q1> |ready> evolves into |Q1> |D1>, and that |Q2> |ready> evolves into |Q2> |D2>. The linear dynamics (e.g., the Schrödinger equation) of quantum physics then implies that (|Q1> + |Q2>) |ready> evolves into |Q1> |D1> + |Q2> |D2>.
47. Because of the measurement problem, I use the term *mixture* carefully. I define a mixture in terms of properties, not in terms of states, because the "measurement state" or "Schrödinger cat state" is a local mixture but a global superposition. A *local mixture* is a situation in which a local observer observes the quantum to have a mixture of definite properties (such as either alive or dead) whereas a global observer observes the quantum to participate in a coherent superposition of correlations between the quantum and some other quantum.

Chapter 10

1. Wojciech Zurek, at Los Alamos National Laboratory, and his colleagues have studied the manner in which quantum phenomena bring our familiar classical world into existence. His work emphasizes the fundamental role of measurement-related processes such as decoherence in creating the world we see around us—a macroscopic world that is, for

the most part, described correctly by classical physics. See Maximilian Schlosshauer, *Decoherence and the Quantum to Classical Transition* (Berlin: Springer Verlag, 2007).
2. There are two recent polls of professional opinion about the measurement problem. The first is detailed in Maximilian Schlosshauer, Johannes Kofler, and Anton Zeilinger, "A Snapshot of Foundational Attitudes toward Quantum Mechanics," *Studies in History and Philosophy of Modern Physics* 44 (2013), 222–230. Thirty-three participants in a quantum foundations conference were asked multiple-choice questions, with multiple answers permitted. Question 5 asked: The measurement problem is (a) a pseudo-problem (27% answered yes), (b) solved by decoherence (15% answered yes), (c) solved or will be solved in another way (39% answered yes), (d) a severe difficulty threatening quantum mechanics (24% answered yes), (e) none of the above (27% answered yes). The second poll is described by Maximilian Schlosshauer, *Elegance and Enigma: The Quantum Interviews* (Berlin: Springer Verlag, 2011). Seventeen recognized quantum foundations experts were asked to provide one-page written responses to each of 17 questions. Question 7 asked whether the measurement problem is a serious roadblock or a dissolvable pseudo-issue. Nine of the 17 said it's an unsolved roadblock, 3 of these 9 noted that decoherence doesn't help solve it, and 1 noted that decoherence solves it in part but that the core of problem remains. Six of the 17 said it's a pseudo-issue and not a roadblock because quantum physics is not about reality and that it's only about knowledge and information; 2 of these 6 noted that new physics offers a possible solution. Of the remaining two experts, one said decoherence solves the measurement problem and the other said the many-worlds interpretation solves the measurement problem.
3. As the preceding note shows, some experts opine that decoherence solves the measurement problem, but many more disagree with this conclusion. The reason decoherence does not solve the measurement problem is that it does not solve the problem of outcomes. For a clear and convincing analysis of this point, see Stephen Adler, "Why Decoherence Has Not Solved the Measurement Problem: A Response to P. E. Anderson," *Studies in History and Philosophy of Modern Physics* 34 (2003), 135–142. Also see Armen Allahverdyan, Roger Balian, and Theo. Nieuwenhuizen, "Understanding Quantum Measurement from the Solution of Dynamical Models," *Physics Reports* 525 (2013), 1–201. The detailed analysis of quantum measurements by Allahverdyan and colleagues concludes (Section 2.7) that decoherence does not solve the measurement problem. For further discussion, see Chapter 11 in this book.
4. Art Hobson, "Two-Photon Interferometry and Quantum State Collapse," *Physical Review A* 88 (2013), 022105.
5. Erwin Schrödinger, "Die gegenwartige Situation in der Qauntenmechanik" [The Present Situation in Quantum Mechanics], *Naturwissenschaften* 23 (1935), 807–812, 823–828, 844–849. Translated by John D. Trimmer at https://www.tuhh.de/rzt/rzt/it/QM/cat.html#note1.
6. Schlosshauer, *Elegance and Enigma*, 142.
7. Quantum expert Rudolf Peierls explains the Copenhagen interpretation in *The Ghost in the Atom*, ed. P. C. W. Davies and J. R. Brown (Cambridge: Cambridge University Press, 1986), Chapter 5 "Rudolf Peierls," 70–82. For brief, authoritative accounts, see Jan Faye, "Copenhagen Interpretation of Quantum Mechanics," in *Stanford Encyclopedia of Philosophy*, 2014, http://plato.stanford.edu/archives/fall2014/entries/qm-copenhagen/; Maximilian Schlosshauer, *Decoherence* (Berlin: Springer-Verlag, 2007), 330–336.
8. Mara Beller, "The Rhetoric of Antirealism and the Copenhagen Spirit," *Philosophy of Science* 63 (1966), 183–204.
9. Louisa Gilder, *The Age of Entanglement: When Quantum Physics Was Reborn* (New York: Alfred A. Knopf, 2008), 5.
10. Physicist and science popularizer Paul Davies interviews David Bohm in *The Ghost in the Atom*, ed. P. C. W. Davies and J. R. Brown (Cambridge: Cambridge University Press, 1986), Chapter 8 "David Bohm," 118–134. For good discussions of the pilot–wave model, see Schlosshauer, *Decoherence*, 354–357; Sheldon Goldstein, "Bohmian Mechanics," in *The*

Stanford Encyclopedia of Philosophy, Spring ed., 2013, http://plato.stanford.edu/archives/spr2013/entries/qm-bohm/; the defining papers are D. Bohm, "A Suggested Interpretation of the Quantum Theory in Terms of 'Hidden' Variables, Parts I and II," *Physical Review* 85 (1952), 166–179, 190–193.

11. John Bell, *Speakable and Unspeakable in Quantum Mechanics* (Cambridge: Cambridge University Press, 1993), Chapter 14. See also John Bell, "Against 'Measurement,'" *Physics World* 1 (1990), 33–40.

12. Ghirardi describes the model in conceptual but occasionally mathematical language in his book for scientists and nonscientists: Giancarlo Ghirardi, *Sneaking a Look at God's Cards* (Princeton, NJ: Princeton University Press, 2005), Chapters 16 and 17. For further discussion, see Schlosshauer, *Decoherence*, 347–349; Bell, *Speakable and Unspeakable in Quantum Mechanics*, Chapter 22. For an extensive review, see Angelo Bassi and Giancarlo Ghirardi, "Dynamical Reduction Models," *Physics Reports* 379 (2003), 257–426. The defining paper is Giancarlo Ghirardi, A. Rimini, and T. Weber, "Unified Dynamics for Microscopic and Macroscopic Systems," *Physical Review D* 34 (1986), 470–491.

13. It's worth point out that, according to Hegerfeldt's analysis (Chapter 5), such an absolutely localized quantum state is impossible because the state would spread instantly to infinite distances, contradicting relativistic causality.

14. Ghirardi, *Sneaking a Look at God's Cards*, 419. Also see Giancarlo Ghirardi, "Collapse Theories," *Stanford Encyclopedia of Philosophy*, 2016, http://plato.stanford.edu/entries/qm-collapse/, which discusses some partly-successful relativistic GRW models.

15. English translation: John von Neumann, *The Mathematical Foundations of Quantum Mechanics* (Princeton, NJ: Princeton University Press, 1955). Analysis of measurement appears in von Neumann's book in Chapters 5 and 6, and the consistency problem referred to earlier appears in Chapter 6.

16. von Neumann, *The Mathematical Foundations of Quantum Mechanics*, 349.

17. Eugene Wigner, "The Problem of Measurement," *American Journal of Physics* 31 (1963), 6–15. Arthur Fine, "Insolubility of the Quantum Measurement Problem," *Physical Review D* 2 (1970), 2783–2787.

18. von Neumann, *The Mathematical Foundations of Quantum Mechanics*, 420.

19. Eugene Wigner, "Remarks on the Mind–Body Question," in *The Scientist Speculates*, ed. I. J. Good (London: Heinemann, 1961), 284–302; reprinted in Eugene Wigner, *Symmetries and Reflections* (Bloomington, IN: Indiana University Press, 1967), 171–184. See also Schlosshauer, *Decoherence and the Quantum-to-Classical Transition*, 364–365.

20. E. Manousakis, "Founding Quantum Theory on the Basis of Consciousness," *Foundations of Physics* 36 (2006), 795–838; R. A. Mould, "Quantum Brain States," *Foundations of Physics* 33 (2003), 591–612. Despite the lack of scientific consensus behind this interpretation, a widely used liberal arts physics textbook for nonscience students—Bruce Rosenblum and Fred Kuttner, *Quantum Enigma: Physics Encounters Consciousness* (Oxford: Oxford University Press, 2006)—leans strongly toward the consciousness interpretation. For example, Chapter 15, titled "The Mystery of Consciousness," states on page 168: "Our concern is with the consciousness central to the quantum enigma—the awareness that appears to affect physical phenomena."

21. Wojciech Zurek, "Decoherence and the Transition from Quantum to Classical," *Physics Today* 44 (October, 1991), 36–44. Zurek's comment (on page 37) was directed toward the many-worlds interpretation, but is applicable to the consciousness interpretation as well.

22. Roger Carpenter and Andrew J. Anderson, "The Death of Schrödinger's Cat and of Consciousness-Based Quantum Wave-Function Collapse," *Annales de la Fondation Louis de Broglie* 31 (2006), 45–52.

23. Quantum foundations expert David Deutsch argues for many worlds in *The Ghost in the Atom*, ed. P. C. W. Davies and J. R. Brown (Cambridge: Cambridge University Press, 1986), Chapter 6 "David Deutsch," 83–105. Many worlds is explained by one of its originators in: Bryce DeWitt, "Quantum Mechanics and Reality: Could the Solution to the Dilemma of

Indeterminism Be a Universe in Which All Possible Outcomes of an Experiment Actually Occur?" *Physics Today* 23 (1970), 30–40. See also the follow-up letters in *Physics Today* 24 (1971), 38–44. For an overview of many worlds and of several many-worlds variants, see Jeffrey Barrett, "Everett's Relative-State Formulation of Quantum Mechanics," in *Stanford Encyclopedia of Philosophy*, 2010, http://plato.stanford.edu/entries/qm-everett/.

24. Guth has written a nontechnical account, for nonscientists and scientists: Alan H. Guth, *The Inflationary Universe: The Quest for a New Theory of Cosmic Origins* (Reading, MA: Addison-Wesley, 1997).

25. Features of the universe explained by inflation include why the universe is the same in all directions, why the cosmic microwave background radiation is distributed evenly, why the universe is flat rather than curved over large distances, and why no magnetic monopoles (magnets with only a north, or only a south, pole) exist.

26. Support for many worlds was stronger two decades ago than it appears to be in 2016. According to physicist Max Tegmark, a "highly unscientific" poll of 90 physicists taken at a 1999 conference on quantum computation resulted in 30 choosing many worlds or something similar. See Max Tegmark and John Archibald Wheeler, "100 Years of Quantum Mysteries," *Scientific American* February (2001), 72–79. But Schlosshauer, *Elegance and Enigma* reported that, of the 17 quantum foundations experts interviewed, only 1 listed many worlds as his favorite, 2 listed it as one of their favorites, and the remaining 14 did not accept many worlds.

27. Bryce DeWitt, "Quantum Physics and Reality," *Physics Today* 23 (September 1970), 30–40.

28. David Deutsch, "Three Experimental Implications of the Everett Interpretation," in *Quantum Concepts of Space and Time*, ed. R. Penrose and C. Isham (Oxford: Clarendon Press, 1986) 204–214; Lev Vaidman, "Many-Worlds Interpretation of Quantum Mechanics," *The Stanford Encyclopedia of Philosophy*, January 17, 2014, http://plato.stanford.edu/entries/qm-manyworlds/.

29. This argument was first suggested in Art Hobson, "Two-Photon Interferometry and Quantum State Collapse," *Physical Review A* 88 (2013), 022105. A complete technical presentation of this argument is presented in Art Hobson, "Resolving Schrödinger's cat," (2016) posted at http://arxiv.org/abs/1607.01298. This "preprint" includes a mathematical proof that the entangled "Schrödinger's cat" superposition is a non-paradoxical coherent superposition of correlations rather than a paradoxical superposition of macroscopic states.

30. Josef Jauch, *Foundations of Quantum Mechanics* (Reading, MA: Addison-Wesley, 1968), 183–191.

31. Marlan Scully, R. Shea, and J. D. McCullen, "State Reduction in Quantum Mechanics: A Calculational Example," *Physics Reports* 43 (1978), 485–498; Marlan Scully, B. G. Englert, and Julian Schwinger, "Spin Coherence and Humpty-Dumpty: III. The Effects of Observation," *Physical Review A* 40 (1989), 1775–1784; Stefan Rinner and Ernst Werner, "On the Role of Entanglement in Schrödinger's Cat Paradox," *Central European Journal of Physics* 6 (2008), 178–183 offer a particularly straightforward and compelling version of this argument; Hobson, "Two-Photon Interferometry and Quantum State Collapse." Also, the proposed solution is loosely related to the so-called *modal interpretation* of quantum physics.

32. Albert Einstein, Boris Podolsky, and Nathan Rosen, "Can Quantum–Mechanical Description of Physical Reality Be Considered Complete?" *Physical Review* 47 (1935), 777–780.

33. If one assumes that Schrödinger's cat and nucleus are in the measurement state, and assumes the cat is in a superposition of the form a|cat alive> + b|cat dead>, one can show easily that either a = 0 or b = 0, implying the cat is not in a superposition. See, for example, Hobson, "Two-Photon Interferometry and Quantum state Collapse."

34. The local states of the electron are obtained by "tracing" the full measurement state over the states of the detector alone (i.e., by "tracing out" the detector). It's well known (see, for example, Schlosshauer, *Decoherence and the Quantum-to-Classical Transition*, 44–46)

that *this gives the correct predictions for the electron (i.e., the same predictions as are given by the full quantum state for the composite system)*. Similar comments apply to the local state corresponding to the detector alone. In other words, *an observer of the electron observes the local state.*

35. The local state of a subsystem is known more commonly known as the *reduced state* of that subsystem.
36. Jauch, *Foundations of Quantum Mechanics*; Scully, Shea, and McCullen, "State Reduction in Quantum Mechanics"; Scully, Englert, and Schwinger, "Spin Coherence and Humpty-Dumpty"; Rinner and Werner, "On the Role of Entanglement in Schrödinger's Cat Paradox"; Hobson, "Two-Photon Interferometry and Quantum State Collapse."
37. Long ago, I had the good fortune to see this routine, live in London, in a 1963 show called *Beyond the Fringe*.
38. Jauch, *Foundations of Quantum Mechanics*, 184–191.
39. This resolution does, however, require that the standard eigenvalue–eigenstate link (namely, the notion that a quantum system has a *definite* value of some observable property if and only if a measurement of that property is *certain* to yield that particular value) be revised. Because, following the measurement, each electron is in a mixture of definite but *unpredictable* (or random) outcomes, it cannot be true that the electrons are in definite states only if that state is predictable. The words "and only if" must be omitted from the standard eigenvalue–eigenstate link. This revision is, however, trivial in the sense that it neither alters nor conflicts with any other quantum principle.
40. Hobson, "Two-Photon Interferometry and Quantum State Collapse." For a more complete version, see Hobson, "Resolving Schrödinger's cat."
41. N. Gisin, "Bell's Inequality Holds for All Nonproduct States," *Physics Letters A* 154 (1991), 201–202. (Note: The title is erroneous; it should read "Bell's Inequality Is Violated for All Nonproduct States.")
42. X. Y. Zou, Lei Wang, and Leonard Mandel, "Induced Coherence and Indistinguishability in Optical Interference," *Physical Review Letters* 67 (1991), 318–321.
43. Nobel Prize winner Serge Haroche advises us that a system should be considered single whenever the binding between its parts is much stronger than the interactions involved in its dynamics. By this criterion, a system and its detector must be considered separate subsystems. Serge Haroche and Jean-Michel Raimond, *Exploring the Quantum: Atoms, Cavities and Photons* (Oxford: Oxford University Press, 2006), 52.
44. For mathematical proof of this claim, see Art Hobson, "Resolving Schrödinger's Cat."
45. L. E. Ballentine and J. P. Jarrett, "Bell's Theorem: Does Quantum Mechanics Contradict Relativity?" *American Journal of Physics* 55 (1987), 696–701.
46. John Bell discovered "Bell's inequality"—the condition for locality—in 1964. He also showed that, in some circumstances, entangled quantum states violate this condition. John Bell, "On the Einstein–Podolsky–Rosen Paradox," *Physics* 1 (1964), 195–200. This paper is reprinted along with 19 foundational papers by Bell, *Speakable and Unspeakable in Quantum Mechanics*.
47. Stuart Freedman and John Clauser, "Experimental Test of Local Hidden-Variable Theories," *Physical Review Letters* 28 (1972), 938–941; Alain Aspect, Jean Dalibard, and Gerard Roger, "Experimental Test of Bell's Inequality Involving Time-Varying Analyzers," *Physical Review Letters* 49 (1982), 1804–1807.
48. The history of entanglement and nonlocality is beautifully recounted in Gilder, *The Age of Entanglement*.
49. Jauch, Foundations of Quantum Mechanics, 184–191.
50. Josef Jauch, "The Problem of Measurement in Quantum Mechanics," in *The Physicist's Conception of Nature*, ed. J. Mehra (Dordrecht: D. Reidel, 1973), 684–686.
51. The "modal interpretation," proposed in 1972 by Bas C. von Fraassen, bears some resemblance to the local state solution. However, it is motivated by purely mathematical considerations concerning the structure of states (the "Schmidt decomposition") in the Hilbert space of the composite system rather than by physical considerations. And it

simply postulates the distinction between the local and global states rather than deriving it from physical principles such as Einstein causality, nonlocality, and experimentally observed mixtures. See O. Lombardi and Dennis Dieks, "Modal Interpretations of Quantum Mechanics," *Stanford Encyclopedia of Philosophy*, 2012, http://plato.stanford.edu/entries/qm-modal/, Dennis Dieks, "Modal Interpretation of Quantum Mechanics, Measurements, and Macroscopic Behavior," *Physical Review A* 49 (1994), 2290–2300; Dennis Dieks, "Quantum Mechanics: An Intelligible Description of Objective Reality?" *Foundations of Physics* 35 (2005), 399–416. For a comprehensive overview of the measurement problem as of 2004, including many interpretations including the modal interpretation, see Maximilian Schlosshauer, "Decoherence, the Measurement Problem, and Interpretations of Quantum Mechanics," *Reviews of Modern Physics* 76 (2004), 1267–1305.
52. Schlosshauer, *Decoherence and the Quantum-to-Classical Transition*, 48–49.
53. A 1/36 probability of 2, a 2/36 probability of 3, and so on. This is because there are 6 × 6 = 36 possible ways the two dice could land. Of these 36 ways, only one way results in a total throw of 2, two ways result in a total throw of 3, and so on.
54. Schlosshauer, *Decoherence and the Quantum-to-Classical Transition*, 41–42.
55. I'm assuming here, for simplicity, that Schrödinger's "radioactive substance" comprises just a single radioactive atom with the desired half-life.
56. The ambiguity occurs only when the corresponding "observable" has equal "eigenvalues"—a case that is dubbed *degenerate*. These eigenvalues are the probabilities of the two outcomes—namely, nondecayed and decayed. So degeneracy occurs only at a single instant: the time at which these two probabilities are equal (i.e., at the half-life).
57. In Hilbert space, the basis vectors of the density operators for the two local systems—the nucleus and the cat—are ambiguous whenever the probabilities of the local states are 50-50. This is because of the "degeneracy" (i.e., equality) of the density operator's two eigenvalues, both of which are 1/2 at this instant. But this ambiguity is of no physical consequence. What is of physical consequence are the *correlations between* the local states of the nucleus and the local states of the cat. The nondecayed nucleus is correlated with the alive cat; the decayed nucleus is correlated with the dead cat. An experimenter designs a detector to correlate with the nucleus in this manner, regardless of which local basis set is chosen to describe the detector and nucleus mathematically. Briefly, it's the physics that's important, not the basis set in which theorists choose to represent the physics.

Chapter 11

1. The best-known translation of this 12th-century poem is by Edward FitzGerald. See *The Rubaiyat of Omar Khayyam*, the First Version, translated by Edward Fitzgerald and illustrated by Edmund J. Sullivan, October 22, 2014. https://archive.org/details/TheRubaiyatOfOmarKhayyam-FirstVersion-Illustrated.
2. In fact, the RTO experiment demonstrates this. The entangled measurement state of the two photons is a single, unmixed, pure state, with zero entropy. So entanglement, although it creates definite outcomes, does not create entropy.
3. Physicists should not be surprised by this because it's well known that an identical problem crops up in classical physics. Newton's laws, applied to the particles presumed to underlie any classical mechanical system, predict an entirely reversible evolution with no sign of the second law. This "problem of irreversibility" was much-discussed during the 19th and early 20th century. It was resolved by Ludwig Boltzmann and others who argued that the analysis must include statistical considerations. One principle that emerged was that, for the vast majority of plausible "initial conditions," classical systems move in such a way that entropy increases on average. Another principle that emerged more recently is that the universal initial condition is the Big Bang, and that this was surely an extremely low-entropy event.
4. The 1922 experiment used a detection screen as shown in Figure 11.1, but did not bend the two streams back into a single stream.

5. In conventional quantum terminology, this is a "spin state" with spin directed into or out of the page, but in this book we simplify matters by avoiding the topic of spin. Similarly, the states designated "plus" and "minus" are "spin up" and "spin down" states.
6. Quoted in Jeremy Bernstein, *Quantum Profiles* (Princeton, NJ: Princeton University Press, Princeton, 1991), 131.
7. Maximilian Schlosshauer, *Decoherence and the Quantum to Classical Transition* (Berlin: Springer Verlag, 2007), Chapters 2 and 3.
8. Ibid., 135.
9. Wojciech Zurek, "Pointer Basis of Quantum Apparatus: Into What Mixture Does the Wave Packet Collapse?" *Physical Review D* 24 (1981), 1516–1525; Wojciech Zurek, "Environment-Induced Superselection Rules," *Physical Review D* 26 (1982), 1862–1880.
10. Wojciech Zurek, "Decoherence and the Transition from Quantum to Classical," *Physics Today* 44 (1991), 36–44; Wojciech Zurek, "Preferred Observables, Predictability, Classicality, and the Environment-Induced Decoherence," in *The Physical Origins of Time Asymmetry*, ed. J. J. Halliwell, J. Perez-Mercader, and Wojciech. Zurek (Cambridge: Cambridge University Press, 1994), Chapter 11, 175–212; Wojciech Zurek, "Decoherence, Einselection, and the Quantum Origins of the Classical," *Reviews of Modern Physics* 75 (2003), 715–771; W. Zurek, "Quantum Origin of Quantum Jumps," *Physical Review A* 76 (2007), 052110; Wojciech Zurek, "Quantum Darwinism," *Nature Physics* 5 (2009), 181–188.
11. Schlosshauer, *Decoherence and the Quantum to Classical Transition*, Chapters 3 and 6 are full of wonderful experimental examples.
12. L. Hackermuller, K. Hornberger, B. Brezger, A. Zeilinger, and M. Arndt, "Decoherence of Matter Waves by Thermal Emission of Radiation," *Nature* 427 (2004), 711–714.
13. The entire foregoing section owes a considerable debt to Schlosshauer, *Decoherence and the Quantum to Classical Transition*, 65–70.
14. Zurek, "Pointer Basis of Quantum Apparatus"; Zurek, "Environment-Induced Superselection Rules"; E. Joos and H. D. Zeh, "The Emergence of Classical Properties through Interaction with the Environment," *Zeitschrift fur Physik B* 59 (1985), 223–243.
15. John Bell, "Against Measurement," *Physics World*, August (1990), 33–40.
16. Stephen L. Adler, "Why Decoherence Has Not Solved the Measurement Problem: A Response to P. W. Anderson," *Studies in History and Philosophy of Modern Physics* 34 (2003), 135–142.
17. P. W. Anderson, "Science: A Dappled World or a Seamless Web," *Studies in History and Philosophy of Modern Physics* 32 (2001), 487–494.
18. Adler begins with a simple quantum system such as an atom, in a superposition state, and shows how environmental decoherence effectively "measures" the quantum. As he shows, the quantum entangles with the environment to eventually (following a series of interactions) produce the measurement state. Adler then points out that, despite decoherence, this measurement state still exhibits the problem of definite outcomes. Thus, he is quite correct in stating that decoherence alone does not resolve the measurement problem. However, we saw in Chapter 10 that the local state analysis does resolve the problem of definite outcomes; Schrödinger's cat is actually in a mixture of either alive or dead, not a superposition of both alive and dead.
19. Maximilian Schlosshauer, "Decoherence, the Measurement Problem, and Interpretations of Quantum Mechanics," *Review of Modern Physics* 78 (2004), 1267–1305; quote, on p 1302.
20. The electron entangles with the quantum vacuum at every one of the other atomic-size regions. Entanglement of just this sort has been verified experimentally and constitutes one more argument for the reality of the collapse process and its consistency with standard quantum physics.
21. Michel Brune, E. Hagley, J. Dreyer, X. Maitre, A. Maali, C. Wunderlich, Jean-Michel Raimond, Serge Haroche, "Observing the Progressive Decoherence of the 'Meter' in a Quantum Measurement." *Physical Review Letters* 77 (1996), 4887–4890. Haroche discusses this experiment in his Nobel speech "Controlling Photons in a Box and Exploring

the Quantum to Classical Boundary," Nobel Lecture, delivered December 8, published July 12, 2013, in *Reviews of Modern Physics* 85, 1083–1102. There are also good descriptions in Serge Haroche, "Entanglement, Decoherence and the Quantum/Classical Boundary," *Physics Today* 51 (1998), 36–42; Phillip Yam, "Bringing Schrödinger's Cat to Life," *Scientific American* June (1997), 124–129; Gary Taubes, "Atomic Mouse Probes the Lifetime of a Quantum Cat," *Science* 274 (1996), 1615.
22. If the detector states are sufficiently different so they *never* make this error, the different states are said to be *orthogonal* or *perpendicular* to each other.
23. Quoted in Ruth Moore, *Niels Bohr: The Man, His Science, and the World They Changed* (New York: Alfred A. Knopf, 1966), 196. Emphasis in Bohr's original.

GLOSSARY

acceleration a change in velocity; a change in either the speed or direction of motion
alpha decay see *radioactive decay*
alpha quantum see *radioactive decay*
analog filling a continuous range of values; the opposite of "digital"
antiparticle a quantum is the "antiparticle" of another quantum if it has the mirror-image properties of the other quantum. As an example, the electron and positron are antiparticles of each other. There are also antiprotons, antineutrons, antiquarks, and so on. See *positron* and *electron–positron annihilation*.
Aspect's experiment a test that showed some experiments violate Bell's locality condition. Thus, nature herself is nonlocal, regardless of the truth or falsehood of quantum physics. Aspect's results agree with the predictions of quantum physics. In his experiment, two entangled photons, A and B, fly in opposite directions to two separate photon detectors. Each detector is equipped with a device that can change quickly the quantum state of that detector's photon. The experiment showed that when one photon's state is changed, *both* photons change their states. Furthermore, the changes are instantaneous or, at any rate, the connection happens faster than light speed. That is, the two photons change their states within a time that is shorter than the time needed for light to connect the two photons.
atom smallest constituent of a chemical element. For example, an iron atom is the smallest piece of iron that still has the chemical properties associated with iron. Atoms are made of protons and neutrons in the nucleus, with electrons on the outside. Because protons and neutrons are made of quarks, atoms are composites of quarks and electrons.
atomic number the number of protons in an atom's nucleus
atomic state one of the possible quantum states that Schrödinger's equation prescribes for the electrons in an atom. Among these possible states are the "energy states" of the atom, states with a particular "allowed" energy. Transitions between these energy states create photons with frequencies that are predictable from the energies of the initial and final states of the atom. See *atom* and *Schrödinger's equation*.
a-tomos fundamental particlelike objects that ancient Greeks such as Democritus believed everything was made of. "A-tomos" means "not able to be cut" or "without parts." Today, the related word *atom* has a somewhat different meaning. Compare *atom*.
basis ambiguity a mathematical objection to the local state resolution of the problem of definite outcomes. The argument is that, if the measurement-state superposition happens to specify equal probabilities for finding a quantum and its detection apparatus in either of two states (such as "decayed nucleus" and "undecayed nucleus"), the mathematics becomes ambiguous because the theory does not specify a specific "basis set" for either system. However, this ambiguity is physically irrelevant, because only the correlations between the apparatus and the observed quantum are physically important, and these correlations are unambiguous and independent of the basis set.

beam splitter a glass plate that partly transmits and partly reflects light beams that strike it.

Bell's locality condition a mathematical condition that entanglement experiments must obey if the experiment obeys the locality principle. As shown by John Bell, locality implies a certain numerical limitation on the degree of correlation between two observers' outcomes. Bell's result was derived from purely probabilistic considerations, having nothing to do with quantum physics, but it has significant implications for quantum physics because many quantum predictions violate Bell's locality condition. Bell's experiments, and quantum theory, show that entangled quanta violate Bell's locality condition. See *locality principle, RTO experiment, Aspect's experiment*.

beta decay see *radioactive decay*

beta quantum see *radioactive decay*

Big Bang quantum event occurring 13.8 billion years ago that created our universe

black hole a region of space from which, because of the strong inward-pulling force of gravity, nothing can escape. Black holes exist in the form of burned-out, collapsed stars, and massive black holes at the centers of galaxies.

boson one of two broad types of quanta. The second type is called a *fermion*. The primary difference between them is that bosons prefer to gather together in a single quantum state whereas fermions prefer isolated solitude. Photons, for example, are bosons, which is why lasers are possible. But, individual electrons are fermions, so they can't cooperate in the way that a laser's photons cooperate.

branch If a quantum is in a superposition of two or more quantum states, each of these individual states is said to be a "branch" of the superposition. If a quantum is in a mixture of two or more properties, each of these properties is a "branch" of the mixture.

Casimir force an attractive force between two electrically neutral flat metal plates surrounded only by empty space. Quantum vacuum fluctuations cause this force, that increases rapidly as the distance between the plates decreases. This phenomenon has been verified experimentally.

charge see *electrical charge*

chemical element see *element*

chemical energy energy resulting from chemical reactions; see *energy*

classical field a physical field that does not exhibit quantum effects because it is not quantized into energy bundles. A field as conceived by Faraday and Maxwell before quantum physics. It is thought that all fundamental physical fields are quantized.

classical physics the principles of physics as understood before quantum physics (1900); the physics of Newton, Faraday, and Maxwell

coherent A superposed quantum is "coherent" in the sense that it is impossible, even in principal, to associate specific experimental outcomes with one or the other branch of the superposition. For example, in the double-slit experiment with no which-slit detector, each quantum is in a coherent superposition of coming through both slits; individual spots on the detection screen cannot be associated with one or the other slit. Compare *incoherent, decoherence, mixture*.

collapse When an electron or other quantum strikes a viewing screen, its quantum state "collapses" instantaneously and nonlocally to become a more compact state. Collapse is also called *state collapse* or *quantum jump*. The term applies to any quantum jump from a more extended to a less extended quantum state. Collapse, and quantum jumping in general, is associated with quantum measurements. See *measurement* and *quantum jump*.

collapse of the quantum state see *collapse*

conducting material a substance within which electrons flow easily; compare *insulating material*

consciousness interpretation the hypothesis that consciousness is fundamental to quantum measurements, implying that human consciousness collapses the "Schrödinger's cat" superposition that quantum physics seems to predict. Critics point out that this interpretation doesn't specify which kinds of humans can collapse quantum superpositions. This implausible hypothesis has been checked experimentally and found to be false. See *interpretations*.

conservation of energy a principle that states the total energy of all participants in any process remains unchanged. Equivalently, the amount of energy in the universe never changes. Energy only changes in form, never in amount. See *energy*.

constructive interference see *interference*

Copenhagen interpretation the standard "textbook" interpretation—and the oldest interpretation—of quantum physics. Laboratory measurements are a central theme. Interprets quantum physics nonrealistically. Quantum states don't represent objective reality, but instead represent only our knowledge of reality, a view, to some extend, that persists today. See *interpretations*.

correlated Two random outcomes are "correlated" if knowledge of one of them alters the probabilities of the outcomes of the other.

current, electric a flow of electrons or other charged quanta

dark energy an invisible form of energy that causes the expanding universe to accelerate. Astronomical observations show that 68% of the universe's energy is in the form of dark energy. Together, dark matter and dark energy account for 95% of the universe's energy. Compare *dark matter*.

dark matter matter that does not experience electromagnetic force and hence cannot glow the way stars glow. Astronomical observations show that 27% of the universe's energy is in the form of dark matter. Together, dark matter and dark energy account for 95% of the universe's energy. Compare *dark energy*.

decoherence the process of converting a quantum superposition into a mixture by entangling the superposition with some other system, such as a measuring apparatus. This process is a fundamental part of the measurement process and entails that measurements are irreversible, thus helping resolve the measurement problem. See *incoherent* and *coherent*.

delayed-choice experiment a double-slit experiment with a which-slit detector with the detector not switched to "on" until after the quantum has come through the slits. John Wheeler suggested this in 1983, and Jean-Francois Roch and colleagues performed it in 2007 using a Mach-Zehnder interferometer rather than the equivalent double-slit setup. Roch arranged the timing so that the detector could be switched on after the quantum (a photon) had entered the interferometer. The last-minute decision had no effect on the experiment's outcome. Even when the detector was turned on after the photon was already in the interferometer, the photon was detected to be in a mixture; when the detector was turned off, the photon was detected to be in a superposition. Wheeler thought this showed the decision altered the photon's state before the photon entered the interferometer, but there's a less radical interpretation: The quantum collapses to the mixture only when it has no opportunity to interfere with itself by passing through the second beam splitter.

destructive interference see *interference*

detection loophole one of several experimental "loopholes" that permit Aspect's experiment and other nonlocality experiments to have interpretations other than nonlocality. The detection loophole arises in experiments in which only a small fraction of the photon pairs created in the experiment are actually detected. If it then happened that the detected pairs were especially prone to violate Bell's locality condition, the results could appear to violate Bell's locality conditions when inspection of all photon pairs would demonstrate that Bell's locality condition was not violated. See also *locality loophole*.

detector effect in a quantum measurement, the detector affects the detected quantum by collapsing it into one or the other of the quantum's superposed states

diffraction a characteristic wave phenomenon. When a wave passes through a narrow opening, it spreads out on the other side of the opening. The spreading is broader for narrower openings. If the opening is less than one wavelength in width, the spreading fills the region behind the slit. If the opening is many wavelengths wide, the wave goes nearly straight through the opening as a narrow "beam," with only slight spreading at the sides.

digital restricted to particular, discrete values; countable; the opposite of *analog*

Dirac equation the differential equation obeyed by relativistic material quanta; similar to Schrödinger's equation

double-slit experiment an experiment first performed by Thomas Young in 1801. Light passing through a pair of thin slits strikes a viewing scree forming an interference pattern on the screen and showing that light is made of waves. Today, this experiment is done not only with light (streams of photons), but also with other kinds of quanta such as electrons, protons, neutrons, atoms, and molecules. All these turn out to be waves—a fundamental point for quantum physics. Furthermore, the experiment can be performed "one quantum at a time," with a stream so dilute that the quanta must pass individually through the slits, without interacting with the other quanta in the stream. An interference pattern still forms on the screen, showing that each quantum comes through both slits, so each quantum is a wave in a field.

elastic energy energy resulting from the ability of a deformed (e.g. stretched or squeezed) object to snap back; see *energy*

electrical charge If an object feels electrical forces when placed into an electrical field, it is said to be "electrically charged" or to "carry electrical charge."

electromagnetic (EM) wave a wave in an electromagnetic field

electromagnetic energy energy resulting from electromagnetic forces; see *energy*

electromagnetic field (EM field) one of the fundamental force fields that fill the universe. An EM field exists wherever an electrically charged object would feel a force. An EM field surrounds every electrically charged object.

electromagnetic radiation a wave in the EM field; also called *electromagnetic wave*; see *electromagnetic spectrum*

electromagnetic spectrum the entire range of electromagnetic waves, including radio, microwave, infrared, visible, ultraviolet, x ray, and gamma ray

electromagnetic wave a wave in an electromagnetic field. EM waves travel at light speed. Wave types include radio, microwave, infrared, visible (or light), ultraviolet, x ray, and gamma ray. Also called *electromagnetic radiation*. See *wave* and *electromagnetic spectrum*.

electron a negatively charged fundamental quantum with a mass about 1800 times smaller than a proton's mass. Atoms contain electrons moving around outside the nucleus. Electrons are leptons—meaning that they experience the electroweak force and gravity, but not the strong force.

electron field see *electron–positron field*

electron shell In a many-electron atom, the quantum states of the electrons form spherical shells centered on the nucleus, with each shell containing several electrons.

electron–positron annihilation When an electron and a positron are brought together, they annihilate each other quickly, typically creating two photons.

electron–positron field the universal quantized field having quanta that are electrons and positrons; also called *electron field, matter field, psi field*.

electron's magnetic moment a property of the electron, related to the electron's magnetic effects. It can be calculated, and also measured, to 12 and, less certainly to 14, figures. Such high accuracy is a phenomenal verification of quantum electrodynamics.

electroweak field theory the unifying theory of electromagnetic and weak forces.

electroweak force field the field responsible for electromagnetic and weak forces. Its quanta are the photon (massless, moves at light speed), W⁺ boson, W⁻ boson, and Z boson (all three have mass and move at less than light speed). See *Standard Model*.

element, chemical a particular kind of atom having a specific number of protons in the nucleus and thus a specific number of electrons and a specific chemical behavior. There are roughly 100 different elements. See *atom*.

EM electromagnetic

EM field see *electromagnetic field*

energy the ability to do work, the ability to move things around, the ability to effect change. There are several forms of energy: kinetic (resulting from motion), gravitational (resulting from gravity), elastic (resulting from the ability of an object to snap back), thermal (resulting from thermal motion or temperature), electromagnetic (resulting from electromagnetic forces), radiant (resulting from radiation), chemical (resulting from chemical reactions), and nuclear (resulting from nuclear reactions).

energy and mass equivalence see *equivalence of energy and mass*

entanglement When two quanta interact (exert forces on each other), their quantum states (which are really just configurations of quantum fields) are partially exchanged, so that the two quanta share their states. We then say the quanta are "entangled," a condition that persists even after the quanta have separated and no longer interact. Entanglement is nonlocal and appears not to diminish with distance. Any number of quanta can be mutually entangled. The protons, neutrons, and electrons in any atom are entangled. In many respects, entangled quanta act as a single unified quantum. Theory and experiment lead to the conclusion that entangled quanta behave nonlocally. Changes in the state of one quantum are transferred instantly to the other quantum, causing real physical changes in the other quantum's state. See *nonlocal, RTO entanglement experiment, Aspect's experiment.*

entropy a property of typical macroscopic systems; a measure of a system's disorganization at the molecular level

environmental decoherence the decoherence process that occurs when a superposed quantum is effectively "measured" by environmental quanta, converting the superposed quantum into a mixture. This is an irreversible, entropy-producing process. See *decoherence.*

epistemological interpretation the notion that quantum physics is only about human knowledge, that quanta and quantum states are not physically real but are merely useful concepts

EPR Albert Einstein, Boris Podolsky, and Nathan Rosen. They published an argument to demonstrate the real and simultaneous existence of precise values of such variables as position and velocity of an electron, contradicting Heisenberg's principle. They concluded that quantum physics is incomplete and that hidden variables must exist that would remove quantum indeterminacy. Later work showed they had neglected quantum entanglement and nonlocality. Their work is the earliest indication of quantum nonlocality.

equivalence of energy and mass a principle that states any object at rest that has a mass of m kilograms must have an energy $E = mc^2$ joules, and any object that has an energy of E joules when at rest must have a mass $m = E/c^2$ kilograms

event horizon an imaginary, spherical surface surrounding a black hole, within which nothing can escape

excited state a quantum state with energy that is greater than the energy of the ground state; compare *ground state*

fermion see *boson*

field A physical entity that exists at every point throughout a region of space, such as a pool of water or a field of grass. Fields can be described quantitatively at every point in the region; for instance, a pool of water has a temperature field, described by the temperature at every point. A gravitational field, a magnetic field, and other types of invisible fundamental fields, fill the universe. They are best described as "properties of space." According to quantum physics, all such universal fields are "quantized." The Standard Model (Table 5.1) describes the known quantized fields: force fields such as the electromagnetic field, electroweak field, and strong force field; matter fields such as the electron field, six quark fields, and six lepton fields; and the Higgs field. See *Standard Model* and *quantized field.*

force an influence on an object that can, if no other forces act, cause the object to accelerate; a push or a pull

frequency the number of waves sent out in each second by a wave source; measured in vibrations per second or "hertz" (Hz).

fundamental forces the small number of different types of forces that can explain all the forces observed to date. Classical physics was able to reduce their number to just two: gravity and electromagnetism. Two others, discovered during the 20th century, are the strong force and the weak force. In 1973, a new theory called the *Standard Model* combined the electromagnetic and the weak forces into a single electroweak force.

gamma ray see *electromagnetic wave*

general theory of relativity Einstein's theory according to which gravity is caused by the curvature of space-time; this theory is a generalization of Einstein's "special theory of relativity."

Ghirardi–Rimini–Weber (GRW) model a revised form of quantum physics postulating that, from time to time, every quantum collapses spontaneously and randomly. This has practically no effect on small systems, but it causes superpositions of large systems to collapse quickly and spontaneously, thus explaining why we never see such macroscopic superpositions. This would solve the measurement problem, but efforts to construct a relativistic version have been unsuccessful, and no experimental evidence for this model has appeared despite recent experiments involving mesoscopic superpositions. See *interpretations*.

global data see *global observer*

global observer an observer who obtains data about both of the quanta in an experiment involving two entangled quanta; such an observer's data is called *global data*. Compare *local observer*.

gluons the quanta of the strong force field. There are eight kinds, all massless, and all move at light speed. See *Standard Model*.

grand unification the yet-unachieved unification of electroweak field theory (of the electromagnetic and weak forces) with strong force field theory (of quarks and gluons)

gravitational energy energy resulting from the gravitational force; see *energy*

ground state the lowest-energy state among the definite-energy quantum states of an atom or other quantum system. A quantum system that is in its ground state has the lowest energy it can have. In hydrogen, the ground state is the state in which the electron's field lies closest around the nucleus. Compare *excited state*.

half-life pertaining to a radioactive nucleus, the time interval during which it has a 50–50 chance (as a result of quantum indeterminacy) of decaying. Equivalently, the half-life is the time during which half of a large sample of identical nuclei will decay.

Haroche's experiment an elaborate experiment demonstrating many quantum principles: superposition, decoherence, measurement, and partial measurement. An atom in a superposition of two Rydberg states is measured by interacting with the electromagnetic field of 0 to 10 photons stored in a high-Q cavity. The state of the atom is detected following the measurement. A second atom is then sent through the same process to observe the effects of decoherence of the photons.

heat a misleading term for *thermal energy*; see *thermal energy*

heating, law of Thermal energy can flow spontaneously from hot to cold, but not from cold to hot. This is one example of an irreversible process. See *irreversible process*.

Heisenberg's principle There is a lower limit on how small a material quantum's indeterminacies in any component of position and the same component of velocity can be. When we alter either of these two indeterminacies, the other indeterminacy must change in such a way that their product remains above this limit. Quantitatively, $\Delta x \times \Delta v \geq h/4\pi m$, where Δx is position indeterminacy, Δv is velocity indeterminacy, h is Planck's constant, and m is the quantum's mass.

hertz the vibration per second, the measurement unit for quantifying wave frequencies, abbreviated Hz

hidden variables unobserved influences that would, if taken into account, remove quantum randomness and make quantum processes predictable. Frequently hypothesized but never experimentally confirmed.

Higgs boson see *Higgs field*

Higgs field a quantized field that fills the universe and confers mass on the fundamental material quanta. It's quantum is the Higgs boson, a material quantum that moves slower than light speed. See *Standard Model*.

high-Q cavity an evacuated region bounded by highly reflective surfaces that can contain photons for a relatively long time before the photons lose their energy. See *Haroche's experiment*.

hydrogen, quantum states of see *quantum states of hydrogen*

hypothesis a tentative, unconfirmed, scientific idea; an educated guess

Hz abbreviation for the hertz; see *hertz*.

improper mixture the mixture of observed properties that arises from the "reduction" of the two-quantum measurement state to predict the properties observed by a single local observer. "Indeterminacy-based mixture" would be a better term, because there is nothing improper about it. A *proper mixture*, on the other hand, arises for nonquantum reasons such as ignorance of the actual state. "Ignorance-based mixture" would be a better term. See *mixture*.

incoherent a quantum in a mixture is "incoherent" in the sense that it is possible, in principle, to associate different branches of the mixture with different experimental outcomes, implying that the different branches are not "unified" or "superposed" into a single entity. For example, in the double-slit experiment with a which-slit detector, each quantum is in an incoherent mixture of coming through one or the other slit. Because, without the detector, each quantum would have been in a coherent superposition of coming through both slits, the detector is said to "decohere" the quantum. Compare *coherent, decoherence*, and *mixture*.

indeterminacy see *quantum indeterminacy*

indeterminate unpredictable

inertia the difficulty-of-accelerating an object; the natural tendency of any massive object to maintain its velocity

inertia, law of the notion that an object will keep moving unless there's something to stop it. Discovered by Galileo, Descartes, and Hobbes. Also called *Newton's first law* because Newton incorporated it into his theory of mechanics.

infrared see *electromagnetic wave*

instantaneous happening all at once, happening in zero time

insulating material a substance within which electrons will not flow, or will flow only with great difficulty; compare *conducting material*

interaction when two objects exert forces on each other

interference a characteristic wave phenomenon; the adding or canceling of two waves. When two waves in the same medium meet, they overlap and pass through each other. Where crests meet crests or valleys meet valleys, the two waves add together constructively to form large crests and large valleys. Where crests meet valleys, the two waves cancel each other destructively.

interferometer See *Mach-Zehnder interferometer*

interpretations This book summarizes several ways of interpreting or altering quantum physics, most of them designed to resolve the measurement problem. For specific examples, see *Copenhagen interpretation, pilot-wave model, Ghirardi–Rimini–Weber model, consciousness interpretation*, and *many-worlds interpretation*.

ion an atom with a net electric charge; an atom with an excess or deficiency of electrons (either more or fewer electrons than protons)

IR infrared electromagnetic radiation

irreversibility problem see *problem of irreversibility*

irreversible process a process during which a system can proceed easily or spontaneously from state A to state B, but can proceed from B to A only with assistance from outside energy sources. Examples include an object falling to the floor and thermal energy flowing from a hot object to a cold object.

isotope Atoms having the same number of protons and also the same number of neutrons are said to belong to the same "isotope." Compare *element*.

Jauch's solution of the measurement problem see *local state solution*

joule the metric unit of energy and of work. One joule is about the amount of work you do in lifting a stick of butter (about 0.1 kilogram) by 1 meter. This increases the butter's gravitational energy by 1 joule.

kelvin a unit of temperature. One degree kelvin means 1 degree above absolute zero. Absolute zero is -273°C.

kinetic energy energy resulting from motion; see *energy*

kinetic theory of warmth Warmth is the random, or disorganized, motion of a substance's molecules.

Large Hadron Collider the high-energy accelerator near Geneva that circulates protons to nearly light speed then smashes them together to create many types of quanta from the energy of the collision

laser light amplification by the stimulated emission of radiation. Lasers use excited atoms in metastable states to create large numbers of photons in identical states, with identical wavelengths, directions of motion, and phases. Such photons form a narrow and intense light beam that can travel great distances with little spreading. The macroscopic light beam can be regarded as a single giant quantum state. See *metastable state*.

lattice regular array. The atoms or molecules constituting most solid materials are arranged in a lattice pattern. Such lattices are responsible, for example, for the symmetric shapes of snowflakes.

law of science a confirmed and reasonably broad scientific idea; same as *principle*

lepton fundamental quanta that experience only the electroweak force (not the strong force) and gravity. They include three kinds of electrons (electrons, muons, taus) and three kinds of neutrinos (electron neutrinos, muon neutrinos, tau neutrinos). See *Standard Model*.

light see *electromagnetic wave*

light speed the speed of a light wave; about 3×10^8 meters per second

local data see *local observer*

locality the notion that objects are directly influenced only by their immediate surroundings.

locality loophole although experiments suggest strongly that nature really is nonlocal, there are several experimental "loopholes" that permit conclusions other than that nature is nonlocal. The experiments suggest that a change occurring at point A can cause an instantaneous change at a distant point B. The locality loophole arises in those experiments in which the time interval between the changes at A and B is sufficiently long to permit information to travel, at or below light speed, between A and B. In this case, the experiment could appear to support nonlocality when the cause-effect relationship actually occurred by local transmission via EM waves. Compare also *detection loophole*.

locality principle Physical processes occurring at one location should have no immediate (instantaneous) effect on the real physical situation at another location. As Einstein put it, "no spooky action at a distance." Classical physics obeys this principle, but quantum physics does not. Experiments and quantum theory both disprove the locality principle. See *Bell's locality condition* and *Aspect's experiment*.

local observer an observer of only one of the quanta in an experiment involving a system comprising two quanta. The observer's data, concerning that one quantum, is called *local data*. Compare *global observer*.

local state solution the author's suggested resolution of the problem of definite outcomes, a resolution first suggested by Josef Jauch in 1968. A careful analysis of the measurement state shows that it is not a superposition of states, but instead a superposition of correlations between states. This theoretical result is entirely consistent with experiment and not at all surprising. This is not offered as one more interpretation or alteration of quantum theory, but rather as a prediction of standard quantum theory.

loophole see *detection loophole* and *locality loophole*

Mach-Zehnder interferometer a device for bringing the reflected and transmitted portions of a light beam back together so they can interfere with each other, and detecting the result.

macroscopic of human size; compare *microscopic*, *mesoscopic*

magnetic moment see *electron's magnetic moment*

many-worlds interpretation a hypothesis that solves the measurement problem by assuming collapse does not occur. Instead, when a quantum superposition is measured, reality splits into zillions of separate realities, in each of which the quantum emerges in a different branch of the superposition. This vanquishes quantum randomness and the measurement problem, but is subject to the "Occam's razor" criticism that the hypothesis is far from simple or efficient. See *interpretations*.

mass the amount of inertia in an object, a measure of the force needed to accelerate an object starting from rest, the difficulty of accelerating an object.

mass and energy equivalence see *equivalence of energy and mass*

matter any physical substance that "weighs," that has weight when situated within a gravitational field such as exists on Earth. Matter always moves slower than light speed. Compare *radiation*.

matter field any quantized field having quanta made of matter; a material quantum field

Maxwell's equations the fundamental equations describing classical and quantum electromagnetic fields

measurement pertaining to in quantum physics, any process in which a quantum phenomenon causes a macroscopic change. Contrary to this word's use in ordinary language, humans needn't be involved in quantum measurements.

measurement problem the problem of making sense of the measurement state, which really amounts to two problems: the problem of definite outcomes, or "Schrödinger's cat," and the problem of irreversibility. The problem of definite outcomes is resolved by the local state solution; the problem of irreversibility is resolved by decoherence. See *local state solution* and *decoherence*.

measurement state an entangled superposition of two quanta, one of which is a macroscopic "which-branch" measuring device. For example, in the double-slit experiment using a which-path detector, the detector and the detected quantum are in this state. Solving the measurement (or Schrödinger's cat) problem is a matter of making sense of this state.

mechanical universe the universe as conceived by classical physics; a predictable, clocklike universe made of independent parts that are, ultimately, particles, all of it running according to Newton's laws.

mesoscopic a little larger than microscopic; having a size in the range 1 micrometer (10^{-6} meter) to 1 millimeter (10^{-3} meter); compare *microscopic, macroscopic*

metastable state a quantum state that does not decay quickly and spontaneously into a lower-energy state. Here, "quickly" means within the lifetime of about 10^{-9} second that is normal for excited states of atoms and molecules before spontaneous decay occurs. Typical metastable states have lifetimes of 1 second or more.

microscopic so small as to be visible only with a microscope. One grain of talcum powder, for example, is about 1 micrometer (10^{-6} meter) to a few micrometers across; this is the rough border between macro and micro. Compare *macroscopic* and *mesoscopic*.

microwave see *electromagnetic wave*

mixture A quantum is in a "mixture" when it exhibits definite but indeterminate properties. For example, a radioactive nucleus is in a mixture of undecayed and decayed when it is either undecayed or decayed but we do not know which one because of ignorance or because nature has not yet determined the outcome. This is different from a "superposition" of undecayed and decayed, in which the nucleus is both undecayed and decayed simultaneously. Thus, a mixture can be ignorance-based (also called a *proper mixture*) or quantum-indeterminacy-based (also called an *improper mixture* or *local mixture*). See *coherent* and *incoherent*.

model a simplified or hypothesized explanation or description of natural phenomena

molecule two or more attached atoms

muon see *lepton*

nano billionth, 10^{-9}; see *nanometer*

nanometer a billionth of a meter, 10^{-9} meter

neutrino a type of fundamental quantum. There are three kinds, each having a small but non-zero mass, and each paired with one of the three electron-like quanta (the electron, muon, and tau). They move at nearly light speed and experience only the gravitational and weak forces. See *lepton* and *Standard Model*.

neutron an uncharged composite quantum made of three quarks. Most kinds of atoms contain one or more neutrons in their nucleus. Neutrons experience the strong force, the electroweak force, and gravity. Compare *proton*.

Newton's law of gravity All pairs of material (i.e., possessing mass) objects attract each other with a force proportional to the product of their two masses and inversely proportional to the distance between them.

Newton's laws Newton's three laws of mechanics (the law of inertia, the law of force and acceleration, the law of equal and opposite forces) and law of gravity; see *Newton's law of gravity*

Newtonian physics physics based on Newton's three laws of mechanics and Newton's law of gravity; see *classical physics, Newton's laws*

Newtonian universe see *mechanical universe*

nonlocal process see *nonlocality*

nonlocality any direct and immediate influence by one object on another object across a distance

nuclear energy energy resulting from nuclear reactions; see *energy*

nucleon a proton or neutron

nucleus the small object at the center of every atom; made of protons and neutrons

objective not observer dependent, the same for all observers; compare *subjective*

Occam's razor the principle of simplicity in science; the notion that scientists should try to find the simplest, or least eccentric, explanation for observed phenomena

ontological interpretation the notion that quantum physics is about reality, that quanta and quantum states are physically real

partial measurement a process, involving a superposed quantum and a measuring apparatus, during which the observed states of the apparatus do not distinguish unambiguously between the superposed states of the quantum. A superposed quantum that is not measured typically causes interference effects that disappear when the quantum is measured. In a partial measurement, such interference effects partially appear. See *Haroche's experiment*.

particle a small object located within a limited spatial region; a small object surrounded by empty space. Newtonian physics conceives the universe to be made of particles. Quanta are not particles, although they sometimes act like particles.

phonon quantized bundle of vibrational energy of a solid lattice structure. Electron–phonon interactions are important in understanding superconductivity. See *superconductor* and *lattice*.

photoelectric effect the ejection of electrons by a metal when light falls on the metal. Einstein showed this is caused by the bundling of light (radiant energy) into quanta that we now call *photons*.

photon a quantum of electromagnetic field energy, a quantum of radiation, a quantum of light

pilot-wave model David Bohm's revised form of quantum physics that postulates a universe made of a quantum field Ψ and classical particles moving in a non-classical manner. The field obeys Schrödinger's equation, but the particles do not. Instead, the field "guides" classical particles in a predictable but highly nonlocal manner. The initial positions of the particles are "hidden variables" within this theory, for these positions have never been observed. Bohm's model shows that a deterministic (having no quantum randomness) hidden-variables theory is at least possible. Although the model's predictions agree with quantum physics at nonrelativistic energies, the model has so far proved impossible to extend to relativistic energies, partly because of its extreme nonlocality. See *interpretations*.

Planck distance the short separation distance, about 10^{-35} meter, between two quanta at which the quantas' masses become so large that the attractive force of gravity overcomes all repulsive forces and the two quanta collapse together.

Planck's constant 6.6×10^{-34} joules per hertz. Planck's constant multiplied by a photon's frequency equals the energy, in joules, of that photon. Planck's constant is a fundamental universal number that specifies the magnitude of quantum effects.

planetary model of the atom a model in which electrons orbit the nucleus in circles or ellipses, the way that planets orbit the sun; see *model*

positron a quantum identical with the electron except for having a positive, rather than negative, charge; the antiquantum of the electron

powers of 10 a system of abbreviation for very large or small numbers. For example, 10^4 means 10,000 (beginning from 1.0, move the decimal four places to the right); 10^{-4} means 0.0001 (move the decimal four places to the left); $3 \times 10^{-4} = 0.0003$.

principle of science a confirmed and reasonably broad scientific idea; same as *scientific law*

probability used in connection with any process (classical or quantum) having uncertain outcomes. Impossible events have probability 0; certain events have probability 1. In quantum physics, the probability of a particular experimental outcome is the outcome's frequency of occurrence (the number of favorable outcomes divided by the number of trials) in a long series of identical trials. Compare *statistics*.

problem of definite outcomes the core of the measurement problem. According to many expert analyses, quantum theory implies measurements do not have definite outcomes but instead yield only superpositions of possible outcomes; yet, experimentally, measurements do have definite outcomes. The author suggests this problem is resolved by the local state solution. See *measurement problem, local state solution, measurement state*.

problem of irreversibility along with the problem of definite outcomes, this is a part of the measurement problem. Schrödinger's equation is "reversible": As long as this equation is obeyed, entropy cannot increase and a quantum system's evolution can be turned around so that the system retraces its previous steps. Thus, the entanglement process (which obeys Schrödinger's equation) and the local state solution alone cannot resolve the measurement problem. The resolution of this dilemma lies in the making of a macroscopic mark during quantum measurements. The environment effectively measures quantum superpositions by entangling with them and collapsing into one branch of the resulting mixture. The process is irreversible because the many dispersed environmental quanta cannot, feasibly, be reversed. The same is true for laboratory measurements, where the macroscopic measuring apparatus plays the "environment" role. Thus, decoherence resolves the irreversibility problem. See *decoherence, reversibility, measurement problem*.

proper mixture see *improper mixture*

proton a positively charged (electrified) composite quantum; made of three quarks. One or more protons reside in the nucleus at the center of every atom. Protons experience the strong force, the electroweak force, and gravity. Compare *neutron*.

pseudoscience any misleading distortion of the scientific process that is presented as science even though it lacks supporting evidence and rational plausibility

psi another name for the matter field of a quantum. This field, when evaluated numerically at a single point x and then squared, gives the probability that the quantum will interact at x. Psi is usually represented by the Greek letter Ψ. See *matter field*.

quantize to restrict a process or entity to only certain allowed energies. A field is said to be "quantized" if it comes in energy bundles that obey the rules of quantum physics. See *quantum*.

quantized field a field that is made of energy bundles called *quanta* that obey the rules of quantum physics. Examples include all the fields of the Standard Model. The universe appears to be made of quantized fields. See *quantum, Standard Model*.

quantum a highly unified, spatially extended, specific quantity or bundle of field energy. The word derives from "quantity." Every quantum is a wave—a disturbance—in a field. Examples include photons, electrons, protons, atoms, and molecules. Plural is *quanta*.

quantum electrodynamics the quantum physics of electrons and photons; a quantum field theory describing electrically charged quanta and their electromagnetic interactions

quantum field theory a theory of quantum fields; usually construed to mean a relativistic quantum field theory—one that includes high-energy, relativistic phenomena such as the creation and destruction of material quanta

quantum indeterminacy the fundamental indeterminacy of many quantum phenomena. Identical causes can result in different outcomes. The future is not encoded in the present and does not exist until it actually happens. Some experiments, such as sending individual identical photons through a beam splitter, have indeterminate outcomes. Unlike the classical randomness of coin flips and the like, quantum indeterminacy does not arise from ignorance. See *Heisenberg's principle*.

quantum jump an instantaneous transition of an entire extended quantum from one configuration or "state" to a different state. Quantum jumps in atoms are instantaneous transitions to a different state of the atom's electrons. These have been observed by several experimental groups and the jumps are apparently instantaneous. To conserve energy, the atom emits or absorbs a photon or some other form of energy. Other examples include nuclear decay, collapse of a quantum onto a viewing screen, or, more generally, the macroscopic detection (and associated collapse) of any quantum.

quantum mechanics an old-fashioned and inappropriate term for *quantum physics*. Quantum physics is not at all "mechanical" or "like a clock" or similar to Newtonian mechanics. See *quantum physics*.

quantum physics roughly, the science of matter and energy on the smallest scales. As a better definition, quantum physics is about the nature and behavior of quanta, the fundamental constituents of the universe.

quantum state a particular configuration, or condition, of a quantum. For examples, see *wave packet* and *quantum states of hydrogen*. Quantum states are often represented by the uppercase Greek letter psi (Ψ). Some experts doubt the reality of quantum states. Their reasons are (1) the measurement problem and (2) the notion that quantum states are merely epistemic, not ontological. See *ontological interpretation, epistemological interpretation*.

quantum state collapse see *collapse*

quantum states of hydrogen the possible quantum states of the single electron in a hydrogen atom, as determined from Schrödinger's equation. These states should be thought of as the possible "standing waves" of the electron field—the states that just fit properly around the nucleus. These states are pictured in Figure 7.3.

quantum vacuum see *vacuum field*

quark one of the fundamental quanta. There are six kinds: up, down, charm, strange, top, and bottom. They experience the strong and the electroweak force. Every proton and neutron is made of three quarks. See *Standard Model*.

qubit the individual computational element, or "bit," of a future quantum computer. Qubits will be designed to be not only in one of the two states (dubbed "0" and "1") of today's classical computers, but also in quantum superpositions of these two states, giving them phenomenal computing power for certain purposes.

radiant energy energy resulting from radiation such as electromagnetic radiation; see *energy*

radiation any physically real entity (any physical substance) that does not "weigh," that has no mass. Radiation always moves at light speed. Compare *matter*. This term can refer to electromagnetic waves or to any other type of "rays" that "radiate" (in the radial direction) outward from a central source

radio see *electromagnetic wave*

radioactive decay a process in which a nucleus quantum jumps spontaneously into a new nuclear state and emits one or more quanta of energy. The energy usually appears as either an *alpha quantum* (two protons attached two neutrons, identical with a helium nucleus) or a *beta quantum* (an electron that was created in, and emitted by, a nucleus). These two processes are called *alpha decay* and *beta decay*.

radioactivity see *radioactive decay*

random unpredictable

randomness see *quantum indeterminacy*

relativity's universal speed limit the widely accepted principle that objects and information cannot be transferred through space at faster than light speed. Violation of this rule would imply physical systems could move backward in time so that events in those systems would violate the second law; thus, you could arrange to kill your grandfather before he meets your grandmother by sending a message backward in time—a contradiction.

resistance, electrical the retarding effect that most materials exert on electric current (flowing electrons)

RTO experiment a "double double-slit experiment" performed in 1990 by John Rarity and Paul Tapster and independently by Zhe-Yu Ou, Zingquan Zou, Lei Wang, and Leonard Mandel. One member of an entangled pair of photons passes through a double-slit setup (or an equivalent Mach-Zehnder interferometer setup) and the other member passes through a second double-slit setup. The correlated impact points of the pairs on the two detection screens demonstrate nonlocality.

Rydberg atom an atom with an outer electron shell that contains a single electron that has been excited into a large outer orbit; see *Haroche's experiment*

Schrödinger's cat Schrödinger's example for demonstrating the measurement problem. A radioactive nucleus is attached to a Geiger counter that is attached to a cat in such a way that, if the Geiger counter clicks, the cat dies. Thus the cat's "alive" or "dead" status acts as a detector for the nondecayed or decayed quantum state of the nucleus. The problem is that quantum physics predicts this process puts the nucleus and cat into the entangled measurement state, which appears (to Schrödinger and others) to be a state in which the cat is both alive and dead. See *problem of definite outcomes*.

Schrödinger's equation the differential equation obeyed by nonrelativistic material quanta. It describes the evolution, or time development, of a system's quantum state Ψ as long as the system is not measured. When a quantum is measured, an unpredictable quantum jump occurs. Relativistic versions of Schrödinger's equation, such as the Dirac equation, are often also called *Schrödinger's equation*. Schrödinger's equation describes matter fields in a manner parallel to Maxwell's equations for electromagnetic fields. See *quantum state, measurement*.

second law of thermodynamics A system that is at least partly organized at the molecular level and is given the opportunity to reorganize is highly likely to proceed to a less organized state. In terms of entropy, the total entropy of all the participants in any macroscopic physical process is overwhelmingly likely to increase or remain unchanged; decreases in total entropy occur only randomly and can be neither sustained nor controlled. See *entropy*.

special theory of relativity Einstein's theory of space, time, and motion according to which space and time are different according to differently moving observers. One prediction of this theory is that neither energy nor information can be transferred at a speed faster than light, because such transfer could violate the principle of causality, according to which effects occur after causes.

spectrum range, assortment; see *electromagnetic spectrum*

speed the distance traveled per unit of time; the distance traveled divided by the time to travel

speed of light see *light speed*

SQUID superconducting quantum interference device. A small ring of superconducting metal with a small insulating segment (Figure 8.3). A superconducting current moving around the ring of this mesoscopic object can be put into specific quantum states and into quantum superpositions. See *superconductor, mesoscopic, superposition*.

Standard Model of Fundamental Fields and Quanta the theory of those quantum fields that are known and at least partially understood today. The theories of the electroweak and strong forces. It includes six kinds of quarks, six kinds of leptons (three electrons and three neutrinos), two kinds of force fields (strong and electroweak), and the Higgs field (and its quantum, the Higgs boson). The quanta of the strong force field are called *gluons*. The quanta of the electroweak force field are the photon, W^+ boson, W^- boson, and Z boson. Summarized in Table 5.1. The Standard Model is known to be incomplete; in particular, it does not include the gravitational force, dark matter, or dark energy.

standing wave a wave that vibrates in place, without moving in any direction. Standing waves typically occur in media that are restricted to a limited region, in which case they must satisfy specific conditions at the boundaries of that region. Examples include a violin string that is fixed in position at both ends; a hydrogen atom's electron field, which much fit properly around the nucleus. See *quantum states of hydrogen* and Figure 7.2.

state collapse see *collapse*

statistics a record of favorable and unfavorable outcomes in a (necessarily finite) real-world series of trials of an experiment such as throwing dice, or a photon's passage through a double slit; compare *probability*

string hypothesis the hypothesis (it's not yet a confirmed "theory") that all the fundamental forces can be reduced to a quantum theory of "strings"—quanta that are not ultimately (when "squeezed" to their smallest size) pointlike, but ultimately like a looped or straight piece of string

string theory see *string hypothesis*

strong field the force field that holds the nucleus together. Its quanta include eight kinds of gluons. See *Standard Model*.

subjective observer dependent, different for different observers; compare *objective*

superconductor a material that conducts electricity with no expenditure of energy (i.e., no electrical resistance). Many metals become superconducting when sufficiently cooled.

supercurrent an electrical current in a superconductor; see *superconductor*

superluminal faster than light

superposition principle If a quantum can be in any one of several different states, then it can be in all these states at the same time. The quantum is said to be in a *superposition* of the states. See *superposition state*.

superposition state a situation in which a quantum is in two or more quantum states at the same time. Examples: a quantum in several different places simultaneously, or moving in several different ways simultaneously, or having several different total energies. See *superpositiion principle*.

system a certain portion of the universe, such as an automobile, a set of pool balls, an electron, a molecule, Earth, or the solar system, to name a few

tau see *lepton*

theory of everything an as-yet undiscovered, unified theory that will describe all the fundamental physical forces and all the fundamental forms of energy. There is a consensus that such a theory can be devised and that it will be a quantum theory.

theory, scientific a logically consistent and experimentally well-verified body of scientific principles that explains a broad range of observed phenomena

thermal energy energy resulting from thermal motion (i.e., resulting from temperature); see *energy*

thermal motion the random motion of atoms and molecules resulting from temperature

thermal radiation the electromagnetic radiation produced by thermal motion. Most thermal radiation from natural processes on Earth is in the infrared region of the spectrum. Sunlight—thermal radiation from the sun—spans the ultraviolet, visible, and infrared regions. See *electromagnetic radiation*.

thermodynamics the general principles of energy

tunnel a quantum starting at point A that is detected later at point B is said to "tunnel" from A to B if this transition would not be allowed classically.

ultraviolet see *electromagnetic wave*

uncertainty see *quantum indeterminacy*

Unruh radiation Quantum field theory predicts an accelerating observer moving through a vacuum will detect quanta that a nonaccelerating observer does not detect. This prediction is unexplainable in terms of particles, but explainable by quantum fields. Unruh radiation has not yet been observed conclusively.

UV ultraviolet

vacuum field a fundamental field that is devoid of quanta. Vacuum fields contain energy and are not "nothing." They exhibit real properties such as the Casimir force and the Lamb shift. The vacuum field exists everywhere; according to quantum physics, a state of true nothingness is not allowed anywhere in the universe.

vacuum fluctuations A quantized field cannot have an energy that is precisely zero, because this would violate Heisenberg's principle. Thus, even in a vacuum, where there are no quanta, all quantum fields must still execute random vibrations called *vacuum fluctuations*.

velocity an object's speed and direction of motion

W^+ boson, W^- boson two kinds of quanta of the electroweak field, both of which have mass; see *Standard Model*

warmth see *kinetic theory of warmth*

wave a disturbance that travels through a "medium," such as a body of water, a rope, or a physical field. For a water wave, the disturbance is a series of valleys and crests on the water's surface. A *sound wave* is a pressure (or density) wave in a gaseous, liquid, or solid material. See *electromagnetic wave*.

wavelength the length of one complete spatial repetition of a wave's shape in space

wave packet a general type of quantum state in which the quantum has a wavy (with crests and valleys) shape within a limited region and is nearly zero outside that region; see Figure 7.1

wave superposition a property of waves. Two waves can be present in the same medium at the same time and can even pass right through each other without disturbing one another. We say that the two waves are "superposed." In a similar manner, quantum states, being waves in fields, can be superposed. See *superposition* and *superposition principle*.

which-path detector in a double-slit experiment, a device that can determine through which slit each quantum passes. Switching on such a device causes the quantum to jump into the "single-slit state" of coming through one or the other slit, but not both. This demonstrates the important point that measurements cause quantum jumps. See *double-slit experiment*.

work work is done whenever a force (a push or pull) acts on an object while the object moves through some distance

x ray see *electromagnetic wave*

Young's experiment a double-slit interference experiment using light, first performed by Thomas Young in 1801. The interference implies light is a wave.

Z boson a quantum of the electroweak field that has mass; see *Standard Model*

Zou's experiment the 1990 experiment of Xingquan Zou, Lei Wang, and Leonard Mandel in which a photon passes through a double-slit setup, and a second arbitrarily distant photon, entangled with the first photon, functions as an optional which-path detector for the first photon. The result is the same as any other double-slit experiment: With the detector off, the first photon is in a superposition of coming through both slits; with the detector on, the first photon collapses to a mixture of coming through one or the other slit. The experiment demonstrates that measurements are nonlocal.

INDEX

acceleration, 41
  and gravity, 245n43
  and inertia, 72
  and mass, 71
  of the universe, 21
  and Unruh radiation, 103–104
Adler, Stephen, 226
alpha decay, 47, 113–114
  in bismuth, 239n12
  and quantum randomness, 113
  and quantum tunneling, 113
  and Schrödinger's cat, 202
  See also radioactive decay
alpha particle. *See* alpha quantum
alpha quantum, 47, 113
  in a radioactive nucleus, 109
  See also radioactive decay
analog quantity, 34–35
Anderson, Andrew, 202–203
Anderson, Carl, 92
Anderson, P. W., 226
antiparticle, 54. *See also* electron-positron annihilation; positron
Arndt, Markus, 162, 223
Aspect, Alain, 176–178
Aspect's experiment, 176–178, 213
atom, 5–6, 19, 63
  and a-tomos, 29
  and the classical worldview, 44, 90
  and the double-slit experiment, 84
  as fields, 60, 89, 105
  forces between, 45–46
  as fundamental constituent of universe, xiii, 18–19, 22
  hydrogen, 8, 136–140
  in interference experiments, 6, 88
  of light, 181
  and molecule, 29
  observation of, 35
  as a particle, 61

planetary model of, 6, 74–75, 122
  quantization of, 33, 73–74
  and quantum indeterminism, 111, 122
  quantum jumps in, 140–143, 148
  quantum states of, 136–140
  radiation from, 8–9, 32, 68, 134–140
  reality of, 143–147
  size of, 58, 122
  spectrum of, 9
  and temperature, 31, 50
  types of, 29
  See also a-tomos
atomic number, 47
atomic state, 8, 140, 148. *See also* quantum state
a-tomos, 28–30
  and Newtonian physics, 42
  See also atom

basis ambiguity, 215–216
beam splitter, 3, 110
  and quantum indeterminacy, 110, 114–115, 167
  and single-photon nonlocality, 180
  and superposition states, 150–153
Becquerel, Henri, 46–47
Bell, John, 36, 126, 175–176, 182–183, 198
Bell's locality condition, 175–182, 250n15.
  See also locality principle
Bentley, Reverend Richard, 60
Berkeley, Bishop George, 70
Berra, Yogi, 22
beta decay, 47–48. *See also* radioactive decay
beta particle. *See* beta quantum
beta quantum, 48. *See also* radioactive decay
Big Bang, 8, 20–22
  energy created by, 240n17
  faster-than-light expansion of, 102, 204
  low entropy of, 58, 240n25, 257n3
  and many-worlds interpretation, 204
  neutrinos from, 237n7

Big Bang (*Cont.*)
  photons from, 182, 228
  quanta created by, 95
  quantum origin of, 123
  radio waves from, 68–69
black hole, 7–8
  at center of our galaxy, 85
  classical physics and, 42
  and Hawking radiation, 245n43
Bohm, David, 126–129, 197–198, 242n6
Bohr, Niels, xiii, 37, 233
  and atomic structure, 74–75
  and the Copenhagen interpretation, 144–146, 154, 195–196
  and Einstein, 109, 137, 139, 248n6
  and Heisenberg, 5
Born, Max, 76, 90, 154, 195–196
  and the interpretation of the quantum state, 240n5
boson, 157, 248n12
  in the standard model, 95–97
  and superconductivity, 244n32
Boswell, James, 70
branch of a superposition, 152–153, 160, 167
  and decoherence, 222–223
  and the GRW model, 199
  and the local state solution, 211
  and the many-worlds interpretation, 203
  and quantum measurement, 165
  in the RTO experiment, 170–173
Butterfield, Herbert, 40

Carpenter, Roger, 202–203
Casimir force, 103
Casimir, Hendrik, 103
charge. *See* electric charge
chemical element. *See* element
chemical energy, 50. *See also* energy
Chopra, Deepak, 12
classical field, 60, 145
classical physics, 2, 38–45, 61
  cannot explain the atom, 122
  in the Copenhagen interpretation, 196
  and hidden variables, 111, 197
  and irreversibility, 257n3
  and quantum indeterminism, 109, 122
Clauser, John, 176–177, 249n18, 250n18
Cleese, John, 207
Cleland, Andrew, 162
clockwork universe. *See* mechanical universe
coherent quantum, 188
  and decoherence, 224
  and the double-slit experiment, 192, 227
  and lasers, 230
collapse of the quantum state, 84–85, 100, 104, 111, 147–148
  and consciousness, 12, 200–203
  and the Copenhagen interpretation, 197

  and decoherence, 221–224
  and the epistemic interpretation, 145, 183
  in the GRW spontaneous collapse theory, 198–199
  in Haroche's experiment, 228–232
  and Heisenberg's principle, 117–118, 122
  in the local state solution, 204–208
  in the many-worlds interpretation, 203–204
  and measurement, 154, 187, 193, 235n2
  and nonlocality, 167–168
  and single-quantum entanglement, 180–181
  and special relativity, 208–211
conducting material, 63, 65. *See also* insulating material
consciousness interpretation, xii, 12–13, 200–203, 254n20. *See also* interpretations of quantum physics
conservation of energy, 52
  and the first law of thermodynamics, 240n23
  Joule's experiments on, 239n14
  and the reality of EM waves, 69–70, 147
constructive interference. *See* interference
Copenhagen interpretation, 144, 154, 195–197. *See also* interpretations of quantum physics
Copernicus, Nicolaus, 40–41, 93
correlated outcomes, 124, 166, 171–173
  in measurements, 211–215, 257n57
  and nonlocality, 175
  and special relativity, 184–186
current, electric, 61, 63
  in electric discharge tubes, 140
  See also superconductivity

dark energy, 19, 22, 49
  evidence for, 20–22
  and the standard model, 94–95, 98
  See also dark matter
dark matter, 19–20, 22, 49
  evidence for, 20–21, 237n9
  and the standard model, 94–95, 98
  See also dark energy
Davisson, Clinton, 75–76, 85–86, 243n16
Davy, Humphry, 61
de Broglie, Louis, 37, 75–76, 77, 84, 243n16
  and Bohm's pilot-wave model, 129, 197, 242n6
  and the Copenhagen interpretation, 248
  the energy of an electron occupies all space, 88–89
decoherence, 218, 221–224
  and the measurement problem, 191, 224–227, 229, 232
  See also coherent quantum
Democritus, 28–30, 77, 100
  and classical physics, 38, 42, 69, 90
Descartes, Rene, 39, 44
destructive interference. *See* interference
detection loophole, 177–178. *See also* locality loophole

Index

detector effect, 184–188
Deutsch, David, 204, 254n23
DeWitt, Bryce, 204
diffraction, 23, 115–120
digital quantity, 18, 31–35. *See also* analog quantity
Dirac, Paul, 12, 37, 76
  and Dirac's equation, 91, 93, 252n43
  quanta interfere only with themselves, 83, 85
  and quantum field theory, 90, 92, 100–101, 141
Dirac equation, 93, 101, 252n43
Dostoevsky, Fyodor, 44
double-slit experiment, 22–28, 80–85, 93, 100, 131
  in Bohm's model, 127–128
  and the detector effect, 184–188
  using electrons, 85–89, 243

Einstein, Albert, 17, 37, 75–76
  and Brownian motion, 238n29
  and collapse of the quantum state, 167, 180
  criticisms of quantum physics, xi, xiii, 11, 109, 152–153
  and the field view of reality, 59–62, 65, 70, 241n9, 241n13
  and hidden variables, 123–124
  and the laser, 155
  his "miraculous" year, 240n18
  and nonlocality, 11, 125–126, 166–167, 212, 236n2
  and the photoelectric effect, 17, 74
  his realist philosophy, 18, 144
  theory of relativity, 23, 38, 42, 52–54, 60, 77
  unified field theory, 30, 46, 97
elastic energy, 50. *See also* energy
electric charge, 6, 9, 62–67, 95
electromagnetic (EM) energy, 32, 50
  quantization of, 73–74, 78–85, 90
  See also energy
electromagnetic (EM) field, 2, 27–28, 38, 59
  Casimir force and, 103
  classical, 60–65
  Dirac's theory of, 90–93
  evidence of quantization of, 80–84
  quantization of, 73–80
  reality of, 69–70, 147
  universality of, 30
  See also electromagnetic spectrum; electromagnetic wave
electromagnetic (EM) radiation, 6, 32, 50, 79
  and measurement, 122
  quantization of, 73–74, 78–85, 90
  and reality of the EM field, 147
  and special relativity, 90
  types of, 67–69
  See also electromagnetic spectrum
electromagnetic (EM) spectrum, 67–69
  of hydrogen, 75, 89, 103, 140
  and quantum field theory, 89, 103
  and quantum jumps, 140–143
electromagnetic (EM) wave, 64–69, 79
  and the Casimir force, 103
  and classical fields, 60–69, 241n13
  reality of, 69–70
  types of, 67–69
  See also electromagnetic spectrum; wave
electron, 6, 63, 95
  and atoms, 63, 100
  and beta decay, 48
  in Bohm's theory, 126–129
  and the chemical elements, 29, 63
  and contact forces, 45–46
  diffraction of, 117–122
  Dirac's theory of, 90–93
  and electric current, 63
  and electromagnetic waves, 65–69
  in electron-positron creation and annihilation, 54, 91–92
  as a fundamental quantum, 29, 73, 95
  in hydrogen atom states, 8, 74, 137, 142
  and light, 134–143
  and magnetic force, 63
  magnetic moment of, 9–10
  mass of, 72
  and matter fields, 60, 75
  quantum field nature of, 75–76, 85–93, 105, 117–122, 132–134
  shell, 68, 229
  size of, 123
  spin of, 91
  in string theory, 100
  in superconductivity, 96–97, 156–158
  See also electron-positron field; electron's magnetic moment
electron field. *See* electron-positron field
electron-positron annihilation, 54, 91–92
electron-positron field, 87–88, 92. *See also* matter field; psi
electron's magnetic moment, 9–10
electroweak field, 94–98, 241n14. *See also* Standard Model
element, chemical, 5, 18, 29
  and isotope, 114
  spectra of, 74
  See also atom
Eliot, T. S., 100
EM. *See* electromagnetic
energy, 8, 38, 49–50
  definition of, 49
  forms of, 22, 49–50
  metric unit of, 50
  of a quantum, 18, 28, 30
  stationary states of, 9, 99, 135–140, 142
  and work, 49
  See also conservation of energy

energy and mass equivalence. *See* equivalence of energy and mass
entanglement, xiii, 12, 124, 168–169
   Aspect's experiment on, 176–178
   and Bell's test for locality, 175–176
   and environmental decoherence, 222, 225, 229
   EPR's discovery of, 124–126
   and irreversibility, 221, 257n2
   and the local state solution, 192, 204–216
   and the many-worlds interpretation, 204
   and measurement, 193, 199, 218–221
   and quantum jumps, 184–188
   and quantum state collapse, 199, 229
   RTO experiment on, 170–174
   Schrödinger's statement on, 148
   universality of, 178–182
   See also nonlocality
entropy, 54–58, 112
   in the Big Bang, 240n25, 257n3
   and irreversibility, 218–221, 225, 227, 257n2
environmental decoherence, 221–229
   in Haroche's experiment, 228–232
   See also decoherence
epistemological or nonrealist interpretation, 131, 146–147, 167, 183
   and the Copenhagen interpretation, 196–197
EPR experiment, 124–126
equivalence of energy and mass, 52–54
event horizon, 85, 245n43
excited state, 139, 142–143, 155, 229. *See also* ground state

Faraday, Michael, 59, 61, 73, 77, 85
   as chemistry apprentice, 61–62
   disagreement with Maxwell, 241n9, 241n13
fermion. *See* boson
Feynman, Richard, 86, 109, 243n10, 244n26
   and the wave vs particle issue, xii, 83, 92, 100–101, 242n2
field, 2, 6–7
   classical, 59–64
   quantum, 18, 27–28, 78–89
   reality of, 69–70
   See also quantum field; Standard Model
force, 19, 38, 41, 43, 45
   four fundamental, 45–49
   See also Standard Model
Freedman, Stuart, 176
frequency, 65–68, 79
   and energy, 67
   and wavelength, 67
fundamental forces, 45–49. *See also* Standard Model

Galileo, 38–40
gamma ray. *See* electromagnetic wave
Geller, Uri, 12
general theory of relativity, 94, 98
   and the Big Bang, 8
   and classical fields, 60
   and dark matter, 20
   and Einstein's unified field theory, 46
   and the Event Horizon Telescope, 85
   and Hawking radiation, 245n43
   and quantum physics, 7, 98–99
Gerlach, Walther, 219
Germer, Lester, 75–76, 85–86, 243n16
Ghirardi, Giancarlo, 198–199
Ghirardi-Rimini-Weber (GRW) model, 198–199. *See also* interpretations of quantum physics
Gilder, Louisa, xiii
Glashow, Sheldon, 96
global observer, 215, 252n47. *See also* local observer
gluon, 95–96
grand unification, 94
gravitational energy, 49–50, 52, 138–139
   in the inflationary theory of the Big Bang, 240n17
   See also energy
Greene, Brian, 12, 46, 184
ground state, 139, 141–143. *See also* excited state
Guth, Alan, 204

Haldane, J. B. S., 41
half-life, 114–115
   of bismuth, 239n12
Haroche, Serge, 159, 228
Haroche's experiment, 228–233
heat. *See* thermal energy
heating, law of, 55, 218. *See also* irreversible process
Hegerfeldt, Gerhard, 101–102, 120
Heisenberg, Werner, xiii, 37, 76, 90, 92, 100, 110
   and collapse of the quantum state, 167, 180
   and the Copenhagen interpretation, 5, 144, 154, 195–196
   and nonlocality, 167
   See also Heisenberg's principle
Heisenberg's principle, 31, 117–123
   and Bohm's deterministic model, 126–129
   and hidden variables, 123–126, 197–198
   and the quantum vacuum, 78, 129–130
Herbert, Nick, 169
Hertz, Heinrich, 65–66, 69
hertz, the, 65
hidden variables, 111, 123–129
   and Bohm's deterministic model, 126–129, 197–198
Higgs, Peter, 72, 97, 244n33
Higgs boson. *See* Higgs field
Higgs field, 72–73, 94–95, 98
   and the origin of mass, 97–98
   See also Standard Model
high-Q cavity, 230. *See also* Haroche's experiment
Hill, Christopher, 1

Index

Huygens, Christian, 22–23, 26, 28
hydrogen, quantum states of. *See* quantum states of hydrogen
hypothesis, 20, 34, 51, 72, 75–76
   string, 99–100
Hz. *See* hertz

improper mixture, 214–215
incoherent mixture, 188, 224, 231. *See also* mixture
indeterminacy. *See* quantum indeterminacy
inertia, 19, 35
   and the Higgs field, 72
   law of, 39–42
   of photons inside superconductors, 97
Infeld, Leopold, 12
infrared, 32, 68. *See* electromagnetic wave
insulating material, 63, 160. *See also* conducting material
interference, 23–26, 64
   of electrons with themselves, 85–89
   of molecules with themselves, 161–162
   of photons with themselves, 80–84, 153–155
   of quantum fields, 134
   in a single slit experiment, 115–116
interferometer. *See* Mach-Zehnder interferometer
interpretations of quantum physics, 196–204
ion, 63, 140, 159
IR. *See* infrared
irreversibility problem. *See* problem of irreversibility
irreversible process, 54–58
   and decoherence, 224–227
   and entanglement, 227–228
   and quantum measurement, 148, 218–221
isotope, 114

Jauch, Josef, 12, 192, 205, 207–208, 211–213
Jauch's solution of the measurement problem. *See* local state solution
Johnson, Samuel, 70
Jonsson, Claus, 86
joule, the, 50

Kajita, Takaaki, 98
Kelvin, Lord, 31
kelvin, the, 57, 96
Khayyam, Omar, 217
kinetic energy, 49, 52, 54. *See also* energy
kinetic theory of warmth, 51

Laplace, Pierre-Simon, 44
Large Hadron Collider, 19, 72–73, 90, 97
laser, 155–158
   in Haroche's experiment, 229–232
   in Wineland's experiment, 159–160
lattice, 156
   and electrical resistance, 157–158

law of entropy. *See* second law of thermodynamics
law of gravity. *See* Newton's laws
law of heating. *See* heating, law of
law of inertia. *See* inertia, law of
law of science, 40–42
Lederman, Leon, 1
Lee, Ka Chung, 179
Leitenstorfer, Alfred, 104
lepton, 94–98, 139, 243n17. *See also* Standard Model
light. *See* electromagnetic wave
light speed, 19, 50, 53
   and definition of the meter, 241n11
   and gluons, 96
   and Hertz's experiment, 65–66
   as the limiting speed, 126
   and Maxwell's equations, 64
   and neutrinos, 237n7
   Roemer's estimate of, 64
Linde, Andrei, 204
local data. *See* local observer
locality, 166–167, 172
   Bell's test for (*see* Bell's locality condition)
   See also locality loophole; nonlocality
locality loophole, 176–178. *See also* detection loophole
locality principle, 125–126, 166–167, 172, 175, 250n16. *See also* Aspect's experiment; Bell's locality condition
local observer, 171–173, 184, 207, 210–215, 252n47. *See also* global observer
local state solution, 192, 204–208, 226–227, 256n51
   rebutting the objections to, 213–216
   why quantum physics requires, 208–213
loophole. *See* detection loophole; locality loophole

Mach-Zehnder interferometer, 2–3, 6, 171–174
   demonstration of quantum superposition, 150–154, 165
macroscopic phenomena, 5, 34, 49
   in classical physics, 38, 76, 114
   in the Copenhagen interpretation, 154
   and entanglement, 179, 182
   in the measurement process, 35–36, 144–148, 154, 184–187, 191
   in quantum physics, 122, 154–157
   and quantum superpositions, 158–165
   See also mesoscopic phenomena; microscopic phenomena; problem of definite outcomes; Schrödinger's cat
magnetic moment. *See* electron's magnetic moment
Mandel, Leonard, 170, 209. *See also* RTO entanglement experiment

many-worlds interpretation, 203–204, 253n2.
    See also interpretations
mass, 19
    and the distinction between matter and
        radiation, 19
    and the fragility of quanta, 35, 37
mass and energy equivalence. See equivalence of
    energy and mass
matter, 17, 19
    amount of in the universe, 22
    See also radiation
matter field, 60, 87–89
    for the electron in hydrogen, 137
    and quantum field theory, 93–95, 127
    See also electron-positron field; psi
Maxwell, James Clerk, 59, 61–66, 70, 77, 85, 89
    differences from Faraday, 62, 241n9, 241n13
    and unification of physics, 93, 97
Maxwell's equations, 62–63, 89–91, 93, 150
McDonald, Arthur, 98
measurement, xii, 35–36, 165, 191–194
    and collapse of the quantum state, 147–148,
        184–188, 228
    Einstein, Podolsky, and Rosen's (EPR) analysis
        of, 124–125
    and interpretations of quantum physics,
        195–204
    and quantum indeterminacy, 121–124
    and quantum jumps, 147–148, 184–188
    and quantum nonlocality, 126, 172
    and the reality of quanta, 143–145
    and Schrödinger's cat, 194–195
    See also measurement problem;
        measurement state
measurement problem, xii, 11–12, 191–194
    and decoherence, 218–221, 224–227
    and entanglement, 169, 184–188
    history of, 211–213
    and interpretations of quantum physics,
        195–204
    and irreversibility, 218–221, 224–227
    local state solution of, 204–211, 213–216
    and macroscopic superposition, 158
    and Schrödinger's cat, 194–195
    See also decoherence; local state solution;
        measurement; measurement state
measurement state, 187, 193–194, 198, 205–208
    and decoherence, 258n18
    and definite outcomes, 225–226, 255n34
    entropy of, 257n2
    and Haroche's experiment, 231
    and nonlocality, 208–213, 252n47
    and objections to the local state solution,
        213–216
    and reversibility, 218–221, 228
    and Schrödinger's cat, 255n33
    See also measurement; measurement problem
mechanical universe, 42–45

Mermin, David, 145
mesoscopic phenomena, 160–165, 222–223. See
    also macroscopic phenomena; microscopic
    phenomena
metastable state, 142–143
    and lasers, 155–156
    and quantum jumps, 142–143
Michelson, Albert, 31
microscopic phenomena, 6, 31–32, 34, 36, 50. See
    also macroscopic phenomena; mesoscopic
    phenomena
microwave. See electromagnetic wave
mixture, 187–188, 193–195, 205, 207–208,
    252n47, 256n39
    and decoherence, 218, 222–228
    and definite properties, 208
    in Haroche's experiment, 229–232
    and resolution of the measurement problem,
        212–215
    and Schrödinger's cat, 258n18
    and special relativity, 210
model, 6, 43, 94
molecule, 5–7, 18, 29
muon. See lepton

nanometer, 66
neutrino, 19, 95, 98, 237n7. See also lepton;
    Standard Model
neutron, 5–6, 46–48, 95–96
    and isotopes, 114
    and the mass of ordinary matter, 72–73
    and quarks, 18, 29, 48, 72–73
    See also proton
Newton, Isaac, 38, 40–45, 59–61, 73
    and light, 22–23, 28
    and particles, 30
    See also Newton's laws
Newton's laws, 20, 31, 37–43, 52
    and the mechanical universe, 42–45
    and reversibility, 257n3
    See also classical physics
nonlocality, 11, 45, 126, 166
    in Aspect's experiment, 176–178
    Bell's analysis of, 126
    Bell's test for, 175–176
    in Bohm's theory, 129
    and the detector effect, 184–188
    Einstein's views about, 11, 124–125
    and entanglement, 166–188
    and the history of the measurement problem,
        211–213
    and the local state solution, 191–192, 204–211
    Mermin's views about, 145–146
    and the modal interpretation, 256n51
    physicists' view about, 252n39
    and quantum state collapse, 167–168
    and reality, 182–183
    and relativistic causality, 126, 183–184

Index

in the RTO entanglement experiment, 170–174
ubiquity of, 178–182
and the unity of the quantum, 126, 239n33
nonlocal process. *See* nonlocality
nonrealist interpretation. *See* epistemological or nonrealist interpretation
nuclear energy, 50. *See also* energy
nucleon, 46–48
and isotopes, 114–115
nucleus, 5–6, 29, 63, 71
and Becquerel's discovery of alpha decay, 46–48
and nuclear energy, 54
and the planetary model of the atom, 74

objectivity in quantum physics, 36, 44, 109, 131, 144, 147
and the Copenhagen interpretation, 196
and the solution of the measurement problem, 204
Occam's razor, 5, 30, 129, 204
O'Connell, Aaron, 162
ontological interpretation of quantum physics, 131, 144. *See also* Copenhagen interpretation
Ou, Zhe-Yu, 170. *See also* RTO entanglement experiment

Pais, Abraham, 144
partial measurement, 222, 231–232. *See also* Haroche's experiment
particle, 5, 11, 17, 22, 59–60
and de Broglie's wave hypothesis, 75–76
and extended quanta, 77–93
in Greek physics, 28–31
inappropriateness of in quantum physics, 104–105
in Newtonian physics, 22–23, 26, 28, 38, 42–45
and quantum fields, 100–104
See also wave-particle duality
Pauli, Wolfgang, xiii, 76, 92, 251n36
phonon, 156, 164, 179–180. *See also* lattice; superconductor
photoelectric effect, 17, 73–74, 227
photon, 2, 235n1
and Aspect's entanglement experiment, 176–177
and Bell's test for locality, 175–176
from the Big Bang, 182
as bosons, 157
creation or destruction of, 54, 90–91
in Dirac's quantum field theory, 90–93
in a double-slit experiment, 80–84
emitted in atomic transitions, 8–9, 139–142
energy of, 79–80, 139–140
entanglement of, 169, 170–174, 177–178

field nature of, 80–84, 155
in Haroche's experiment, 228–232
and the Higgs field, 72
and the human eye, 235n5
inertia of within superconductors, 96–97
and instantaneous quantum jumps, 142–143, 148
and lasers, 155–156
and the Mach-Zehnder interferometer, 3–5, 150–154
and the origin of mass, 96–98
and the photoelectric effect, 73–74
in Planck's hypothesis, 34, 242n3
and quantum indeterminacy, 110–111, 114–115
as a ripple in a field, 6–7, 79
in a shop window, 1–7
and single-quantum entanglement, 180–181
in a single-slit experiment, 117
size of, 84–85
speed of, 19, 237n8
in the Standard Model, 95–97
in string theory, 100
superposition of, 150–154
unity of, 29, 110
pilot-wave model, 127–129, 197–198, 243n6. *See also* interpretations of quantum physics
Planck, Max, 17, 31–35, 76, 79–80, 217
Planck distance, 99, 123, 246n13
Planck's constant, 80, 120, 140
planetary model of the atom, 5–6, 74–75, 122, 139, 246n12
Plato, 22–23
Podolsky, Boris, 124, 183, 212
positron, 54, 91–93, 244n26
powers of ten, 50
principle of science, 4, 40
probability, 112–114, 117
in the Copenhagen interpretation, 197, 240n5
and Heisenberg's principle, 120
and the meaning of quantum states, 133–139
and quantum measurement, 164–165
waves, 87
See also statistics
problem of definite outcomes, 191–193, 200
and the local state solution, 204, 207–208, 212–216
See also measurement problem; measurement state
problem of irreversibility, 191, 210, 218–221, 226–227, 257n3. *See also* decoherence; measurement problem
proper mixture. *See* improper mixture
proton, 5–6, 8
and the chemical elements, 29, 47
compared with electron, 122
and electric currents, 63
identity of all, 88

proton (Cont.)
  and isotopes, 114
  in the Large Hadron Collider
  and the mass of ordinary matter, 72–73
  and quarks, 29, 48, 71–72, 95–96
  and radioactive decay, 47–48, 122–123
  in the Standard Model, 93–96
  and the strong force, 46–47, 96
  See also neutron
pseudo-science, 12, 13, 144
psi, 87, 89, 92, 133–134, 240n5
Pusey, Matthew, 146–147

quantization, 6, 33–34, 37, 79
  of Dirac's and Maxwell's equations, 91–93
  and entanglement, 166
  and indeterminacy, 115–117
  of lattice vibrations, 156
  of matter, 85–89
  and nonlocality, 167
  of radiation, 80–95
  of superconducting current, 160–161
  See also quantum, the
quantum, the, xi, 17–22, 77–105
quantum electrodynamics, 9
quantum field, xii, 2, 11, 28, 30, 78–80, 93
  calculation of mass of proton and neutron, 72
  creation and annihilation in, 54
  differences from classical fields, 60
  theory of, 89–93
  throughout all space, 31
  See also quantum; Standard Model
quantum field theory, 89–93
quantum indeterminacy, 109–130. See also Heisenberg's principle; randomness
quantum jump, 140–143, 147–148
  in Bohm's model, 197–198
  and entanglement, 166, 184–188, 193
  in GRWo spontaneous collapse model, 198–199
  and measurement, 154, 184–188
quantum mechanics, 76. See also quantum physics
quantum physics, xi–xii, 17–18
  accuracy of, 7–10
  applications of, 10
  basic idea of, 34, 37
  fundamental issues of, 11
  and general relativity, 7
  and reality, 6
  universality of, 5, 34
quantum state collapse. See collapse of the quantum state
quantum states, 131–148
  changes over time, 147–148
  of the hydrogen atom, 134–140
  quantum jumps between, 140–143
  reality of, 143–147

stationary (standing wave) states, 135–137
wave packet, 132–134
See also collapse of the quantum state; epistemological interpretation; ontological interpretation
quantum states of hydrogen, 134–140
quantum vacuum. See vacuum field
quark, 18, 29, 47–48
  and the mass of protons and neutrons, 71–72
  and the origin of mass, 71–73
  and the strong force, 47–48, 71
  See also Standard Model
qubit, 158, 163–164

radiant energy, 50
  and Einstein's proof of equivalence of energy and mass, 53
  and the photoelectric effect, 74
  See also energy
radiation, 6, 9
  alpha and beta (see alpha decay; beta decay)
  from the Big Bang, 21–22, 204, 255n25
  from the chemical elements, 74
  classical view of, 75
  compared with matter, 85–89, 92
  contrasted with matter, 19, 73, 75–76
  and creation of matter, 54
  in the double-slit experiment, 80–85
  energy of a photon of, 79
  the eye's sensitivity to, 68
  Hawking, 245n43
  and the Higgs field, 73
  and lasers, 155
  mass of, 19
  and the photoelectric effect, 74
  Planck's analysis of, 31–32, 73
  and prisms, 141
  in radioactive decay (see alpha decay; beta decay)
  and reality, 147
  as ripples in the vacuum field, 79
  temperature of, 57
  Unruh, 103–104
radio. See electromagnetic wave
radioactive decay, 46–48
  indeterminacy of, 11, 45, 111, 113–115
  mass change due to, 54
  See also alpha decay; beta decay
radioactivity. See radioactive decay
random, 2
  thermal motion, 31–32, 51, 56, 68
randomness, 2–5, 11, 31, 44, 109–110
  avoidance of in the many-worlds interpretation, 203–204
  avoidance of in the pilot-wave model, 197–198
  and beam splitter experiments, 150–152
  and definite states, 256n39

and the detector effect, 184–188
evidence of, 110–113
and Heisenberg's principle, 117–123
and hidden variables, 123–130
and improper mixtures, 214–215
in interactions with atoms, 84
of an orbital electron, 138–139
probability and, 112–113
and quantum state collapse, 148, 167
in the quantum vacuum, 78, 102, 104
radioactive decay and, 113–115
and statistical predictability, 115–117
in the Stern-Gerlach experiment, 219–221
of transitions of electrons within atoms, 141–143
See also quantum indeterminacy
Rarity, John, 170. *See also* RTO entanglement experiment
relativity's universal speed limit, 102, 125–126, 153, 212
and nonlocality, 183–184
resistance, electrical, 96, 156–158
Riek, Claudius, 104
Rimini, Alberto, 198–199
Roemer, Olaf, 64
Rosen, Nathan, 124, 183, 212
RTO entanglement experiment, 170–174
and an "atom of light," 181
nonlocality of, 175–176
and resolution of the measurement problem, 209–211
and the unity of the quantum, 181, 183
and universal entanglement, 227–228
Rubin, Vera, 20
Rutherford, Ernest, 122, 143
Rydberg, Johannes, 229
Rydberg atom, 229–231, 246n12. *See also* Haroche's experiment

Sagan, Carl, 147, 177
Salam, Abdus, 96
Schrödinger, Erwin, 37, 76, 89–90, 132–133, 137, 148, 178
disagreement with the Copenhagen interpretation, 248n6
dissatisfaction with entanglement, 212
his interpretation of Schrödinger's cat, 194–195
and "psi" as symbol of the quantum state, 243n22
See also Schrödinger's cat; Schrödinger's equation
Schrödinger's cat, xii, 12, 36, 158
and basis ambiguity, 216
and the detector effect, 188
and entanglement, 169
and Haroche's experiment, 231
and improper mixtures, 214–215

and interpretations of quantum physics, 195–204
and the irreversibility problem, 218
Jauch's resolution of, 212–213
and the local state solution, 204–208
and the measurement problem, 191–192, 210, 218
non-paradoxical nature of, xii, 205, 211, 255n29
as Schrödinger told it, 194–195
and the Stern-Gerlach experiment, 220
See also problem of definite outcomes
Schrödinger's equation, 89–91
in Bohm's pilot-wave model, 126–129, 197–198
contrasted with quantum jumps, 184
Dirac's more general formulation of, 91–93, 252n43
and dynamical evolution of the quantum state, 147–148
and quantized energies of the hydrogen atom, 140–141
and quantum superposition, 150
and reversibility, 219–221
in single-electron diffraction, 117–120
and standing waves, 134–140
and stationary states of the hydrogen atom, 136–140
and wave-particle duality, 93
Schwinger, Julian, 92, 100
second law of thermodynamics, 57
and the Big Bang, 58
and decoherence, 224–227
and the definition of energy, 239n13
and the first law of thermodynamics, 240n23
and the growth of a leaf, 57–58
and measurement devices, 148
in the measurement problem, 218–221
and Newton's classical physics, 257n3
and time reversal, 102, 112
See also entropy
special theory of relativity, 11, 38
and Bohm's pilot-wave theory, 129, 197–198, 242n6
and Dirac's equation, 141
and Einstein's Nobel Prize, 17
and entanglement, 168
and equivalence of energy and mass, 52–54
in the history of physics, 94
and the local state solution, 205, 208–210, 214
and particles in quantum physics, 101–102
and the positron, 244n26
and quantum field theory, 83, 89–93, 100–101
and quantum state collapse, 183–184
and Schrödinger's equation, 90
and Unruh radiation, 103–104
spectrum. *See* electromagnetic spectrum
speed. *See* velocity

speed of light. *See* light speed
SQUID, 160–161, 163–164. *See also* superconductor; superposition
Standard Model of Fundamental Fields and Quanta, 93–98
   beyond the, 98–100
standing wave, 135–137
   and the quantum vacuum, 103
   and SQUIDs, 160–161
   See also quantum states of hydrogen
state collapse. *See* collapse of the quantum state
stationary states of energy. *See* energy: stationary states of
statistics, 112–114
   and half life, 114–115
   and interference pattern, 86
   quantum physics and, 115–117
   See also probability
Stern, Otto, 219
string hypothesis, 46, 99–100
string theory. *See* string hypothesis
strong force, 46–48, 71–73, 96, 98
   as a field, 71
   and gluons, 96
   in the history of physics, 94
   and neutrons, 47
   and the nuclear surface, 245n7
   and quarks, 47
   See also Standard Model; weak force
subjectivity in science, 36, 145
   and the consciousness interpretation, 201
   and the Copenhagen interpretation, 196–197
subjective idealism, 70, 145
   See also epistemological or nonrealist interpretation; objectivity in quantum physics
superconductivity, 96, 156–160, 244n32
   and massive photons, 96–97
   See also SQUID
supercurrent, 158, 160–161. *See also* superconductor
superluminal. *See* relativity's universal speed limit
superposition, 4, 131, 149–150
   absence of in Bohm's model, 197
   and the consciousness interpretation, 200–201
   and the Copenhagen interpretation, 158
   and entanglement, 169–174, 179–180
   evidence for, 150–154
   of existing and not existing, 236n3
   and the GRW spontaneous collapse hypothesis, 198–199
   in Haroche's experiment, 228–230
   of large molecules, 6, 161–162
   and the local state solution, 205–208
   macroscopic or mesoscopic, 158–165
   and the many-worlds interpretation, 203–204
   and the measurement state, 187–188, 191–195, 210–211, 214, 255n29
   and Newtonian physics, 150
   principle, 149
   and quantum computers, 158
   and quantum state collapse, 167–168, 221–224
   and Schrödinger's cat, xii, 158, 191–195, 212, 218, 252n47
   in a shop window, 235n2
   in a SQUID, 160–161
   and waves, 6, 149–150, 152–154
system, 43
Szarek, Stanislaw, 182

Tan, Sze, 180
Tapster, Paul, 170. *See also* RTO entanglement experiment
tau. *See* lepton
Teller, Edward, 111
theory, scientific, 5, 8, 20, 51, 72
   versus "hypothesis," 99–100
   versus "model," 6
theory of everything, 8, 30, 97
thermal energy, 50–52
   and the definition of "energy," 239n13
   and electrical resistance, 156
   versus "heat," 239n15
   and infrared radiation, 68
   and irreversible processes, 54–56
   Joule and, 239n14
   and the law of heating, 55
   in neon signs, 236n12
   See also energy
thermal motion, 32, 50
thermal radiation, 32
   and infrared radiation, 68
   See also electromagnetic radiation
thermodynamics, 31, 38, 50
   statistical nature of, 56
   See also second law of thermodynamics
Tomonaga, Sin-Itiro, 92, 100
Tonomura, Akira, 86, 243n15
tunneling, 113, 161

ultraviolet, 9, 68. *See* electromagnetic wave
uncertainty. *See* quantum indeterminacy
Unruh, William, 103
Unruh radiation, 103–104
   and Hawking radiation, 245n43
UV. *See* electromagnetic wave; ultraviolet

vacuum field, 22, 30–31, 78–79
   and the Casimir force, 103
   and dark energy, 237n16
   and radiation from atoms, 139
   and Unruh radiation, 103–104
vacuum fluctuations, 31, 103
Valery, Paul, 42

velocity, 119
von Neumann, John, 12, 200–201

W^+ and W^- bosons, 95–96. *See also* Standard Model
Wang, Lei, 170. *See also* RTO entanglement experiment
warmth. *See* kinetic theory of warmth
wave, 6, 11, 23–28. *See also* electromagnetic wave
wavelength, 9, 65
　and energy, 67
　and frequency, 67
　See also frequency
wave packet, 132–134
　entanglement of, 168–169
　of photons in a laser, 155
　superposition of, 149
wave superposition, 149–150. *See also* superposition
weak force, 46–48
　and beta decay, 47–48
　and electroweak unification, 96–98, 241n14
　in the history of physics, 94
　and quarks and leptons, 94–95
　in the Standard Model, 95–96
Weber, Tullio, 198–199

Weinberg, Steven, 11, 89, 96, 238n24, 241n14, 242n2
Wheeler, John, 220, 226
which-path detector, 185–188
　and decoherence, 188, 221–222, 224
　and entanglement, 192–193, 205
　and the measurement state, 211
　in Zou's experiment, 208–209
　See also double-slit experiment
Wigner, Eugene, 12–13, 201, 247n10
Wilczek, Frank, 11, 72, 83, 242n2
Wineland, David, 159, 228
work, 49–50, 52

x ray. *See* electromagnetic wave

Young, Thomas, 24, 26–28, 238n21
Young's experiment, 24–28, 238n22
　beauty of, 238n21

Z boson, 95–97. *See also* Standard Model
Zeilinger, Anton, 6, 11, 88, 161, 171, 178, 223
Zou, X. Y., 170, 209. *See also* RTO entanglement experiment
Zou's experiment, 208–209
Zurek, Wojciech, xii, 201, 223, 252n1